PHARMACOLOGY

Richard M. Kostrzewa, Ph.D.
*Professor of Pharmacology, James H. Quillen College of Medicine
at East Tennesse State University*

Sulzburger & Graham Publishing, Ltd.
New York

Dedication:
To a former student and friend, Dr. Ellen Kunkel Bagden, who lovingly watches from heaven over her husband Alan and two children, Alex and Eric.

About the Author:
Dr. Richard M. Kostrzewa received a bachelor's degree in chemistry and a master's degree in pharmacology from the Philadelphia College of Pharmacy & Science. His Ph.D. degree in pharmacology was obtained at the University of Pennsylvania. Currently a Professor of Pharmacology at the James H. Quillen College of Medicine at East Tennessee State University, Dr. Kostrzewa has held positions at the Louisiana State University Medical Center, Tulane University Medical Center and Veterans Administration Medical Center in New Orleans. He has taught in Medical, Dental, Nursing and undergraduate courses of Pharmacology for 25 years. He is the author of nearly 100 peer-reviewed research articles on neuropharmacology studies related to schizophrenia, tardive dyskinesia, Parkinsonism and attention deficit hyperactivity disorder. One paper on a common neurotoxin has been featured as a Citation Classic, being among the 400 most-cited articles in the life sciences. He and his wife, Florence, are proud parents of ten children.

Editor-in-Chief, Publisher:
Neil C. Blond, Esq.

Graphic Illustration:
Jeremy Borenstein

Copy Editor:
David Gregory

© 1995 Sulzburger & Graham Publishing, Ltd.
P.O. Box 20058
Park West Station
New York, NY 10025

All rights reserved. No part of this book may be reproduced by any electronic or other means without written permission from the publisher.

ISBN 0-945819-48-X
Printed in The United States of America

Contents

1. PHARMACOLOGY PRINCIPLES 7
   - Routes of Drug Administration 7
   - Drug Distribution 15
   - Drug Deposition 16
   - Drug Metabolism 16
   - Relationship Between Bioavailability and Effect 17
   - Drug Excretion 17
   - Pharmacokinetics 19
   - Pharmacodynamics 25
   - Specific Terms 26
   - How Can Dose-Response Curves Be Used to Compare Drugs? 28
   - Therapeutic Doses 29
   - Drug Safet; Therapeutic Index (TI) 29
   - How Does Agonist Binding At Receptors Produce An Effect? 30
   - Variability in Drug Responses 30
   - Drug Interactions 31
   - Drug Metabolism 31
   - Enzyme Induction and Inhibition 34
   - Factors Influencing Biotransformation 37

2. THE AUTONOMIC, PARASYMPATHETIC AND SYMPATHETIC NERVOUS SYSTEMS 39
   - The Autonomic Nerves Nervous System (ANS) 39
     - Overview 39
     - Divisions of the ANS 39
     - Receptors in the ANS 40
     - Neurotransmitters and Metabolism 41
       - Ach 41
       - NE and EPI 42
     - Neurotransmitter Actions 43
     - SecondMessengers in the ANS 45
     - Metabolic Effects of the SNS 46
   - The Parasympathetic Nervous System (PSNS) 47
     - Overview 47
     - Muscarinic Receptor Agonists = Cholinomimetics 48
       - Choline Esters 48
       - Natural Alkaloids 50
       - Synthetic Alkaloid 50
       - Toxicity of Cholinometics 51
     - Cholinesterase Inhibitors (ChE-Is) 51
       - Pharmacokinetics 51
       - Classification of ChE-Is 52
       - Types of ChE-Is and Actions 52

    ChE-I Insecticide Poisoning 57
   Muscarinic Receptor Antagonists (M-Blockers) 57
  The Sympathetic Nervous System (SNS) 58
   Overview 58
   Catecholamine Adrenoceptor Agonists 59
   Direct-Acting Noncatecholamine Adrenoceptor Agonists 61
   Indirect-Acting Noncatecholamine Adrenoceptor Agonists 62
   Adrenoceptor Antagonists 65
   Adrenolytics 71
   Ganglionic Stimulants and Blockers 72

3. Somatic Nerves: Skeletal Muscle 77
  Skeletal Muscle Relaxants 77
   Nondepolarizing Drugs ("Pachycurares") 77
   Depolarizing Drugs ("Leptocurares") 79
  Spasmolytics 80

4. Diuretics 83

5. Cardiovascular Pharmacology 93
  Cardiovascular Pharmacology 93
  Treatment of Hypertension 93
   Antihypertensives 93
   Summary of Antihypertensive Drugs 101
   Summary on Uses of Antihypertensive Drugs 102
   Anti-Anginal Drugs 103
  Antiarrhythmics 107
  Treatment of Congestive Heart Failure 113
   Digital Glycosides 114
  Antihyperlipidemics 119
   Treatment of Digitalis Toxicity 119
   General Approach to Treatment of Heart Failure 119
   Other Drugs for Treating Congestive Heart Failure 119
   Non-Drug Measures for Controlling Hyperlipidemia 121
   Drugs Used for Hyperlipidemia 121
  Drugs that Alter Blood Coagulation/Hemostasis/Clot Dissolution 125

6. Treatment of Allergy 131
  The Allergic Response 131
  Histamine H_1-Blockers 134
  Mast Cell Ca^{++}-Channel Blockers 137
  Pharmacological Antagonists to Histamine and Autacoids 137
  Antiasthmatics 137
  Treatment Approaches 139

7. Gastrointestinal Drugs 141
  Treatment of Ulcers 141

Treatment of Constipation: Laxatives and Cathartics 147
    Laxatives 147
    Cathartics 149
    Enemas 150
Treatment of Diarrhea 150
Emetics and Antiemetics 152

8. Pain and Inflammation 155
Background on Pain 155
Eicosanoids and the Inflammatory Process 155
Drugs for Integumental Pain or Inflammation 160
    Analgesic/Antipyretic Drug Lacking Antiinflammatory Activity 160
    Analgesics with Antiinflammatory Activity 161
        Aspirin 161
        Nonsteroidal Antiinflammatory Drugs (NSAIDs) 166
        Remittive Agents for Rheumatic Diseases 168
        Steroidal Antiinflammatory Drugs (SAIDs) 171
        Anti-Gout Drugs 172
    Treatment of Headache 177
    Local Anesthetics 178
Drugs for Visceral Pain: Narcotic Analgesics 181

9. Pharmacology of the CNS 187
General Anesthetics 187
    Stages of Anesthesia 188
    MAC of Anesthetics 189
    Properties of Inhaled Anesthetic Drugs 191
Anxiolytics 193
    Benzodiazepines 193
    Treatment of Anxiety 197
    Treatment of Insomnia 197
Barbiturates 198
    Non-Barbiturate Sedative-Hypnotics 200
Ethanol 201
Methyl Alcohol 204
Other Alcohols 205
    Estimate of Blood Alcohol Levels 205
Classical CNS Stimulants 206
CNS Toxins 208
Treatment of Neurologic and Psychiatric Disorders 208
    Epilepsy 208
        Anticonvulsants 211
    Antiparkinsonian Drugs 218
    Antipsychotics (Neuroleptics) 221
    Treatment of Depression 225

  First Generation Antidepressants 228
  Second Generation Drugs 232
  Electroconvulsive Therapy (ECT) 234
 Treatment of Mania 235
Drugs of Abuse 235
 Specific Drugs of Abuse and Withdrawal Treatments 236

10. ENDOCRINE PHARMACOLOGY 241
 Hypothalamic-Pituitary Axis 241
  Prolactin 243
 Oxytocin and Uterine Drugs 243
  Uterine Stimulants 243
  Uterine Relaxants 246
 Thyroid and Antithyroid Drugs 246
  Hypothyroidism 248
  Hyperthyroidism (High T_3 and T_4 levels) 249
 Adrenocortical Hormones: Cortisol, Aldosterone and Androgens 250
  Glucocorticoids 250
 Androgens and Anabolic Steroids 255
  Androgens 257
 Estrogens and Progestins 258
  Estrogen 261
  Progestins 264
  Contraceptives 265
  Estrogen and Progesterone Antagonists 267
 Parathyroid Hormone (PTH) and Ca^{++} Regulation 267
  Vitamin D and Ca^{++} Regulation 268
  Other substances affecting Ca^{++} Activity 270
 Insulin, Glucagon and Oral Hypoglycemic Agents 276
  Insulin 276
  Exogenous Insulin 276
  Oral Hypoglycemic Agents 279
  Glucagon 281

11. ANTIMICROBIAL CHEMOTHERAPY 283
 Basic Principles 283
 Antifolate Bacteriostatic and Bactericidal Drugs 287
  Sulfonamides 287
  Trimethoprim 289
 Quinolones and Fluoroquinolones 291
 Urinary Tract Antiseptics 292
 Beta-Lactam Antibiotics 293
  Penicillins 293
  Cephalosporins 297
  Other β-Lactam Antibiotics 299
 Aminoglycosides and Aminocyclitols 300

Tetracyclines 302
Chloramphenicol 304
Macrolide Antibiotics 306
    Erythromycin 306
    Clarithromycin and Azithromycin 306
Miscellaneous Antibiotics 307
Treatment of Tuberculosis (Mycobacterium tuberculosis, M. leprae, M. avium) ... 307
Antifungal Drugs 312
Antiviral Drugs 317
Antiparasitics 322
    Anthelmintics for Intestinal Nematodes (Roundworms) 322
        Drugs for Intestinal Nematodes 323
    Anthelmintics for Blood and Tissue Nematodes (Filariasis) 325
        Drugs for Blood and Tissue Nematodes 326
    Anthelmintics for Cestodes (Tapeworms) 328
    Anthelmintics for Trematodes (Flatworms) 329
    Antiprotozoal Drugs 335
        Drugs Used to Treat Malaria 343
Antiseptics and Disinfectants 347

12. Cancer Chemotherapy 351
    Development of Cancer 351
    Approach to Treatment of Cancers 351
    Most Common DNA Sites Alkylated 353
    Bifunctional versus Monofunctional Alkylating Agents 353
    Development of Resistance to Alkylating Agents 355
    Cell cycle Nonspecificity of Alkylating Agents 355
    Specific Alkylating Agents 356
    Antimetabolites 363
    Purine Analogs 365
    Pyrimidine Antagonists 365
    Vinca Alkaloids 367
    Podophyllotoxins 367
    Antibiotics 367
    Miscellaneous Agents 369

13. Toxicology Principles 373
    Toxic Agents 374
    Treatment of Heavy Metal Poisoning 374
    Air Pollutants 380
    Solvents 381
    Insecticides 382
    Herbicides 383
    Fumigants 384

Index 387

CHAPTER 1
PHARMACOLOGY PRINCIPLES

Pharmacology is the study of drugs on living systems. It is important to realize that chemicals that have profound effects on isolated cells may have no therapeutic effect *in vivo*. Basically, a chemical must be adequetely absorbed and distributed to the desired tissue. Also, the chemical must have sufficient resistance to metabolic inactivation by enzymes and not be too rapidly excreted. Similarly, the chemical cannot be too extensively bound to proteins or other constituents, which would render the compound inactivated. After gaining access and/or entry to the target cells, the active chemical species must have sufficient intrinsic activity to alter cellular processes that are essential for inducing an effect. Finally, the chemical must have minimal toxicity. Only after meeting these criteria can a chemical be considered for classification as a drug. All of these different processes are explained in this first section on "Principles."

ROUTES OF DRUG ADMINISTRATION
There are advantages and disadvantages to each of the different routes by which drugs can be administered. These are briefly described.

  A. Enteral (intestinal) routes

   1. **Oral**. When swallowed, a drug immediately passes through the esophagus to the stomach. Some molecularly small or acidic drugs can be absorbed from the stomach. However, most other drugs will not be appreciably absorbed from the stomach, for several reasons. First, the surface area of the stomach is relatively small vs. small intestine. Second, fluids quickly pass from the stomach through the pylorus to the small intestine. Third, drugs with a high pK_a (basic or alkaline drugs) will become ion-trapped in the stomach and cannot be appreciably absorbed.

   Therefore, most absorption of orally administered drugs occurs from the small intestine. After absorption, drugs enter the portal circulation and must pass through the liver before entering the general circulation. The liver is the principle organ involved in drug metabolism, containing a multitude of enzymes that biotransform (i.e., activate or inactivate) drugs. The term, *first pass metabolism*, indicates that a high percentage of drug is activated or inactivated the first time it passes through the liver after absorption from the GI tract. Some drugs with high first pass metabolism can still be given by an oral route if dosage adjustment is made. However, for others, a different route of administration is required.

   Among individuals the (a) rate and (b) total amount of drug absorption from the intestine is highly variable. This is a factor in the variability of response to a drug in a population (i.e., group of people).

   2. **Sublingual**. If a tablet or wafer is retained in the mouth under the tongue, substantial absorption can occur. An advantage of this route is that the drug does not pass through the liver before entering the arterial circulation for distribution throughout the body. Drugs administered by this route are characteristically associated with a rapid onset of action. Most anti-angina drugs are given by this route.

   The sublingual route is ideal for potent lipophilic drugs, or small peptides which would be rapidly inactivated by acid and pepsin in the stomach.

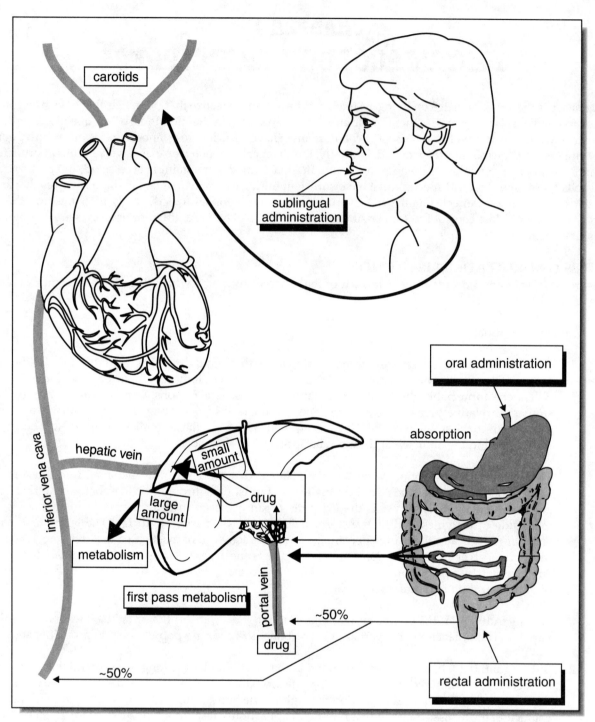

Figure 1.1

3. **Rectal**. Rectal suppositories are frequently used for drug administration, particularly in infants or elderly patients who are unable to swallow pills or capsules, or might spit out unpleasant-tasting medicines. A major disadvantage is that the suppository might be entrapped in feces in the rectum (impairing absorption), be defecated or exert a cathartic effect before adequate absorption occurs.

   About half of rectally-administered drugs are not subjected to first pass metabolism. Drugs that would be adversely affected by stomach acid or intestinal enzymes are sometimes given rectally. This route is especially important for administering *anti-emetic* drugs which would be eliminated in vomitus if taken orally.

B. Parenteral (not intestinal) routes

   1. **Intravenous (i.v.)**. This route directly introduces the drug into the systemic circulation. The pumping action of the heart quickly distributes the drug throughout the body. Characteristically, drugs administered by this route have a rapid onset.

      The i.v. route is especially valuable for drugs needed to stimulate a failing heart; or for terminating convulsions. The i.v. route is also preferred for comatose patients; for hospital patients in which drug effects are titrated; or when assurance is needed that all of the indicated dose has been administered.

      It is important that i.v. drugs be soluble in water solutions. Otherwise, particles of drug will produce pulmonary embolism. Drugs given i.v. are not subjected to first pass metabolism.

      Disadvantages of this route include (a) greater risk of infection including toxic shock, (b) abrupt onset of adverse effects and (c) inability to interrupt drug delivery when an adverse effect has occurred after a bolus injection. Extravasation of the drug is particularly significant if the drug is irritating.

   2. **Intramuscular (i.m.)**. The rich blood supply to muscle ensures rapid absorption of aqueous-based drugs. Drugs administered by this route are not subjected to first pass metabolism. The i.m. route is especially valuable for delivery of drugs that would be metabolized by pepsin in the stomach (e.g., insulin).

      To produce a prolonged drug action, i.m. injection of lipophilic drugs produces a depot which is gradually absorbed. Irritant drugs should not be given i.m.

   3. **Subcutaneous (s.c.)**. As per the i.m. route, drugs administered s.c. are rapidly absorbed and are not subjected to first pass metabolism. This route is important when neither oral nor i.v. routes are desired. Irritant drugs should not be given s.c.

   4. **Inhalation**. Particularly for delivery of drug to the bronchi or lung, inhalation therapy is an advantage. Many antiasthmatic drugs are administered by this route, ensuring a rapid bronchodilatory action.

      For volatile substances, like anesthetic agents, this route is preferred. The large surface area of the lung ensures rapid absorption of drug into the bloodstream. The onset of action is nearly equivalent to that of i.v. injection of drug.

   5. **Intranasal**. Nasal decongestants can be taken as nasal mists.

6. **Dermal.** Highly lipid-soluble drugs can be absorbed intact through the skin. Dermal patches have made this route popular. Nicotine patches are used for smokers who wish to terminate the habit. Scopolamine patches are used to prevent motion sickness.

   This route is also used for producing focal effects (e.g., benzocaine applied to an aching knee).

7. **Topical.** To produce effects on the body surface, the topical route is logical. Minoxidil is applied to the scalp to grow hair. Keratolytics are applied to warts to remove them. Antimicrobials are applied directly onto infected areas of the dermis. Alcohol sponge baths are used to remove body heat and reduce fever.

   Intraocular (eye drops) and intra-aural (ear drops) routes would constitute topical applications.

8. **Intravaginal.** Anti-fungal drugs are often administered by this route for vaginal fungal infections. This route is also used to deliver drugs used for dysmenorrhea, with the target site being the uterus.

9. **Other** specialized routes are used.

   Intra-articular. Drugs like corticosteroids can be given directly into inflamed joints.
   Subdural, subarachnoid, epidermal. For childbirth, local anesthetics may be administered by these routes.
   Intrathecal. For delivery of drugs to the central nervous system, this route may be used. This is becoming more popular for treating intractable cancer pain with morphine.
   Intracardiac. Epinephrine is sometimes injected directly into a stopped heart.
   Intra-arterial. For producing a focal effect while limiting systemic effects, this route may be used. Floxuridine is administered directly into the hepatic artery to kill colon cancer cells that have metastasized to the liver.

Problems with dosing. Patients do not always take medication at the recommended doses or at the scheduled times. This is likely to occur for medicines taken several times a day. For disorders in which there are few

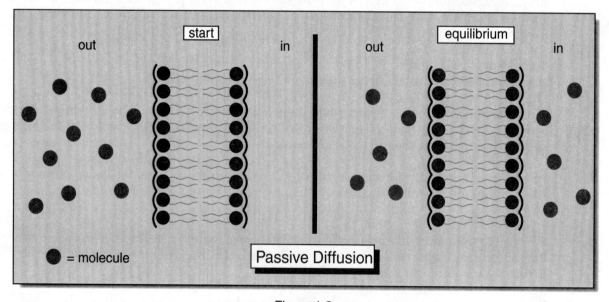

Figure 1.2

PHARMACOLOGY PRINCIPLES 11

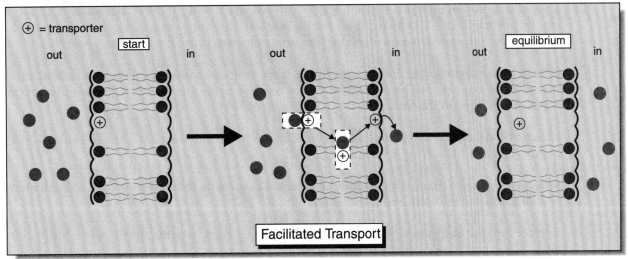

Figure 1.3

symptoms this problem is greatest. For example, a patient taking an antibiotic for a non-symptomatic infection simply forgets to take the medicine at the recommended times. Patients with ulcers, however, have pain as a reminder to take their medicine. *Compliance* is the term used to indicate the ability of patients to follow instructions for taking their medication. Compliance rates may be low, high or in-between.

Drug absorption. In order to understand drug absorption it is important to realize that molecules pass through membranes by at least 4 different processes:

1. **Passive diffusion.** Small molecules and lipid-soluble molecules simply diffuse through lipid-membranes to gain entry to cells. This process continues, in theory, until there are equal numbers of molecules on each side of the membrane. Therefore, molecules move down the concentration gradient, from higher to lower conc. Energy is not required.

2. **Carrier-mediated diffusion = facilitated transport.** Small molecules, lipid-soluble molecules or charged molecules bind to transporters in the membrane which carry the drug to the other side of the membrane and release it. There are many different types of transporters, but each type is selective for specific kinds of molecules. The transporters can be saturated, so this process may not be as rapid as passive diffusion. As for passive diffusion, molecules move down the conc. gradient. Energy is not required.

3. **Active transport.** Small molecules, lipid-soluble molecules or charged molecules bind to transporters in the membrane which carry the drug to the other side of the membrane and release it. These transporters are selective for specific molecules. The transporters can be saturated. However, drugs can move from a low conc. to a higher conc. Energy is required.

4. **Pinocytosis.** *Vesicles* form in some membranes and *engulf drugs* that are then transported to the inside surface membrane and released. Large molecules, like proteins can be incorporated into cells by this mechanism.

Drug absorption must occur by one or more of the above processes. For most drugs taken orally, passive diffusion is the ordinary means by which drugs are absorbed. Absorption rate is dependent on several processes:

1. **Rate of dissolution of tablet.** If a tablet is compressed so hard that it never breaks apart, little of it can be absorbed. If 1-2 hr is required for dissolution, the drug cannot be absorbed rapidly. If dissolution of the tablet is instantaneous, the drug is immediately available for absorption.

2. **Rate of blood perfusion of the absorptive site.** If the drug is absorbed from the small intestine, but the patient is jogging, then blood is redistributed away from the gut, to skeletal muscle. Absorption will likely be slow in this instance.

3. **Effect of food on the drug.** Some foods interfere with drug absorption. The Ca^{++} in milk binds tetracycline and impairs its absorption.

   When a drug is an irritant, it is generally advisable that the drug be taken simultaneous with meals, so that foods will partially buffer the drug.

4. **Surface area for absorption.** If a drug could only be absorbed from the stomach, it is likely that a large percentage would never be absorbed, since the stomach has a relatively small surface area.

5. **Time of contact with the absorptive surface.** Again, if a drug could only be absorbed from the stomach, there would be little time in which the drug could be absorbed if the stomach emptied rapidly.

   This principle is used when some poisons are taken by mouth. A common approach is to accelerate the transit of drug through the gastrointestinal (GI) tract with a cathartic, so that the drug exits through the anus as quickly as possible—thereby reducing the time available for absorption.

6. **Membrane permeability to the drug.** As indicated above, lipid-soluble drugs will readily pass through most biological membranes (these are lipid membranes). Charged drugs may not pass through the membrane to any appreciable extent.

7. **Effect of stomach acid on the drug.** If acid inactivates a drug (e.g., the anti-ulcer drug omeprazole), then virtually all of the drug would be inactivated prior to absorption. However, if this drug is stable in the lower intestine, then an enteric-coated prep (that dissolves only at alkaline pH) can be used to protect this kind of drug.

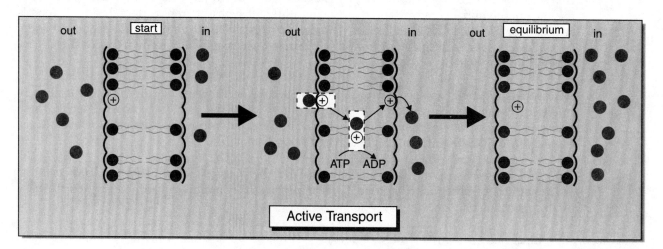

Figure 1.4

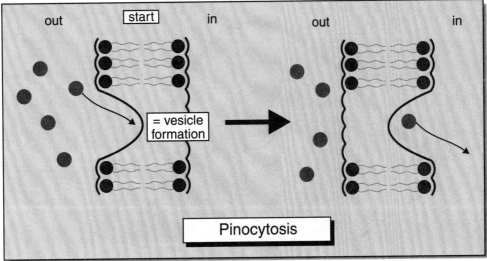

Figure 1.5

If unstable at alkaline pH, drugs would be inactivated upon entering the alkaline small intestine. Such could be expected if epinephrine were inadvertently taken orally.

8. **Effect of digestive enzymes on the drug.** If a drug is metabolically inactivated by digestive enzymes, then an oral route cannot be used. This is the case for insulin, a protein that would be digested by pepsin in the stomach.

9. **Effect of pH on ionization of the drug.** Parietal cells in the stomach release hydrochloric acid. If a drug is converted to an ionized form in the presence of acid, then such a drug would not be appreciably absorbed in the stomach. The basis for this process is discussed below.

    Absorption of acidic and basic drugs. Most drugs are slightly acidic ($pK_a \sim 4.5$) or basic ($pK_a \sim 8.4$). As noted above, the lipid-soluble form easily traverses membranes; charged, water-soluble, molecules do not easily cross membranes. Therefore, the pH of the surrounding fluid has a great influence on absorption of such drugs. This is illustrated below.

    Acidic drug ($pK_a = 4.5$): $HA \leftrightarrow H^+ + A^-$
    At pH 4.5, 50% of drug exists in undissociated form as HA.
    At pH 3.5 (more acidic), the reaction is driven leftward, and 90 parts per 100 (90%) exist as HA.
    At pH 1.5 (stomach pH), 999 parts per 1,000 (99.9%) is HA.
    At pH 7.4 (plasma pH), 1 part per 1,000 (0.1%) is HA.

    These values are from the *Henderson-Hasselbalch equation* (below). The amount of nonionized (undissociated drug, HA) is dependent on the value of the pK_a of the drug and pH of the surrounding fluid. Concentrations of each form of drug (nonionized, HA; ionized, A-) can be calculated by knowing the pH and pK_a. For example, at a pH of 1.5:

$$pH - pK_a = \log \frac{[base]}{[acid]}$$

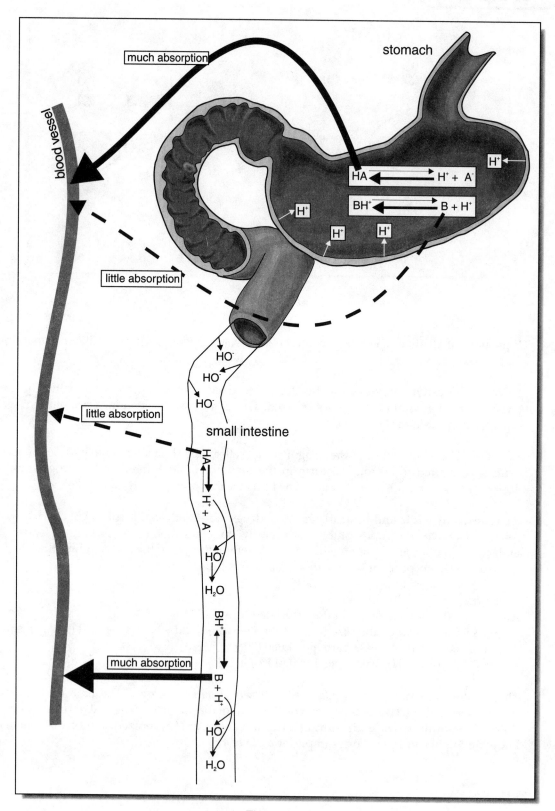

Figure 1.6

PHARMACOLOGY PRINCIPLES

$$1.5 - 4.5 = -3 = \log \frac{1}{1{,}000}$$ (Note: log of -10 = -1; log of -100 = -2; etc.)

Therefore, in the stomach the HA form of drug (undissociated form) would predominate. Accordingly, *an acidic drug would be well-absorbed from the stomach* **(Figure 1.6)**.

Basic drug ($pK_a = 8.4$): $BH^+ \leftrightarrow B + H^+$
At pH 8.4 (intestinal pH), 50% of drug exists in undissociated form as B. *A basic drug would be well-absorbed from the intestine* **(Figure 1.6)**.
At pH 7.4 (plasma pH), 1 part in 10 exists as B.
At pH 1.5 (stomach pH), only 1 part per 10,000,000 exists as B. This drug would not be well absorbed from the stomach **(Figure 1.6)**.

DRUG DISTRIBUTION

1. **Blood flow**. The amount of blood flowing to a tissue has a major influence on the amount of drug distributed to that tissue.

2. **Plasma binding**. The majority of drugs will at least partially bind to plasma albumin. This represents *an inactive form* of the drug. However, this bound form can dissociate, releasing the free (active) form of drug **(Figure 1.7)**. For these reasons, plasma bound drug is a *depot* or reservoir for drugs.

   Plasma-bound drugs compete with one another for the plasma binding sites. These drugs tend to displace one another, with more free (active) drugs being released, thereby producing excessive amounts of active drugs **(Figure 1.7)**. This interaction often produces drug toxicity.

   Some drugs are strongly bound in certain tissues. For example, heavy metal drugs, such as those containing platinum or gold, are bound to metallothionein protein in liver and kidneys.

   Highly *lipid-soluble drugs* tend to localize in tissues such as brain and adipose tissue which have a high lipid content.

3. **Drug redistribution**. Drugs which inicially locate in a tissue and then mobilize and relocate in another tissue are said to "redistribute." The action of rapid-acting *thiobarbiturate* anesthetics is terminated via redistribution to adipose tissue/muscle, etc. from brain. This process is more important than metabolism in terminating the action of such drugs.

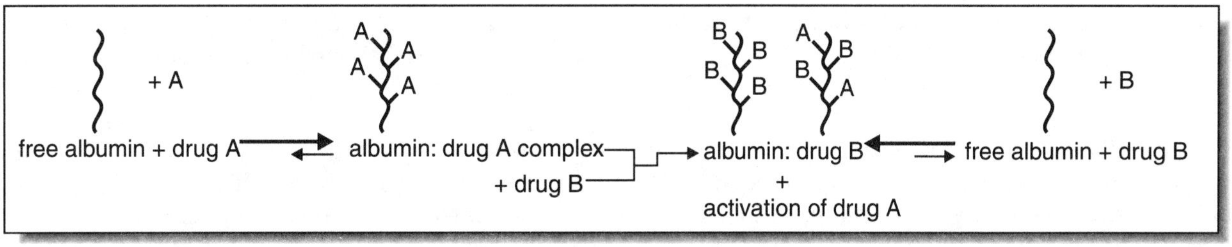

Figure 1.7

4. **Barriers.** There are specific biological barriers that tend to exclude a drug from certain tissues. The *blood-brain barrier* represents a series of tight junctions between endothelial cells lining the capillary network in the brain. The enzymatic make-up of these cells also constitutes part of the barrier. The purpose is to exclude certain natural and many unnatural substances from the brain. Epinephrine, for example, cannot cross the barrier intact—it would be excluded by tight junctions; any that would permeate the barrier would be degraded by enzymes. Highly lipid-soluble drugs are not excluded by the blood-brain barrier; polar drugs are excluded.

Another barrier, much like the blood-brain barrier in its capacity to exclude drugs, is the *placental barrier*. These two barriers are the most restrictive to drug distribution. There are several minor barriers in the body and these are described later.

DRUG DEPOSITION

Some drugs are sequestered in specific tissues for long periods of time. *Cisplatin*, an anti-neoplastic drug, binds to metallothionein in liver and remains at that site for months. *Gold compounds*, used for treatment of rheumatoid arthritis, are found in joints for months. Lead becomes sequestered and inactivated in bone. DDT is stored for years in adipose tissue. These depots can be beneficial or hazardous.

DRUG METABOLISM

The liver has the major role in metabolism of drugs, employing a group of similar enzymes largely present as *smooth endoplasmic reticulum*. The kidney and other tissues also have such enzymes, but to a much lesser extent. These enzymes, when combined with cyanide, absorb UV light at a wavelength of about 450 nm. Consequently, this group of isozymes is classified as cytochrome P_{450} enzymes (cyt P_{450}).

Some drugs like *barbiturates* have the ability to induce these enzymes. After repeated administration of barbiturates, the rate of barbiturate metabolism may be increased approximately twofold. These induced enzymes will also more rapidly metabolize many other drugs which have no chemical similarity to barbiturates. It is important to be aware of which drugs potently induce cyt P_{450} enzymes, so that proper dosage adjustments may be made for (a) long-term treatment and (b) when other drugs are also given to the patient.

Some drugs compete for metabolism. The consequence is a reduced rate of biotransformation of one or both drugs, with a net change in the peak concentration and duration of action of one or both drugs. This effect is observed when the antianxiety drug diazepam and the anti-ulcer drug cimetidine are co-administered.

Drug dosage reductions must be made when administering drugs to patients with liver impairment (less enzymes). Geriatric patients also have a reduced ability to metabolize drugs, so dosage adjustments might have to be made in these patients as well.

The different mechanisms involved in drug metabolism are described in the chapter on drug metabolism.

Almost universally, the process of metabolism produces substances that are more water-soluble and therefore, more readily excreted by the kidneys into urine. The metabolites may be more active, equally active or less active than the parent molecule. The multitude of chemical reactions involved in the metabolism or biotransformation of drugs is illustrated later in this chapter.

Bioavailability is a measure of the drug that is available to the systemic circulation after administration. This value is a measure of both absorption and immediate metabolism. For example, if a drug is 100% absorbed and not metabolized, 100% of the drug will reach the systemic circulation intact, so bioavailability is 100%.

If a drug is 50% absorbed and not metabolized, bioavailability is 50%.

PHARMACOLOGY PRINCIPLES

If a drug is 100% absorbed but first-pass metabolism is 50%, bioavailability is 50%.

RELATIONSHIP BETWEEN BIOAVAILABILITY AND EFFECT

Despite this apparently clear relationship, drug companies conduct detailed pharmacokinetic studies to determine the optimal dosing of a drug. In the example below, the same amount of drug is administered i.v. or orally. If the minimum effective concentration for a drug effect is represented by line A (i.e., this is the minimal concentration of drug that produces an action), then the slowly absorbed drug will act longer than the i.v. or rapidly-absorbed drug. If the threshold is at line B, then all 3 drugs have about the same duration of action. However, if line C represents the plasma conc. of drug that produces adverse effects, then only the slowly-absorbed drug avoids this problem **(Figure 1.8)**.

DRUG EXCRETION

Some drugs are excreted unchanged in

1. **Lung**. Anesthetic gases are exhaled largely in unchanged form. This is the major route for excreting gaseous anesthetics, but a minor route for most other drugs.

2. **Kidney**. Most drugs are excreted primarily by the kidney into urine. Parent drug as well as metabolites will be excreted by this route. A few drugs are almost exclusively eliminated unchanged via this route (e.g., lithium, mannitol).

Most drugs have a relatively low molecular weight and easily pass through the capillary endothelium in the glomerulus **(Figure 1.9)**. Only free, non-bound drug will be so filtered. Drugs may be subsequently reabsorbed in the nephron.

Renal elimination of acidic drugs can be hastened by alkalinizing the urine with sodium bicarbonate ($NaHCO_3$), which has the effect of producing the ionized form of the drug in the lumen of the nephron. Being ionized, the drug is unable to pass through membranes and cannot be reabsorbed **(Figure 1.10)**. Similarly, renal elimination of alkaline drugs can be hastened by acidifying the urine with ammonium chloride (NH_4Cl) **(Figure 1.10)**.

Some drugs are secreted into the nephron by *carrier-mediated transport* processes. These sites are located on the plasma side of the nephron. This is the major route of elimination of penicillin. By co-administering probenecid, a drug that is excreted via the same carrier transport system, the rate of

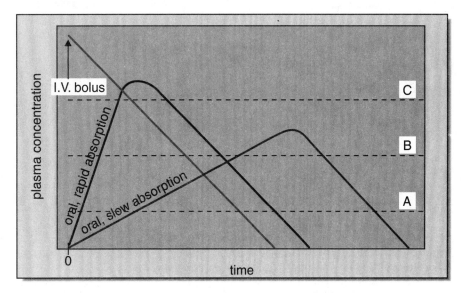

Figure 1.8

penicillin excretion can be slowed **(Figure 1.9)**. Similar competition for these carrier-mediated sites is an important consideration for elimination of many other drugs.

Carrier-mediated transport sites are also located on the luminal surface of the nephron. These sites are important for reabsorption of many substances like glucose. Some drugs can also bind and compete for natural substrates at these sites. Uricosurics are anti-gout drugs which act by binding to these carrier sites and thereby blocking the reabsorption of uric acid **(Figure 1.9)**. Virtually all drugs that are weak acids or are metabolized to weak acids, interfere with uric acid secretion and tend to produce hyperuricemia and gout. Children, up to the age of about 6 years, have a limited capacity for secreting drugs into the nephron by carrier-mediated transport sites.

Lithium is a drug that is reabsorbed from the kidneys through sodium (Na^+) channels. By administering excess Na^+, the rate of lithium excretion can be increased.

Because of the major role that the kidney has on drug elimination, it is important to realize that drug dosage reductions are often required when there is renal impairment. Otherwise, the reduced rate of excretion will result in higher plasma levels of drugs and an increased incidence of toxicity.

3. **Feces**. Anthelmintics are intended to act only within the gut. Consequently, they are excreted almost exclusively by this route.

Most drugs are partially excreted in feces, either because they were not absorbed or because they gain entry through the liver to the bile and are then secreted with bile into the duodenum.

An important process known as *enterohepatic circulation* must be taken into account for many drugs which (1) are absorbed from the GI tract, (2) excreted intact or partly metabolized by the liver into bile, (3) secreted with bile into the GI tract, reabsorbed, etc.

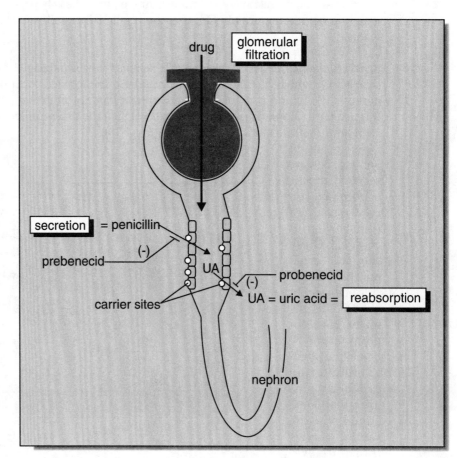

Figure 1.9

(Figure 1.11). This cycling of the drug through the GI tract and liver generally prolongs the action of drugs for days. This is observed with the laxative phenolphthalein. Many drugs that are metabolized to glucuronides in the liver are excreted into bile, then secreted into the gut where bacterial β-glucuronidase cleaves the glucuronide, producing free drug which is more lipophilic and more readily absorbed (Figure 1.11).

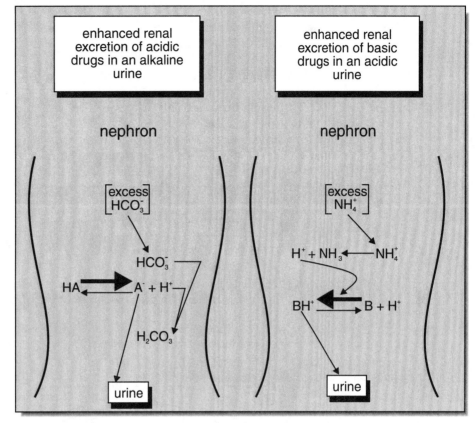

Figure 1.10

4. **Milk**. Although this is a minor route for drug elimination, it becomes of significance to mothers who nurse infants. Many drugs with high lipid solubility and relatively long half lives are likely to be excreted by this route. This includes antidepressants and antipsychotics. Also, alkaline drugs tend to become concentrated in milk (pH is 6.5) versus plasma (pH is 7.4), since more of the drug would be ionized and "ion trapped" in milk.

5. **Blood (menstrual fluid), sweat and tears**. Although trace amounts of drugs are found in these body fluids, drug elimination via these fluids is insignificant.

PHARMACOKINETICS

The previous sections provide general principles, descriptions of those factors that are relevant to drugs gaining entry into the body and persisting for a sufficiently long enough time to exert an effect. The following sections intended to mathematically apply some of the information on drug distribution (pharmacokinetics) and drug actions (pharmacodynamics).

A. Pharmacokinetic Factors

Pharmacokinetic calculations are based primarily on three factors, known as the volume of distribution (V_d), clearance (CL) and half-life ($t_{1/2}$) or β ($0.693/t_{1/2}$) of a drug.

1. **Volume of Distribution (V_d).** This is a hypothetical value of the volume it would take to totally dissolve an administered drug, while attaining the concentration of drug in blood.

$$V_d = \frac{\text{total amount of drug administered}}{\text{plasma drug concentration}}$$

   a. Example 1, for a drug restricted to the vascular compartment. If all of an i.v. drug stays within the vascular system, its V_d = 5 liters (i.e., total blood volume). It is important to note that the V_d may be expressed on the basis of (a) plasma volume or (b) liters (L) per kg. Mannitol is an example of a drug restricted to the vascular compartment.

$$V_d = \frac{200 \text{ mg (test dose)}}{40 \text{ mg/liter of plasma}} = 5 \text{ L}$$

   b. Example 2, for a drug that distributes to other fluids. When some drug distributes to other fluids (e.g., interstitial fluid), less remains in plasma. An example of this is aspirin.

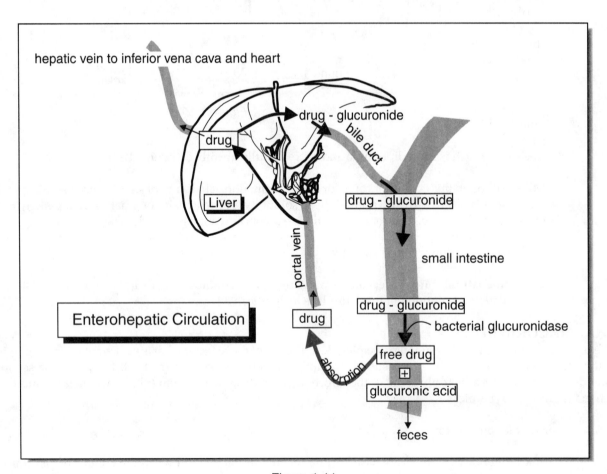

Figure 1.11

PHARMACOLOGY PRINCIPLES

$$V_d = \frac{650 \text{ mg (analgesic dose)}}{60 \text{ mg/liter of plasma}} = 11 \text{ L}$$

c. Example 3, for a drug that distributes to total body water (intra- and extracellular). When some drug distributes to fluids and tissues, less remains in plasma. An example is ethanol.

$$V_d = \frac{10 \text{ g (1 drink)}}{200 \text{ mg/liter of plasma}} = 50 \text{ L}$$

d. Example 4, for a drug that is bound by protein. Many drugs are bound to proteins in plasma and elsewhere. An example is the antidepressant, fluoxetine (Prozac).

$$V_d = \frac{50 \text{ mg}}{20 \text{ ug/liter of plasma}} = 2500 \text{ L}$$

There are standard tables with values for V_d. A value of
  5-10 L indicates distribution primarily in plasma or within the vascular system;
  10-20 L indicates distribution in extracellular fluid;
  20-50 L indicates distribution in total body water and
  >50 L indicates drug binding to tissues.

Many different factors can alter the V_d, including sex, age and disease.

2. **Clearance (CL).** This is a hypothetical value of the volume of blood (CL_B), plasma (CL_P) or other fluid that can be cleared of a drug per unit time. For a drug that is primarily metabolized by the liver, the CL is usually related to the amount of blood that can be delivered to the liver per min. For a drug that is primarily excreted unchanged by the kidney, the CL is usually related to the amount of blood that can be delivered to the kidney per minute.

For a drug that is metabolized and/or excreted by several tissues, CL_{Total} is the sum of CL_{Liver}, CL_{Kidney}, $CL_{Other\ tissues}$.

$$CL_{Total} = CL_{Liver} + CL_{Kidney} + CL_{Other}$$

3. **Half-life ($t_{1/2}$).** This represents the time needed to eliminate half the drug.

B. Quantitative Pharmacokinetics

The above statements related to V_d, CL and $t_{1/2}$ are used in the following equations to estimate the optimal dosing of drugs for individual patients. In any curve, as per the one shown in **Figure 1.12**, there are definite statements that can be made on how the drug is being handled by a patient. If you were to have a curve similar to the one in **Figure 1.12** you could predict such things as (a) the plasma level of drug that will be achieved with the next dose(s) of drug, (b) how often the drug should be given to this patient, (c) how quickly this patient absorbs the drug and (d) excretes it.

1. **For a drug administered as an intravascular bolus.** The drug in **Figure 1.12** is assumed to distribute evenly throughout the body. This is a one compartment model. The concentration of drug in plasma (C_p) is measured at 2 points (•).

The time of drug administration is recorded as t_0 (Time Zero) and only 2 blood determinations are made, at t_1 (C_1) and t_2 (C_2). From these data it is possible to determine, by extrapolation, the drug concentration (C_p^o) at time zero, the elimination rate constant

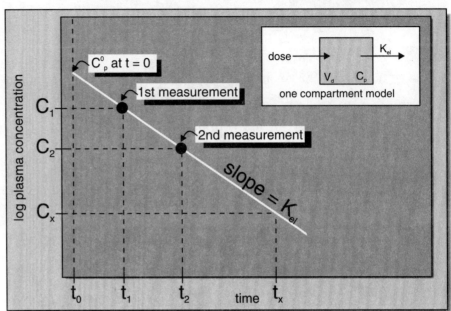

Figure 1.12

(slope = K_{el}) and the half life ($t_{½}$) of the drug, as well as the concentration of drug at any time. The following equations are used.

$$K_{el} = slope = \frac{\ln C_1 - \ln C_2}{t_2 - t_1} \qquad t_{½} = \frac{\ln 2}{K_{el}} = \frac{0.693}{K_{el}}$$

The figure can be used to estimate the concentration of drug at any point in time, simply by drawing intercepts on the figure as shown. At time x (t_x), the concentration of drug will be C_x. The equation below can be used instead:

$\ln C_x = \ln C^o - kt_x$ where C^o is the conc. of drug at t = 0.

The V_d, CL and $t_{½}$ are related by the following equations:

$$V_d = \frac{Dose}{C_p^o} \qquad CL = V_d \times K_{el} \qquad t_{½} = \frac{V_d \times 0.693}{CL}$$

Generally, if the V_d is large, the $t_{½}$ tends to be long since the drug is largely distributed to extravascular sites, and not easily available for hepatic metabolism. These interrelations afford each physician the means to learn much about the way drugs are handled by his/her patient, while using as few as two sampling times.

2. **For a drug administered extravascularly.** The drug in the **Figure 1.13** is assumed to distribute evenly throughout the body (one compartment) and the concentration of drug in plasma (C_p) is measured at several points. It is more difficult to predict plasma concs. (C_p) of drug by simple extrapolation of the line in **Figure 13**. Note that there are two sample times (•) during absorption (rise of curve,

starting at t_1) and two sample times during elimination (decline in curve from maximal plasma concentration (C_{max}). The C_{max} is attained when there is as much drug entering the body as leaving it.

C_p^o is obtained by extrapolation of the K_{el} to t_0 or with an equation.

In order to determine the absorption rate constant (K_{abs}), it is necessary to (1) back-extrapolate the K_{el} to t_0, then (2) subtract the points on the bottom curve from the top extrapolated line. The plasma concentrations (C_p) at t_1 and t_2 represent the C_1' and C_2' on the new line with its slope being K_{abs}. The following equations can be used:

$$K_{el} = \text{slope} = \frac{\ln C_3 - \ln C_4}{t_4 - t_3} \qquad K_{abs} = \frac{\ln C_1' - \ln C_2'}{t_2 - t_1}$$

Sometimes it is difficult to sample at the best times. However, it is possible to calculate the area under the concentration-time curve (bottom curve) and use this value in the equations. The area under the curve (AUC) is an estimate of the total amount of drug that was in the body. AUC is related to V_d and CL by the equation:

$$AUC = \frac{\text{Dose}}{CL} = \frac{\text{Dose}}{V_d \times K_{el}}$$

For a drug administered orally, the 'dose' is actually the amount of drug that is bioavailable = Dose x f, where f is the fraction absorbed and present after first-pass metabolism.
The V_d can be obtained from the standard tables, thereby simplifying equation solving. The V_d, K_{el} and $t_{1/2}$ are able to be determined for each patient. By knowing the AUC the calculations are valid, even for more complicated models that will be described.

3. **Two-compartment model.** In this situation a drug distributes to another compartment (e.g., intracellular fluid). This analysis requires added sampling times and becomes more difficult mathematically, but follows the same principles as above.

In **Figure 1.14** the drug is administered as an i.v. bolus. Blood samples are taken at 4 times. After back-extrapolating the second part of the curve, the hypothetical

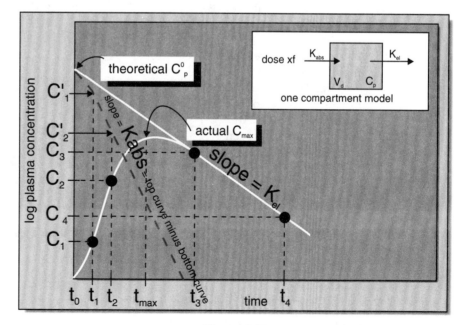

Figure 1.13

plasma concentration at t_0 is obtained (B). The slope of this line is β, the elimination constant. When the extrapolated line is subtracted from the top line, the rate of distribution of drug to the second compartment is obtained, as well as the hypothetical concentration of drug in this compartment at t_0 (A). The slope of the new line is α, the distribution constant. C_1 and C_2 must be converted to C_1' and C_2' on this new line. The following equations can be used:

$$C_P^0 = A + B \qquad \beta = \frac{\ln C_3 - \ln C_4}{t_4 - t_3} \qquad \alpha = \frac{\ln C_1' - \ln C_2'}{t_2 - t_1} \qquad t_{1/2} = \frac{0.693}{\beta}$$

V_1 and V_2 are the volumes of the respective vascular and peripheral compartments and are related by the equations:

$$V_d = \frac{\text{Dose}}{\text{AUC} \times \beta} = \frac{\text{Dose}}{\text{AUC} \times K_{el}} = V_1 + V_2 \qquad V_1 = \frac{\text{Dose}}{C_P^0}$$

4. Pharmacokinetic calculations can also be used to determine multi-dose schedules for drugs that achieve therapeutic concentrations only after several cumulative doses have been given. Although it might seem that a multiple dose (loading dose) should simply be given initially, there is the attendant risk that high drug concentrations may produce adverse effects.

   In **Figure 1.15**, the cumulative effects of multi-doses of a drug administered i.v. are shown. There are several rules of thumb for regularly-timed drug administrations:

   a. From the first drug dose it takes 5 half lives ($5t_{1/2}$) to achieve steady state.
   b. After the last dose it takes $5t_{1/2}$ to fully eliminate a drug.
   c. When the dosing interval (τ) is equal to the $t_{1/2}$, then the loading dose (L) should be twice the maintenance dose.
   d. The dosing rate should be equivalent to the clearance rate after steady state (ss) has been attained: Dosing rate = $CL \times C_{ss}$.

   Mean plasma concentration =

   $$\overline{C}_p = \frac{\text{Dose}}{V_d \times K_{el} \times \tau}$$

C. Factors Affecting Pharmacokinetics of Drugs

   Many factors affect the way that the body will absorb,

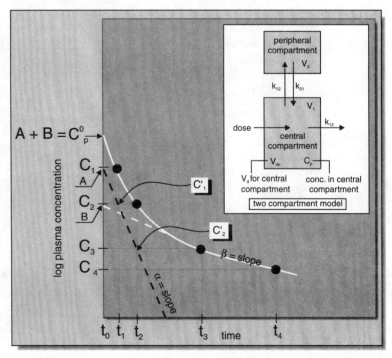

Figure 1.14

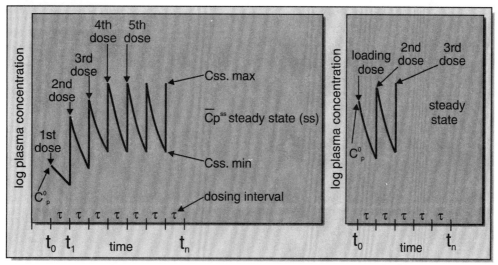

Figure 1.15

distribute, metabolize and excrete drugs. All influence the pharmacokinetic disposition—this is illustrated in **Figure 1.16**.

PHARMACODYNAMICS

The previous section on pharmacokinetics illustrates predictive methods a physician uses to estimate the plasma level of drugs and to determine how each patient absorbs and eliminates particular drugs. The present section discusses how a drug produces an effect after reaching the target site of action.

Drugs exert effects through one of several possible mechanisms:

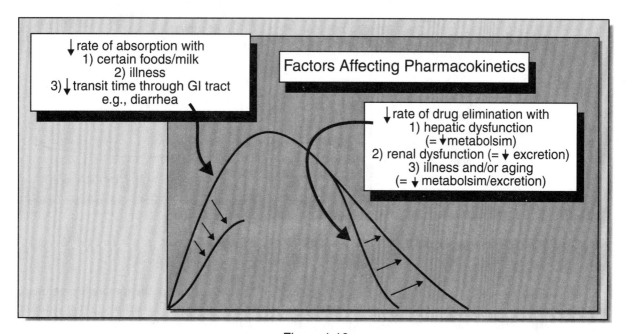

Figure 1.16

1. **Non-specific physical interactions**.
   a. By increasing the osmolality of blood, osmotic diuretics draw water from extracellular spaces, reduce edema and produce diuresis.

   b. Part of the action of anesthetics may be related to non-specific disordering of neuronal membranes.

2. **Non-specific chemical interactions**.
   a. Antacids exert an antiulcer action by chemically neutralizing stomach acid.

   b. By binding heavy metals, chelators inactivate them.

3. **Specific binding to cell receptors**.
   a. This is the major action of drugs. Most drugs either enhance or inhibit normal activities of tissue cells in the body, by acting on specific binding sites (receptors) on the outer cell membrane or inside the cell. A series of intracellular actions occurs as a consequence of such binding.

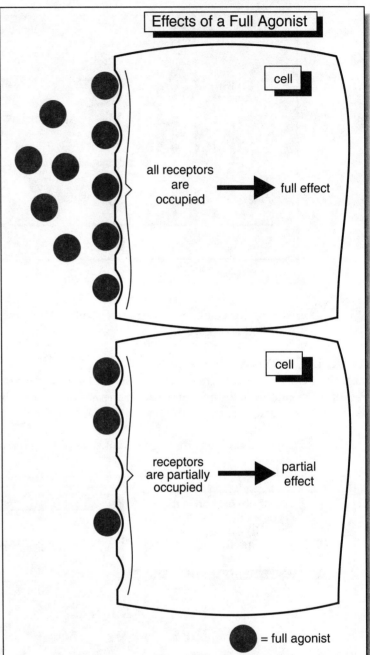

Figure 1.17

SPECIFIC TERMS

1. *Receptors* are protein binding sites that are linked to cellular biochemical processes that *initiate a response*.

2. *Agonists* are drugs that bind to specific receptors. Full agonists *produce* the maximal possible *effect* when all receptors are occupied (full receptor occupancy) **(Figure 1.17)**. Partial agonists produce a partial effect with full receptor occupancy **(Figure 1.18)**. (For simplicity, the concept of spare

receptors is not included in the discussion. This is based on the fact that some receptors are extra; when these spare receptors are inactivated, agonists can still elicit a full response.)

3. *Intrinsic activity* refers to the inherent ability of a drug to elicit a response. Full agonists have greater intrinsic activity than partial agonists. This property is related to the ability of a drug to interact with the receptor in a way that induces the necessary cellular changes.

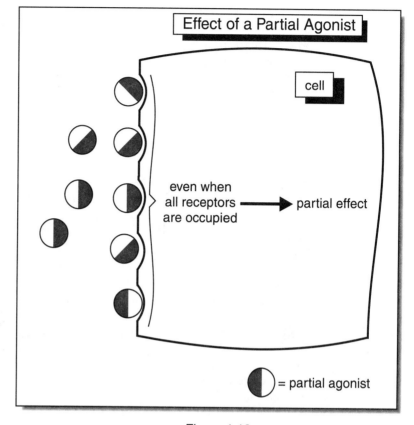

Figure 1.18

The drug with the highest intrinsic activity is not necessarily the drug with the lowest effective dose. Intrinsic activity is determined on an isolated tissue *in vitro*. In this situation the drug has no difficulty in reaching its receptor. However, "dose" relates to whole body *in vivo* systems. In this situation the drug with high intrinsic activity may be poorly absorbed or rapidly metabolized and rapidly excreted, or it may have difficulty in reaching the target site (e.g., unable to cross barriers). All of these pharmacokinetic factors must be considered.

4. *Antagonists* are drugs that oppose the action of agonists:

   a. *Physiological antagonists* are drugs that produce biological effects opposite to the agonist. For example, histamine reduces blood pressure, while norepinephrine elevates blood pressure. Therefore, norepinephrine is a physiological antagonist of histamine. A physiological antagonist acts at a different kind of receptor than the agonist.

   b. *Chemical antagonists* are those that inactivate the agonist. For example the chelator penicillamine will bind the gold in gold-containing drugs and inactivate the drug.

   c. *Receptor antagonists* are those drugs able to occupy the receptor, yet produce *no effect* **(Figure 1.19)**. An antagonist has no intrinsic activity. Antagonists are used in medicine to block the agonist binding site **(Figure 1.20)**. Receptor antagonists represent the most common type of pharmacological antagonist.

   Competitive antagonists are displaced by high concentrations of agonists. This type of antagonism is associated with weak hydrogen bonding. Non-competitive antagonists are not displaced by high concentrations of agonists. This type of antagonism may represent covalent bonding to the receptor.

d. *Inverse agonists* are drugs that occupy the receptor and produce effects *opposite* to that of agonists **(Figure 1.21)**. There are only a few systems in which this occurs. Diazepam is a drug that has anti-anxiety effects. Some drugs bind to the same receptor to actually produce anxiety.

5. Effects of agonists and antagonists on receptors: Dose-effect relationship. Drugs that act on receptors produce dose-dependent effects. This may be expressed as a graded dose response curve or a quantal dose response curve.

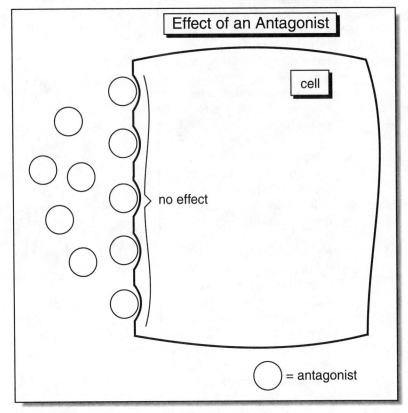

Figure 1.19

a. **Graded dose-response curve**. By gradually increasing the dose of norepinephrine (NE), an increasingly greater elevation in blood pressure (BP) is produced until the maximal response is attained **(Figure 1.22)**. Because of the difficulty in working with a curve that is a hyperbola, a semi-log plot is usually used, with the dose given in log units **(Figure 1.23)**. Note how this transformation converts most of the curve to a straight line. A graded dose-effect curve could be obtained from a single subject. The dose at which 50% of the maximum response is attained is known as the ED_{50}.

b. **Quantal dose-response curve**. By gradually increasing the dose of a sedative, more people will fall asleep. In this example, people are either asleep or not asleep. Basically, the effect is 'all or none'. Another example would be the effect of an insecticide (toxin) in producing a lethal effect on pests: bugs are either dead or not dead. You could not construct this curve with one subject; many subjects are needed. You can see from **Figure 1.24** that a quantal dose-response curve looks the same as a graded dose-response curve. However, the ED_{50} indicates the dose at which *50% of subjects respond*.

HOW CAN DOSE-RESPONSE CURVES BE USED TO COMPARE DRUGS?

The ED_{50} is a convenient way for comparing the potency of drugs. *Potency* is an indicator of the sensitivity of a system to a drug (i.e., affinity of receptor for agonist). In **Figure 1.25**, drug A is more potent than drug B, as indicated by the lower ED_{50} value (i.e., a lower dose of drug A produces a 50% response).

Efficacy is an indicator of the "power" of a drug, as reflected by the maximal attainable response. In **Figure 1.25**, drug B has greater efficacy than drug A.

The response to a partial agonist, as stated above, is not as great as that of a full agonist **(Figure 1.26)**. In this example, the full and partial agonists have similar potency. In reality, the partial agonist may be more potent or less potent than the full agonist.

An antagonist will attenuate the response of an agonist. The effect of a *competitive antagonist* is characterized by a mere "shift to the right" in the dose-response curve **(Figure 1.27)**. The maximal response does not change, but the ED_{50} is higher. In other words, if more agonist is added, the antagonist will be displaced from the receptor so that all receptors can be occupied by the agonist and elicit a full response.

The effect of a *non-competitive antagonist* is characterized by a reduction in the maximal response but no change in the ED_{50} **(Figure 1.28)**. In essence, some of the receptors are permanently inactivated by the antagonist, and no amount of agonist can displace this (reduced maximal effect). At the same time those receptors not occupied by the non-competitive antagonist have the same affinity for the agonist (no change in ED_{50}). Since there cannot be full receptor occupancy, a full effect cannot be obtained.

THERAPEUTIC DOSES

The therapeutic dose of a drug may fall almost anywhere on the dose-effect curve, according to the safety of the drug, health status of the patient and desired effect. The potency and safety of a drug have a lot to do with the therapeutic dose. Some drugs are so safe that toxicity is not produced by even a high dose (e.g., penicillin). In this case other factors determine the dose. Some drugs are so toxic that even small overdoses can produce severe adverse effects (e.g., digoxin). In this case safety largely determines the therapeutic dose.

DRUG SAFETY; THERAPEUTIC INDEX (TI)

$$TI = \frac{LD_{50}}{ED_{50}}$$

The TI is one measure of a drug's safety, being a ratio of the amount of drug needed to kill vs. the amount of drug needed to cure 50% of subjects. The higher the value, the safer the drug. Actually, since this value is determined with animal data, the TI may not be the most appropriate index of human safety data for a drug. There are many other safety indexes.

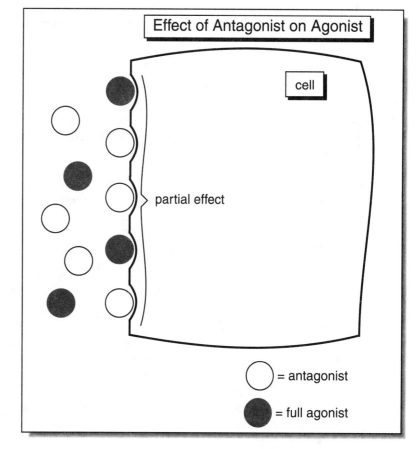

Figure 1.20

30 PHARMACOLOGY

HOW DOES AGONIST BINDING AT RECEPTORS PRODUCE AN EFFECT?

The interaction of agonist with a receptor is associated with one of several kinds of changes, depending the coupling of the receptor with other processes in the cell.

A. Some receptors change *ion permeability*, which can alter (a) resting membrane potential, the threshold for nerve excitation or the calcium ion conc. in cells. Any of these changes influence cell activity.

B. Some receptors are coupled to *second messenger systems* (e.g., cyclic adenosine-3',5'-adenosine monophosphate = cAMP; inositol triphosphate, diacyl glycerol, etc.) that activate cellular processes, resulting in an effect.

C. Some intracellular receptors influence *gene expression* (e.g., glucocorticoids influence transcription).

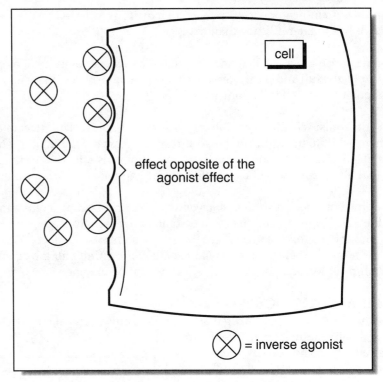

Figure 1.21

VARIABILITY IN DRUG RESPONSES

1. *Tolerance* represents the development of resistance to drug effects. For example, if nicotine is given repeatedly, its interaction with the receptor eventually ceases to induce some cellular effects.

2. *Tachyphylaxis* represents the rapid development of resistance to drug effects. For example, tyramine acts by releasing cytoplasmic stores of norepinephrine in sympathetic nerves. If given repeatedly i.v. at short intervals (minutes), it quickly depletes norepinephrine, so that no effect can be elicited.

3. Idiosyncratic reaction is an abnormally exaggerated response to a drug. This might be related to genetic differences in enzymes or other proteins with which the drug interacts. For example, a small dose of the mus-

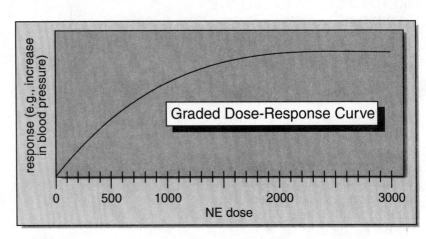

Figure 1.22

cle relaxant, succinylcholine, produces excess muscle relaxation in people that lack enzymes for inactivating the drug.

4. A drug allergic response is one produced in allergic individuals. An example is the IgE-mediated anaphylactic shock (marked fall in BP, bronchoconstriction and difficulty in breathing) produced by penicillin in an allergic person. This response could also be classified as an idiosyncratic reaction or as a *hypersensitivity reaction*.

DRUG INTERACTIONS

Drugs act at many sites in the body. For this reason it is not unusual that many different drugs interact with one another. Agonist-antagonist interactions were described above.

One drug may inhibit or enhance the action of other drugs according to how processes such as absorption, distribution, metabolism, excretion, etc. are altered.

Some drugs also increase the effect of others. *Potentiation* is an abnormally large response, sometimes produced when 2 or more drugs are co-administered. An example is the total cancer cell kill that is more easily achieved when anti-metabolites are used in conjunction with alkylating antineoplastic agents. Usually drugs that act by different cellular mechanisms are more likely to produce potentiation. Conversely, drugs acting at the same receptor or by a similar mechanism are more likely to produce an *additive effect*.

DRUG METABOLISM

The body enzymatically biotransforms drugs to more-polar water-soluble compounds, attacking the most labile groups on the drug molecule. This is described below.

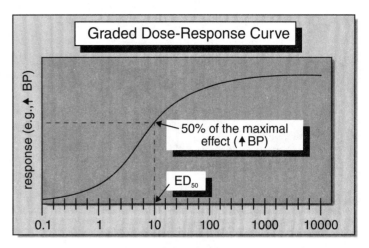

Figure 1.23

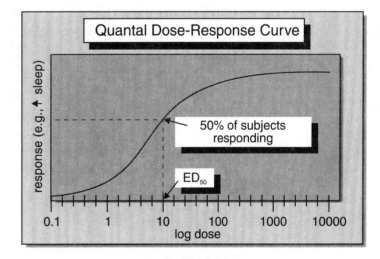

Figure 1.24

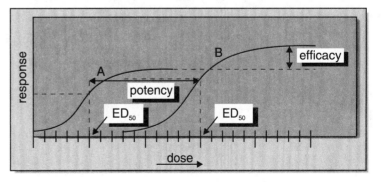

Figure 1.25

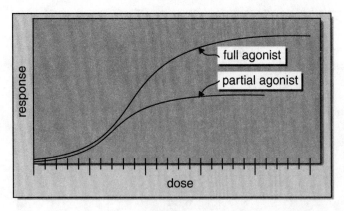

Figure 1.26

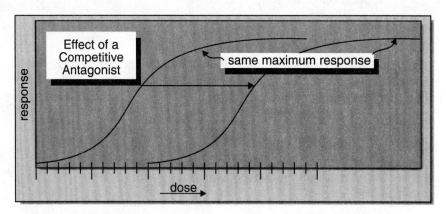

Figure 1.27

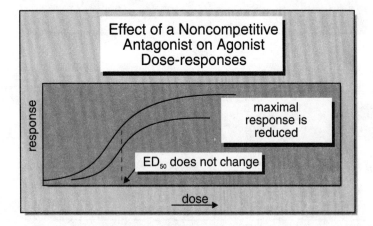

Figure 1.28

1. Tissues involved in drug metabolism are: Liver, Lung, Kidneys and Other, in that order.

2. Outcome of biotransformation: increased water solubility.
   Activation: An inactive substance (prodrug) is converted to an active substance (drug).
   Inactivation: An active drug is converted to a less active drug or inactive drug.

3. Reactions scheme:

   drug →(phase 1 reaction)→ polar metabolite (active or Inactive) →(phase 2 reaction)→ conjugate (inactive)

   drug →(phase 2 reaction)→ conjugate (inactive)

4. Types o-*f phase 1 reactions: oxidation
                                    reduction
                                    hydrolysis

A. Phase 1 Oxidation Reactions

These are mediated by cytochrome P_{450} enzymes that are located on microsomes on the smooth endoplasmic reticulum (SER). Cytochrome P_{450} refers to the class of roughly 2 dozen isozymes that have high affinity for carbon monoxide, forming a complex that optimally absorbs light with a wavelength of about 450 nm. In oxidation reactions, oxygen is added to the drug molecule.

$$\text{drug } H_2 + O_2 \xrightarrow[Mg^{++}]{\text{cyt } P_{450},\ NADPH \to NADP} \text{drug } OH + H_2O$$

1. Attack on drug carbon atoms:

   a. Aliphatic Oxidation: $RCH_3 \longrightarrow RCH_2OH$

   b. Aromatic Oxidation: Ar–R → [epoxide intermediate] → R–Ar–OH

   epoxide

The short-lived epoxide is toxic and/or carcinogenic.

2. Attack on drug nitrogen atoms:

   a. N-Oxidation: $RNH_2 \longrightarrow RN=O$

   b. N-Dealkylation: $RR'NCH_3 \longrightarrow RR'NH + HCHO$

   c. Deamination: $RCH_3NH_2 \longrightarrow RCHO + NH_3$

3. Attack on drug oxygen atoms:

   a. O-Dealkylation: $ROCH_3 \longrightarrow ROH + HCHO$

4. Attack on drug sulfur atoms:

   a. Sulfoxidation: $RR'S \longrightarrow RR'S=O$

   b. S-Dealkylation: $RSCH_3 \longrightarrow RSH + HCHO$

   c. Desulfuration: $RSH \longrightarrow ROH$

5. Attack on drug halogen:

   a. Dehalogenation: $RR'CHX \longrightarrow [RR'CHX] \longrightarrow RR'CHO + HX$

B. Phase 1 Reduction Reactions

   1. Carbonyl reduction: $\underset{RCR}{\overset{O}{\|}} \longrightarrow \underset{RCHR}{\overset{OH}{|}}$

   2. Azo reduction: $RN=NR' \longrightarrow RNH_2 + R'NH_2$

   3. Nitro reduction: $RNO_2 \longrightarrow RNH_2$

C. Phase 1 Hydrolysis Reactions

   1. Esters: $RCO\text{-}OR' \xrightarrow{\text{esterase}} RCOOH + R'OH$

   2. Amides: $RCO\text{-}NHR' \xrightarrow{\text{amidase}} RCOOH + R'NH_2$

   (with $H^+ + HO^-$)

Hydrolysis occurs mainly via soluble enzymes in blood. Some ethnic groups are genetically predisposed to low plasma cholinesterase (ChE) levels, such that the actions of drugs normally metabolized by ChE would be potentiated. The muscle relaxant, succinylcholine, is one such drug.

ENZYME INDUCTION AND INHIBITION

New synthesis of additional cyt P_{450} enzymes is sometimes stimulated by a drug (the inducer), regardless of whether the drug is a substrate (e.g., phenobarbital). This process is known as enzyme induction.

1. When a drug is a substrate for the induced enzyme, the drug will be metabolized at a faster rate (metabolic tolerance). The increased metabolism can produce sub-threshold and ineffective levels of this or other drugs, unless dosage adjustments are made.

2. After enzyme induction has occurred, the abrupt withdrawal of the inducer will have the effect of reducing the amount of cyt P_{450} enzymes, thereby slowing metabolism of other drugs that are still being administered. This predisposes to toxic effects.

3. Some drugs compete for biotransformation by cyt P_{450} enzymes, thereby mutually inhibiting metabolism (e.g., cimetidine and diazepam). Inadvertent high levels of one of the drugs, resulting from reduced metabolism, can produce toxicity.

Phase 2 reactions are characterized by the formation of conjugates (complexes) between the drug and one of several groups such as:

1. Glucuronic acid
2. Sulfate (-SO_4)
3. Acetate (CH_3CO-)
4. Glycine/Glutamate/Glutathione
5. Methyl (CH_3-)

Phase 2 reactions almost always inactivate drugs, but prominent exceptions are morphine glucuronide (active) and minoxidil sulfate (active).

A. Glucuronide Conjugation:

Newborns, especially premature infants, have low activity of glucuronyl transferase (GT). Accordingly, infants are not able to efficiently inactivate drugs and are at particular risk to drug toxicities.

$$UDP-\alpha-D-\text{glucuronic acid (UDPGA)} + \text{drug} \xrightarrow{GT} \text{drug glucuronide}$$

$$RNH_2 \longrightarrow RNH-\text{glucuronide}$$

$$ROH \longrightarrow RO-\text{glucuronide}$$

$$RSH \longrightarrow RS-\text{glucuronide}$$

$$RCOOH \longrightarrow RCO-\text{glucuronide}$$

Infants are likewise unable to efficiently glucuronidate the bile pigment bilirubin, formed by the breakdown of fetal hemoglobin. When the amount of bilirubin exceeds the binding capacity of plasma proteins, the resulting high amount of free bilirubin will deposit in the brain (kernicterus) and produce irreversible damage.

Drug glucuronide metabolites, formed in the liver, are largely excreted into bile which is secreted into the duodenum during digestion. The jejunum, as well as anaerobic intestinal bacteria, possess high amounts of glucuronidases which cleave glucuronides, thereby reactivating drugs. If these are absorbed, the enterohepatic circulation will recycle the drug, prolonging its duration of action.

B. Sulfate Conjugation

Sulfate conjugation occurs by the following scheme:

$$3'-D-\text{phosphoadenosine}-5'-\text{phosphosulfate (PAPS)} + \text{drug} \xrightarrow{\text{sulfotransferase}} \text{Drug } SO_4$$

$$ArOH \longrightarrow ArSO_4$$

$$ArNH_2 \longrightarrow ArSO_4$$

C. Glycine, Glutamate or Glutathione (GSH) Conjugation occurs as follows:
drug—CO OH $\longrightarrow$ drug—CO SCoA + H_2NCH_2COOH $\longrightarrow$ drug—CO $NHCH_2COOH$ + CoASH

The reaction of a drug with GSH is a potentially deleterious one. If a drug is taken in large overdose, GSH can be depleted from liver cells, placing them at risk to damage by normal oxidative reactions. Acetaminophen (Tylenol) toxicity is associated with GSH depletion, liver destruction and death. If other -SH donors are administered, such as cysteamine or acetylcysteine, then acetaminophen toxicity can be prevented.

$$\text{acetaminophen} + O_2 \xrightarrow{\text{cyt } P_{450}} \text{metabolite} \xrightarrow[\text{or other —SH donor}]{\text{GSH}} \text{inactive product}$$

D. Acetate Conjugation

This involves the enzyme N-acetyltransferase (NAT):

$$\text{drug—}NH_2 + CH_3CO\,SCoA \xrightarrow{\text{NAT}} \text{drug—}NH\,COCH_3 + \text{CoASH}$$
$$\text{drug—}NHNH_2 + CH_3CO\,SCoA \longrightarrow \text{drug—}NHNH\,COCH_3 + \text{CoASH}$$

A high percentage of people are slow acetylators. The duration of action of those drugs primarily inactivated by acetylation would be greatly prolonged in such patients.

E. Methylation utilizes the cofactor S-adenosylmethionine (SAM), a methyl donor:

$$\text{drug} \xrightarrow{\text{SAM}} \text{drug methoxy metabolite}$$

$$ROH \longrightarrow RO\,CH_3$$
$$RNH_2 \longrightarrow RNH\,CH_3$$
$$RSH \longrightarrow RS\,CH_3$$

F. Other Biotransformations. Drugs are substrates for many kinds of enzymes:

1. **Monoamine Oxidase (MAO)**

$$\text{drug—}CH_2\,NH_2 \xrightarrow[2H_2O]{\text{MAO}} RCHO + NH_2 + H_2O_2$$

MAOs are located on the outer mitochondrion membrane. MAO_A is present in most cells. MAO_B is primarily located in astroglia and in the brain and this enzyme has a major role in inactivating the CNS neurotransmitter, dopamine.

2. **Diamine Oxidase**

$$H_2N\,\text{drug}\,CH_2NH_2 \longrightarrow H_2N\,\text{drug}\,CHO + NH_3$$

3. **Alcohol Dehydrogenase** (deH_2ase) and **Aldehydehyde DeH_2ase**:

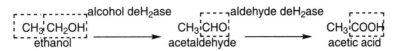

4. **Transulfuration**

$$CN^- + S_2O_3^{-2} \xrightarrow{\text{sulfotransferase}} SCN^- + SO_3^{-2}$$
cyanide + thiosulfate → thiocyanate + sulfite

$$CN^- + HSCH_2COCOOH \longrightarrow SCN^- + CH_3COCOOH$$
CN⁻ + β-mercaptopyruvic acid → thiocyanate + pyruvic acid

FACTORS INFLUENCING BIOTRANSFORMATION

1. Age: Liver size and enzyme content is less in old age.

2. Genetics: Some ethnic groups have certain phase 1 or phase 2 reactions that are faster or slower, compared to the overall population.

3. Nutrition: Some foods increase cyt P_{450} enzyme activity (e.g., cabbage family). Malnutrition, hypovitaminosis, deficiencies in essential fatty acids, high sugar, etc. reduce cyt P_{450} activity.

4. Disease: Hepatitis, cirrhosis, fatty liver, pulmonary disease, and many other health conditions diminish cyt P_{450} activity.

5. Sex: Males, through the influence of testosterone, metabolize drugs more rapidly than females.

6. Endocrines: Hypothyroidism is associated with reduced cyt P_{450} activity. Pregnancy, elevated glucocorticoid levels etc. have an influence on cyt P_{450} activity.

Many other factors can modify the metabolism of drugs.

CHAPTER 2
THE AUTONOMIC, PARASYMPATHETIC AND SYMPATHETIC NERVOUS SYSTEMS

THE AUTONOMIC NERVES NERVOUS SYSTEM (ANS)

A. OVERVIEW

The ANS is an involuntary nervous system that regulates the heart, smooth muscle and glandular systems, in an attempt to maintain internal homeostasis. According to the external situation, the body is geared toward energy conservation (parasympathetic nervous system activity: slow heart rate, food digestion, gluconeogenesis, etc.) or energy utilization (sympathetic nervous system activity: increased cardiac output, increased blood flow to skeletal muscles, increased respiration, glycogenolysis, etc.).

It is noted that SNS stimulation directly increases heart rate, but that baroreceptor reflexes (related to elevated blood pressure) usually override the direct SNS effect and produce a net decrease in heart rate.

B. DIVISIONS OF THE ANS

1. **Parasympathetic Nervous System (PSNS).** The PSNS arises via cranial nerves III, VII, IX, X and S_1-S_4 sacral nerves **(Figure 2.1)**. This is the "craniosacral" division of the ANS. These *preganglionic nerves*, as well as the *postganglionic nerves*, are cholinergic. Except for a few specialized ganglia (e.g., ciliary ganglion), the preganglionic fibers make synaptic ganglionic contact with postganglionic fibers in the end organs per se (i.e., terminal ganglia). Accordingly, preganglionic fibers are long and postganglionic fibers are short. Typically, only few preganglionic fibers fire at one time, so physiological responses are discrete.

2. **Sympathetic Nervous System (SNS).** The SNS arises via intermediolateral cells of the spinal cord in thoracic nerves T_1-T_{12} and lumbar nerves L_1-L_2. This is the *thoracolumbar* division of the ANS **(Figure 2.2)**. These preganglionic fibers make synaptic contact with postganglionic fibers predominately in the chain of ganglia alongside the spinal column ("paravertebral ganglia," i.e., next to the vertebral column). In some cases the SNS preganglionic fibers make synaptic contact with postganglionic fibers in prevertebral ganglia (i.e., in front on the vertebral column; e.g., mesenteric ganglia) or terminal ganglia (i.e., in the end organs of urinary and reproductive systems).

   Preganglionic fibers are usually short and postganglionic fibers are long. Postganglionic fibers have a widespread distribution which often includes several organs. Therefore, when the SNS is activated there tends to be a *generalized response* which involves physiological reactions throughout the whole body. This response prepares the body for a "fight or flight" reaction, wherein the body is prepared to use energy and do muscular work to guard against danger or threat.

   Preganglionic nerves are *cholinergic* (i.e., they make the neurotransmitter acetylcholine [Ach]), while postganglionic nerves are usually *noradrenergic* (i.e., they make the neurotransmitter noradrenaline, same as norepinephrine [NE]). The major exception is SNS postganglionic cholinergic nerves to sweat glands. Also, a few SNS postganglionic fibers to the kidneys release DA as the neurotransmitter. The adrenal medulla is a specialized ganglion of the SNS, with the postglionic cells lacking axons.

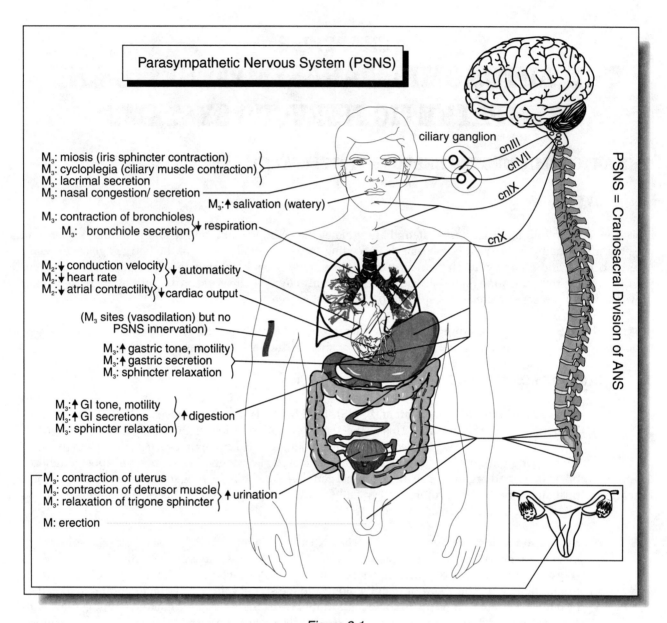

Figure 2.1

These cells make epinephrine (Epi) and release it into the blood, so ultimately Epi is distributed throughout the body. Thus, Epi is a neurohormone.

C. RECEPTORS IN THE ANS

The preganglionic cholinergic nerves for both the PSNS and SNS act at N_N nicotinic sites on the postganglionic nerves (N_N is the Nicotinic site on Nerve) **(Figure 2.3)**. Postganglionic parasympathetic cholinergic nerves release Ach, which acts at muscarinic (M) sites on end organs (smooth/cardiac muscle, glands). The majority of postganglionic sympathetic nerves release NE, which acts at α- and β-adrenoceptors on end organs (smooth/cardiac muscle, glands). A few of the sympathetic postganglionic nerves release Ach, which acts at M sites on sweat glands.

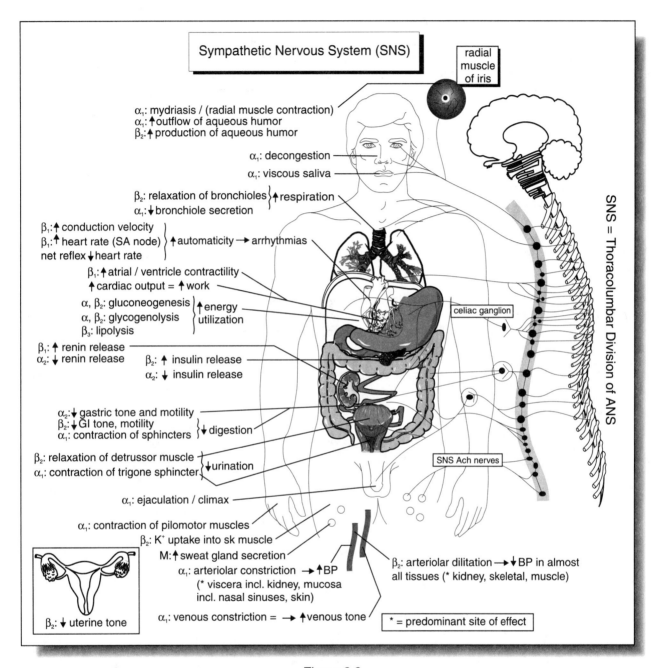

Figure 2.2

D. NEUROTRANSMITTERS AND METABOLISM

1. ACH (FIGURE 2.4)

In cholinergic nerves Ach is synthesized in a one step process involving the enzyme *choline acetyltransferase* (*CAT*), which condenses choline with activated acetate (i.e., acetylCoA). Ach is stored in granules (also known as vesicles) as a stable complex with proteins and ATP. Approximately 10,000 molecules of Ach are in each granule.

When cholinergic nerves are stimulated, granules in the nerve endings fuse with the outer nerve membrane and release Ach into the synapse. After diffusing across the synaptic cleft, Ach acts at postsynaptic M and/or N receptors. Most synaptic Ach is catabolized by *acetylcholinesterase* (AchE) to acetate and choline, with most of the choline being reaccumulated by the cholinergic nerve and eventually reconverted to Ach. Only negligible amounts of Ach normally diffuse from the synapse, and this Ach would be catabolized by nonspecific cholinesterase (i.e., pseudocholinestrase or pseudo-ChE).

2. NE AND EPI (FIGURE 2.5)

In SNS postganglionic nerves NE is synthesized as follows. The essential amino acid tyrosine is first converted to L-dihydroxyphenylalanine (L-DOPA) by *tyrosine hydroxylase* (T-OH), the rate-limiting step. Next, *dopa decarboxylase* removes a -COOH group to form dopamine (DA). These steps occur in the cytoplasm. Some DA is catabolized by *monoamine oxidase* (MAO) to dihydroxyphenylacetic acid (DOPAC) and eventually further catabolized by catechol-O-methyltransferase (COMT) to homovanillic acid (HVA), inactive metabolites. However, most DA is actively accumulated by synaptic granules and converted to NE, through the action of *dopamine-β-hydroxylase* (DBH).

In nerve endings more than 90% of the NE is stored in granules as a stable complex with proteins and ATP, in the ratio of 1 ATP molecule to 4 NE molecules. Approximately 10,000 molecules of NE are in each granule. There is a small amount of cytoplasmic NE which can be catabolized by mitochondrial MAO to inactive dihydroxy-

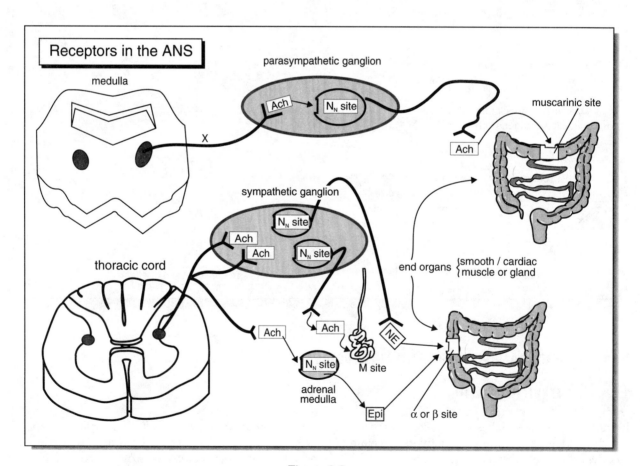

Figure 2.3

mandelic acid (DHM) and ultimately by COMT to vanillylmandelic acid (VMA), the principal inactive metabolites found in urine.

When SNS nerves are stimulated, granules in the nerve endings fuse with the outer nerve membrane and release NE into the synapse. After diffusing across the synaptic cleft NE acts at postsynaptic α– and β-adrenoceptors. NE can also act at presynaptic autoreceptors which modulate additional release of NE. Presynaptic β_1 receptors are facilitatory, while presynaptic α_2 receptors are inhibitory.

About 15% of synaptic NE diffuses away or is catabolized. About 85% of synaptic NE is reaccumulated by presynaptic nerves by an active uptake process, so that NE can be sequestered again in granules. The reuptake process is the major means of inactivating synaptic NE, and it will be shown later that drugs which interfere with this process have profound effects.

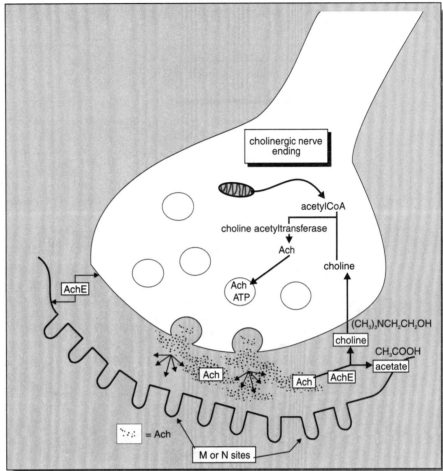

Figure 2.4

In adrenal medullary cells **(Figure 2.6)** epi is formed preferentially, so that these cells contain epi:NE in the ratio of about 4:1. Synthesis of epi proceeds through NE formation, as described in SNS nerve endings. Epi formation occurs when NE exits the granule into the cytoplasm, to be acted upon by *phenylethanolamine-N-methyltransferase* (PNMT). Some of this epi is catabolized by MAO (ultimately to VMA) but most is reaccumulated by granules and stored in a stable complex with ATP and proteins, as in SNS nerve endings. The illustration shows 3 epi, 1 NE and 1 ATP molecule in a complex; the ratio is variable, but overall there will be about 4 epi to 1 NE. When adrenal medullary cells are stimulated, epi is released from the storage granules, as described for NE nerves. However, being a neurohormone that acts at distal sites, most of this epi diffuses into the circulation. Only small amounts are reaccumulated by active reuptake processes.

E. NEUROTRANSMITTER ACTIONS

The granules of nerves that contain Ach, NE or Epi also contain peptides that are released as co-transmitters. These peptides may have actions of their own, but it seems that their primary action is to modulate the response to Ach, NE or epi. Also, the granules may contain proteins (e.g., chromogranins like DBH) that are co-released when each granule releases the approximate 10,000 molecules of Ach, NE or Epi.

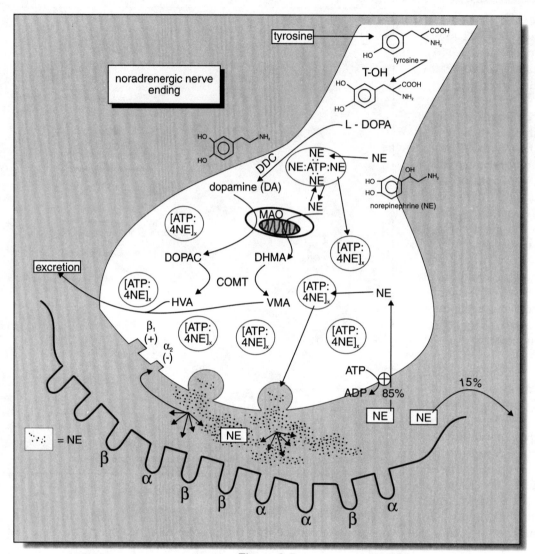

Figure 2.5

1. **Ganglia**. Ach is released by preganglionic nerves of both the PSNS and SNS. In each case the postganglionic receptor is nicotinic (N_N site) **(Figure 2.3)**.

2. **PSNS postganglionic nerves**. These nerves release Ach onto receptors that are muscarinic (M sites) at the end organs **(Figure 2.3)**. Receptor localization:
    M_1 on nerves in CNS and ganglia
    M_2 on cardiac muscle
    M_3 on smooth muscle and glands

3. **SNS postganglionic nerves**. The majority of these nerves release NE onto end organ receptors that are classified as α– or β-adrenoceptors. A few SNS postganglionic nerves, primarily those to sweat glands, release Ach onto muscarinic receptors (M sites) **(Figure 2.3)**.

a. **Responses to NE and Epi**. SNS activity involves (a) release of NE at end organs and (b) release of epi into the circulation, from where epi eventually exerts actions at end-organs.

NE receptors: $\alpha_1, \alpha_2, \beta_1, \beta_3$
Epi receptors: $\alpha_1, \alpha_2, \beta_1, \beta_2,$ (β_3 minor)

b. Receptor localization: α_1 and β_2 on smooth muscle and glands
                                  β_1 on heart muscle and presynaptic NE nerves
                                  β_3 on lipocytes
                                  α_2 on presynaptic NE nerves

4. **Effects of PSNS and SNS activity**. Effects of PSNS and SNS activity are illustrated in **Figures 2.1 and 2.2** respectively. The receptor associated with each response is likewise indicated.

F.   SECOND MESSENGERS IN THE ANS

When agonists act at α_1, M_1, and M_3 receptors, the consequent G protein activation produces phospholipase C (PLC) activation. Phosphatidylinositol-4,5-bisphosphate (PI), a substrate for PLC, is then hydrolyzed to (a) inositol triphosphate (IP_3) which promotes Ca^{++} release from intracellular storage sites, and (b) DAG which activates protein kinase C (PKC). Both messengers induce the physiological response **(Figure 2.34)**.

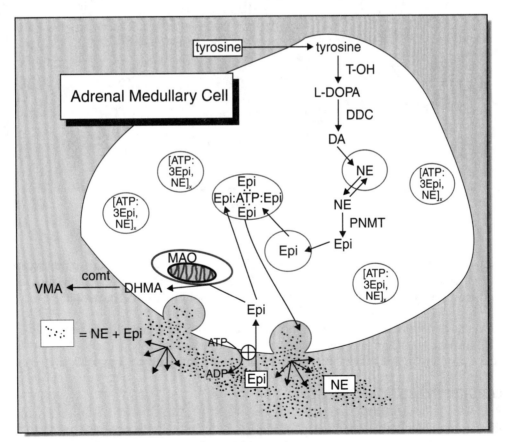

Figure 2.6

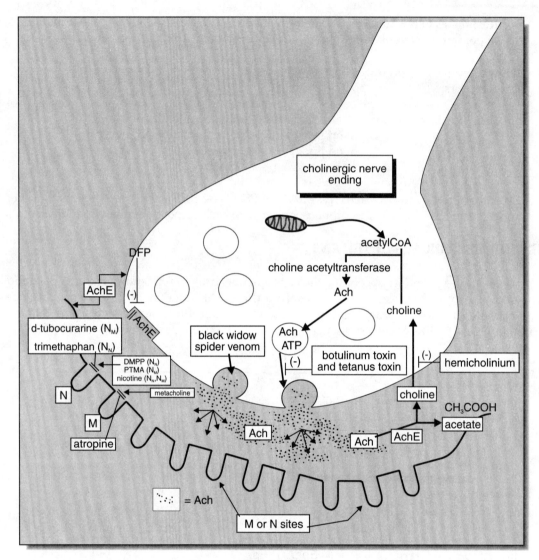

Figure 2.7

When agonists act at β-adrenoceptors, the consequent G protein (Gs) activation produces adenylyl cyclase (AC) activation. This enzyme cleaves ATP to cAMP, a second messenger that activates protein kinase A (PKA), an enzyme that induces the physiological response via protein phosphorylation **(Figure 2.35)**.

When agonists act at α_2 and M_2 receptors, the consequent activation of an inhibitory protein (G_i) results in an inhibition of AC activity. The change in cellular content of the second messenger, cAMP, alters the activity of cAMP-dependent protein kinases which accounts for the physiological effect **(Figure 2.36)**.

G. METABOLIC EFFECTS OF THE SNS

1. **Elevation of blood glucose (Figure 2.30).** Epi activity at the β_2 adrenoceptor results in the sequential activation of adenylyl cyclase and the cascade of reactions that results in glycogenolysis and elevated blood glucose. These are needed for epi actions that expend energy. This scheme of events

occurs in liver and muscle, with the exception being that muscle lacks G-6-phosphatase so that lactic acid is produced here, not glucose *per se*.

2. Reduction of blood insulin **(Figure 2.31)**.

3. Elevation of blood lipids **(Figure 2.32)**.

4. Hypokalemia **(Figure 2.33)**.

THE PARASYMPATHETIC NERVOUS SYSTEM (PSNS)

A. OVERVIEW

Chemical sustances can facilitate cholinergic actions in the following ways. Black widow spider venom promotes granule release of Ach. Acetylcholinesterase inhibitors (AChE-Is), like diisofluorophosphate (DFP), inhibit catabolism of Ach, prolonging its duration in the synapse and enhancing its effect. Direct acting muscarinic receptor agonists, like methacholine, mimic actions of Ach as do the direct acting nicotinic (N) receptor agonists, dimethylphenylpiperazinium (DMPP) (N_N site), phenyltrimethylammonium (PTMA) (N_M site) and nicotine (N_N and N_M sites)**(Figure 2.7)**.

Chemical substances can attenuate cholinergic actions in the following ways. Botulinum toxin and tetanus toxin inhibit fusion of granules with the nerve membrane, thereby inhibiting release of Ach from the nerve. Hemicholinium competitively inhibits uptake of choline into cholinergic nerves and this leads to depletion of Ach in granules. Atropine blocks muscarinic (M) receptors, while trimethaphan and d-tubocurarine (d-Tb) block N_N and N_M receptors, respectively **(Figure 2.7)**.

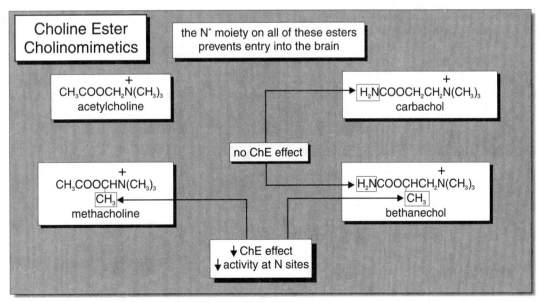

Figure 2.8

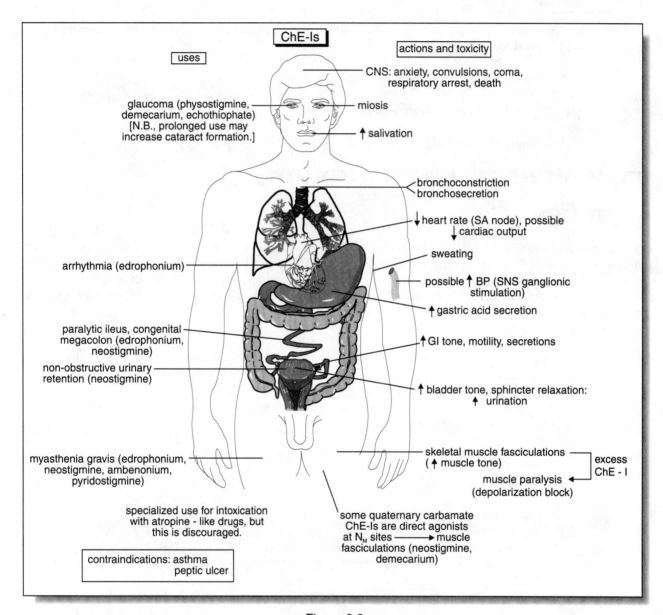

Figure 2.9

B. MUSCARINIC RECEPTOR AGONISTS = CHOLINOMIMETICS

These drugs are (a) agonists at muscarinic (M) receptors, (b) have some agonist activity at nicotinic (N) receptors and (c) are partially or fully resistant to hydrolysis by cholinesterase (ChE).

1. **CHOLINE ESTERS ARE SEEN IN FIGURE 2.8.**

    A. *Acetylcholine* (Ach; Miochol) is so rapidly hydrolyzed by ChE, that it is useless as a drug except in ophthalmology where topical application produces rapid complete miosis, as for cataract surgery. If it given i.v. in high amount, Ach effects would be similar to parasympathetic nerve (PSN) stimulation with

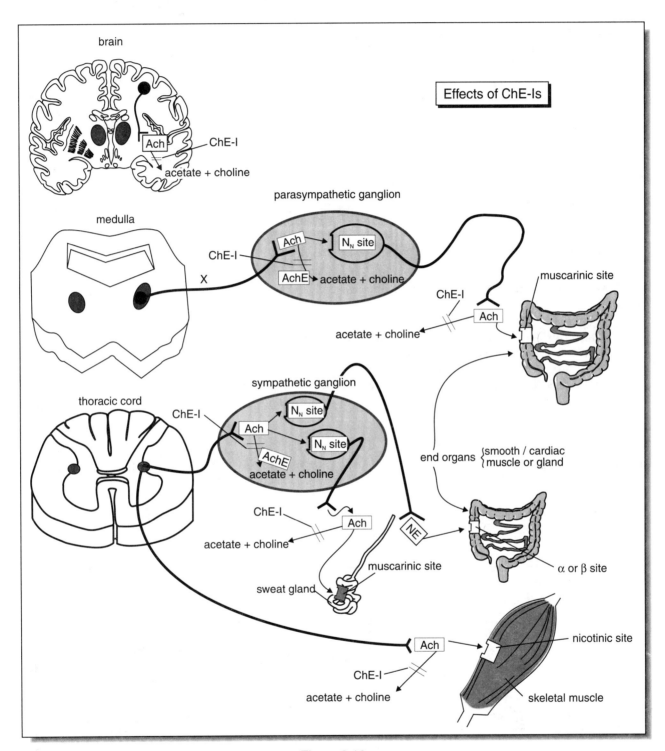

Figure 2.10

added vasodilation (vessels have M receptors despite lack of PSN input) and sweat gland secretion (M receptors for sympathetic cholinergic input to the glands).

B. *Carbachol* (Carcholine) is used primarily as a second-line drug for narrow and open angle glaucoma. M effects of carbachol are greater at the iris, urinary bladder and gastrointestinal (GI) tract versus Ach.

C. *Bethanechol* (Urecholine) s.c. is a drug of choice for neurogenic and postoperative urinary retention and gastric atony. There are minimal cardiovascular actions. M effects are like those of carbachol; N effects are absent. Side effects are similar to those after i.v. Ach. Contraindications are peptic ulcer, asthma and hyperthyroidism.

D. *Methacholine* (Methacholine), unlike bethanechol, lacks partial selectivity for urinary and GI effects, although there is selectivity for M sites.

2. NATURAL ALKALOIDS

These bulky alkaloid molecules are not hydrolyzed by ChE and are able to enter the brain and cause CNS excitation.

A. *Muscarine* is the active principle in some poison mushrooms and causes predictable effects based on selective activity at M sites.

B. *Pilocarpine* (Pilocar) has activity like muscarine, but has especially prominent effects on salivary, sweat and gastric glands. It is a drug of choice for open-angle and closed-angle glaucoma.

3. SYNTHETIC ALKALOID

Oxotremorine has activity like muscarine, but produces prominent tremor and ataxia, resembling Parkinsonism.

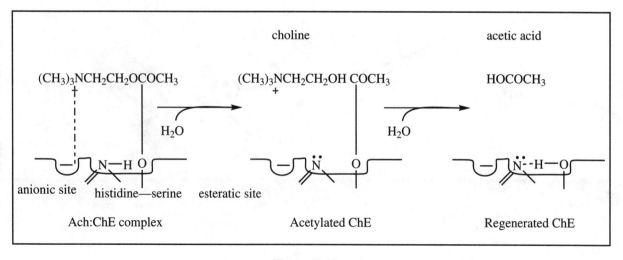

Figure 2.11a

4. TOXICITY OF CHOLINOMETICS

After systemic administration of any of the compounds, toxicity would resemble excess PSN stimulation. The alkaloids would additionally have central effects. Atropine-like drugs are antidotes!

C. CHOLINESTERASE INHIBITORS (CHE-Is) FIGURE 2.9

These agents (drugs, insecticides and nerve gases) bind competitively or non-competitively with non-specific ChE (*pseudo-ChE* is butyryl-ChE or BuChE) in blood and liver, and with specific ChE (*acetyl-ChE* is AChE) at synapses, thereby inhibiting the major means for inactivating Ach. Consequently, synaptic actions of Ach become prolonged in (a) brain, (b) PSNS ganglia, (c) PSNS end organs, (d) SNS ganglia, (e) post-ganglionic SNS cholinergic sites (primarily sweat glands) and (f) skeletal muscle **(Figure 2.10)**. The net effect resembles excess stimulation of all cholinergic nerves in the body, along with some sympathetic nerves (the latter being a reflection of SNS ganglionic stimulation).

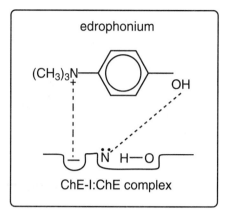

Figure 2.11b

1. PHARMACOKINETICS

A. Absorption

1. Quaternary ChE-Is are poorly absorbed from the skin, lungs and conjunctiva. Orally, much higher doses would be needed versus a parenteral route.

2. Tertiary carbamate ChE-Is are well absorbed from all sites, except skin.

3. Tertiary organophosphate ChE-Is, including thiophosphate insecticides (malathion, parathion), are well absorbed from all sites including skin.

B. Distribution to:

> CNS: Tertiary ChE-Is enter the CNS, not quaternary ChE-Is.
> Eye: After oral administration, only tertiary ChE-Is affect the eye.

C. Metabolism

With the exception of echothiophate, organophosphate ChE-Is are unstable in aqueous solutions. Such insecticides have limited lives when sprayed on crops. Carbamate aqueous solutions are stable.

Both AChE and BuChE interact with all ChE-Is.

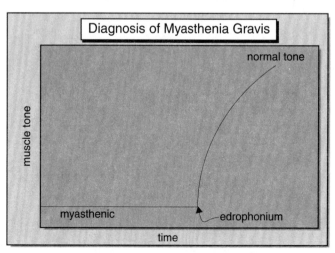

Figure 2.12

Thiophosphates ($\overset{S}{\underset{O}{=P}}$) are metabolized to the

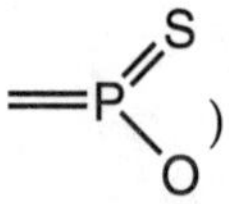

active species ($\overset{O}{\underset{O}{=P}}$) in insects and vertebrates.

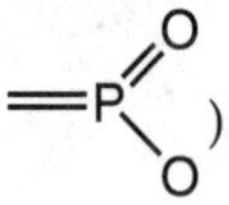

D. Duration

This is related to stability of the ChE:ChE-I complex, not to metabolism *per se*.

2. **CLASSIFICATION OF CHE-IS IN THIS CHAPTER**

Tertiary ChE-Is include physostigmine, carbaryl (an insecticide) and all organophosphates except echothiophate which is quaternary. All other agents are quaternary.

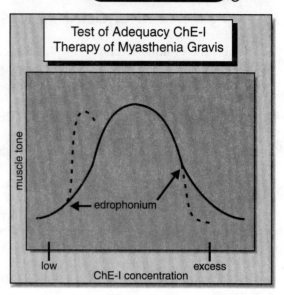

Figure 2.13

Ach Metabolism occurs in about 1/150 microseconds, as shown in **Figure 2.11a**.

3. **TYPES OF CHE-IS AND ACTIONS**

A. Simple alcohol with quaternary nitrogen group (ammonium group) **(Figure 2.11b)**.

Edrophonium (Tensilon) is a competitive inhibitor of ChE. By occupying the electrophilic and electrophobic ChE sites, edrophonium blocks access of Ach to the enzyme. This drug has a short duration of action (5-15 minutes) and is used to diagnose myasthenia gravis and to test the adequacy of ChE-I therapy.

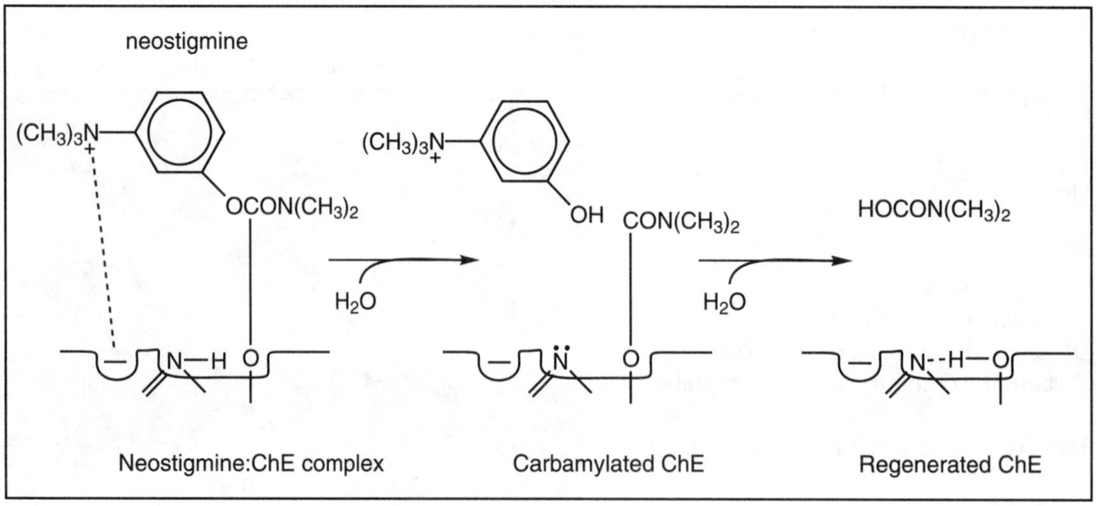

Figure 2.14

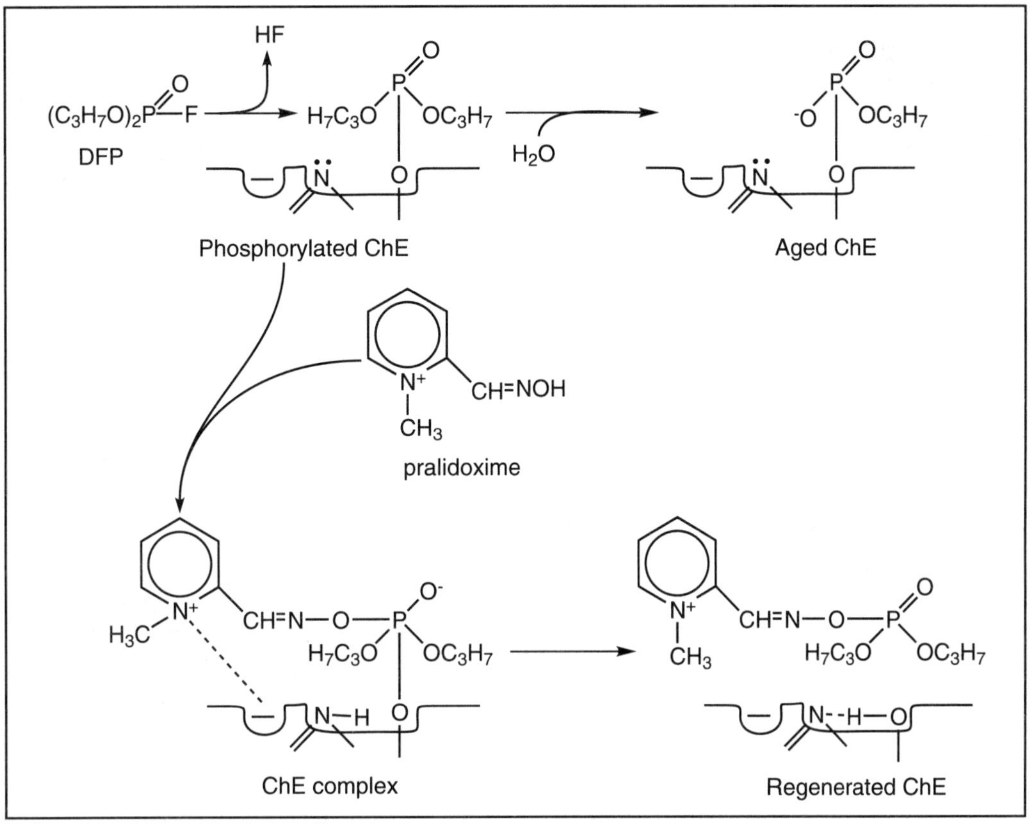

Figure 2.15

1. **Tensilon test for diagnosis of myasthenia**. Myasthenia gravis is an autoimmune disease in which antibodies are made against N_M receptors, thereby reducing sites at which Ach can act and bring about muscle contraction. Muscle tone is low. Edrophonium competitively inactivates ChE, increasing synaptic levels of Ach at the neuromuscular junction (NMJ), which enhances Ach interaction with N_M sites. Consequently, muscle tone becomes beneficially increased **(Figure 2.12)**.

2. **Tensilon test of adequacy of ChE-I therapy of myasthenia**. When ChE-I levels are too low, low muscle tone persists. In this situation a test dose of edrophonium will add to ChE inhibition, approaching optimal ChE inhibition, and muscle tone will increase. However, in the situation where ChE-I concentration is in excess, depolarization block of skeletal muscle occurs from excess synaptic Ach, and muscle tone declines. A test dose of edrophonium in this condition will produce even greater ChE inhibition, resulting in greater depolarization block of skeletal muscle, and a greater decline in muscle tone. Therefore, according to the response, edrophonium can determine whether muscle tone is low because of (a) inadequate ChE inhibition or (b) excess ChE inhibition **(Figure 2.13)**. (Note, in the latter case there may be excess salivation, diarrhea, urinary urgency, etc.—due to excess Ach levels at PSNS end organs.)

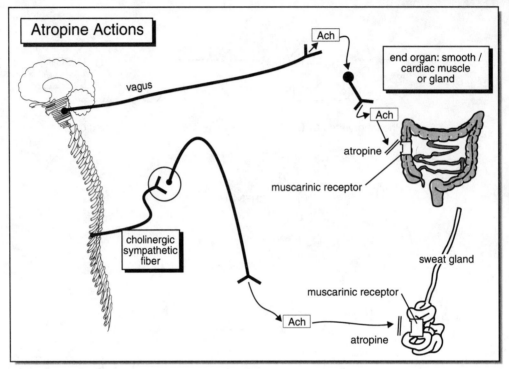

Figure 2.16

B. Carbamate ester of alcohols with a quaternary nitrogen group. These are competitive inhibitors of ChE.

Neostigmine (Prostigmin)—duration, less than 2 hours **(Figure 2.14)**.
Pyridostigmine (Mestinon)—duration, 3-6 hours.
Demecarium

C. Carbamate ester of alcohols with a tertiary nitrogen group (Physostigmine [Eserine]). Unlike the quaternary analogs, the tertiary nitrogen analogs enter the CNS. Many insecticides are tertiary nitrogen carbamates.

D. Organophosphates. These inhibitors phosphorylate ChE. With time ("aging") the covalent bond becomes even stronger, resulting in irreversible inactivation of ChE. New ChE must then be made, in order for recovery to occur. This is a process requiring several weeks. Many insecticides and nerve gases are organophosphates.

1. **Therapeutic agents**
   a. Echothiophate (Phospholine) and isoflurophate (Floropryl; DFP) are occasionally used topically for glaucoma and some ocular surgical procedures. Atropine prevents many systemic effects that might otherwise occur as the drug is absorbed from the eye.

   b. *Pralidoxime* (Protopam; 2-PAM) is a strong nucleophile, able to reactivate ChE from organophosphate ChE-Is if aging has not occurred **(Figure 2.15)**. This quaternary agent will not enter the CNS. It must be given parenterally, and in combination with atropine which blocks muscarinic sites in both the CNS and periphery. Since the ChE-Is have long durations of action the atropine and 2-PAM injections must be given repeatedly at intervals.

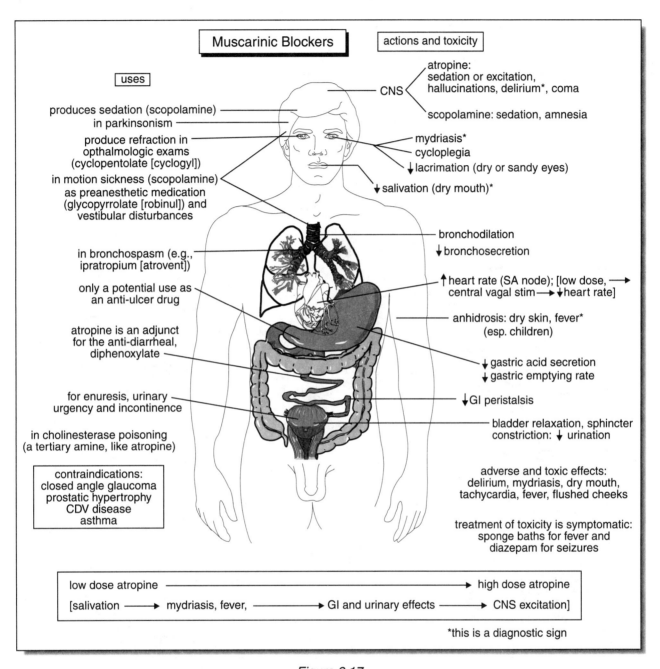

Figure 2.17

2. **Organophosphate nerve gases**
   a. Sarin, soman and tabun are highly lipid-soluble agents that are absorbed through the skin and any other body site. These produce (a) effects that resemble profound PSNS stimulation, (b) muscle fasciculations and muscle paralysis (i.e., respiratory paralysis) and (c) CNS excitation and convulsions. Death is likely in the absence of an antidote.

PHARMACOLOGY

b. Nerve gas antidote. This is available in syringes containing:
   i. atropine (2 mg) for its anti-muscarinic effect;
   ii. pralidoxime (5 mg) for regenerating ChE and
   iii. diazepam (10 mg) for its anticonvulsant effect.

3. Thiophosphate insecticides.

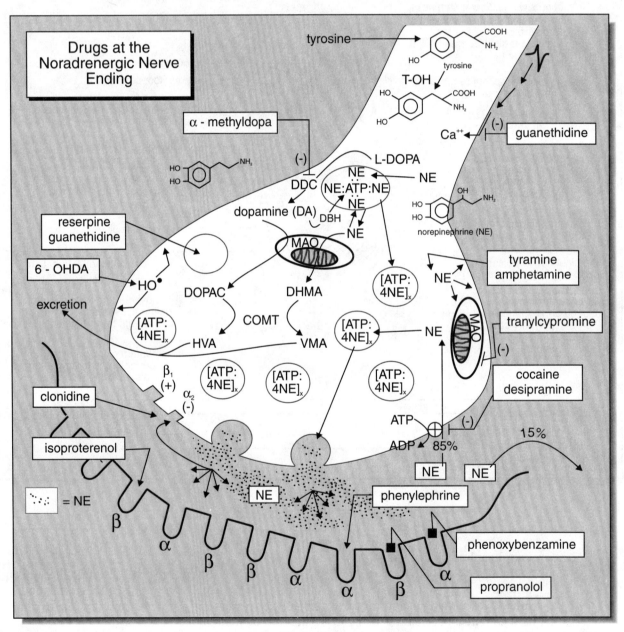

Figure 2.18

i. Tetraethyl pyrophosphate (TEPP).
ii. Malathion and Parathion. These highly lipid soluble organophosphates are metabolized in insects to , which confers ChE-I activity. These insecticides readily cross membranes and produce convulsions and respiratory paralysis in insects.

Vertebrates have enzymes that also activate these substances. Mammals and birds, but not insects and fish, have enzymes that effectively inactivate malathion. Parathion, however, is not inactivated in vertebrates, and is accordingly not available for general use.

4. **CHE-I INSECTICIDE POISONING**

   A. Diagnosis

   Symptoms reflect excessive muscarinic stimulation. Miosis is commonly observed. Blood levels of ChE will be reduced to less than 50% of normal.

   B. Treatment

   Remove clothing and shower the patient. If there was oral consumption, give a lavage and/or emetic. After vomiting has occured, give a cathartic. Administer atropine at intervals. If the ChE-I was an organophosphate and exposure was within 24 hours, infuse 1-2 g of 2-PAM and repeat at intervals. Diazepam, at intervals, may be needed to control convulsions. Positive respiratory assistance may be needed.

D. **MUSCARINIC RECEPTOR ANTAGONISTS (M-BLOCKERS)**

   1. **Other names.** Anticholinergics, antimuscarinics, parasympatholytics, muscarinic antagonists, atropine-like drugs.

   2. **Mechanisms and actions.** By binding at muscarinic (M) sites these drugs block the synaptic actions of acetylcholine (Ach) in the CNS, smooth muscle, cardiac muscle and exocrine glands **(Figure 2.16)**. PSNS postganglionic nerves and SNS postganglionic cholinergic nerves are mainly affected outside the CNS. By knowing the effects of PSNS stimulation, actions of M-blockers can be deduced **(Figure 2.17)**.

   3. **Pharmacokinetics**
      Absorption: Tertiary amines like atropine and scopolamine are well-absorbed. Quaternary amines like methantheline (Banthine) and propantheline (Pro-Banthine) are poorly absorbed.
      Distribution: After systemic administration, tertiary amines affect all M sites in the body. Quaternary amines do not affect M sites in the CNS. Both classes are given topically to affect the eye.
      Metabolism: Hydrolysis and conjugation reactions result in short half lives of these drugs (hours).
      Excretion: More than half the parent drug is excreted unmetabolized in urine.

   4. **Source and agents.** Atropine and scopolamine are the prototypic drugs of this class. Sources include belladonna (deadly nightshade), Jimson weed, thorn apple (stink weed) and tomato leaves. Belladonna means beautiful woman: rosy (flushed) cheeks (vasodilation), warm skin (decreased sweating) and big eyes (dilated pupils). Chemically, the agents are tropic (as in the name atropine) acid derivatives.

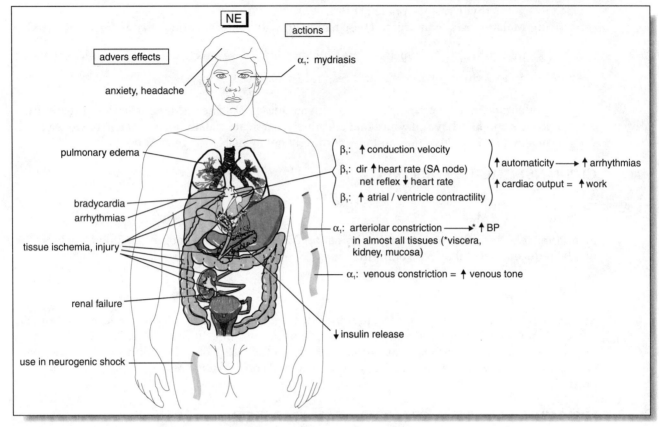

Figure 2.19

5. **Other drugs with prominent M-blocking avtivity**. Older H_1-blockers, antidepressants, antipsychotics.

THE SYMPATHETIC NERVOUS SYSTEM (SNS)

A.   OVERVIEW (FIGURE 2.18)

Chemical substances can facilitate adrenergic actions in the following ways. Indirect acting sympathomimetics include tyramine and amphetamine which displace NE from cytoplasmic binding sites, thereby promoting NE release **(Figure 2.27)**. A sympathomimetic effect occurs, due to action of released NE at adrenoceptors **(Figure 2.27)**. Cocaine and desipramine (and other tyicyclic antidepressants) inhibit reuptake of NE from the synapse into the nerve ending. Since this process is the principle means for inactivating NE, the duration of NE action is prolonged. Direct acting adrenoceptor agonists include drugs like phenylephrine (α_1), isoproterenol (β_1, β_2) and terbutaline (β_2).

Chemical substances can attenuate adrenergic actions in the following ways. Reserpine and guanethidine deplete granules of NE stores. Guanethidine also inhibits excitation-secretion coupling, so that the nerve action potential does not effectively promote NE secretion. The effect is analogous to a local anesthetic action. α-Methyldopa is a competitive inhibitor of DOPA decarboxylase. It inhibits formation of DA from L-DOPA, leading eventually to absence of NE in granules. α-Methyldopa per se is converted to α-methyl-DA and α-

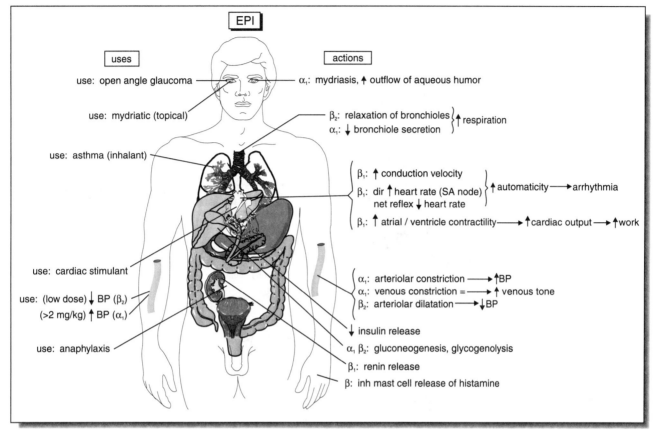

Figure 2.20

methyl-NE, which act as false transmitters in dopaminergic and noradrenergic nerves, respectively. Tranylcypromine is an MAO-inhibitor, and although it initially elevates NE content, it ultimately inactivates NE nerves (possibly by promoting false transmitter formation or through a guanethidine-like effect on excitation-secretion coupling). 6-Hydroxydopamine (6-OHDA) is a laboratory research drug that generates highly reactive free radicals that actually destroy NE nerves. Clonidine is an α_2-agonist that inhibits release of NE from the nerve. α- and β receptor blockers include phenoxybenzamine and propranolol, respectively.

A note of cardiovascular (CDV) effects. When the SNS is activated, NE is released from sympathetic nerves and epi is released from the adrenal medulla. The acute CDV effect is a reflection of action of (a) NE and epi at β_1-adrenoceptors in the heart, (b) NE and epi at α_1-adrenoceptors in certain vascular beds (viscera, mucosa, skin) and (c) epi at β_2-adrenoceptors primarily in vascular smooth muscle. Actions at α_2-adrenoceptors are not so important in the acute CDV actions of NE and epi. The net effect is:
 a. increased mean blood pressure and decreased perfusion of viscera (α_1);
 b. increased perfusion of skeletal muscle (β_2) and
 c. activation of the baroreceptor reflex, producing increased PSNS activity and decreased SNS activity to the heart causing bradycardia.

B. CATECHOLAMINE ADRENOCEPTOR AGONISTS

1. **Effects of i.v. NE (Levophed) (α, β_1).** NE acts at α_1, α_2 and β_1 receptors, not at β_2 receptors. Clinically, NE is typically infused and its effects are related to actions at α_1 and β_1 receptors. Also, because it

PHARMACOLOGY

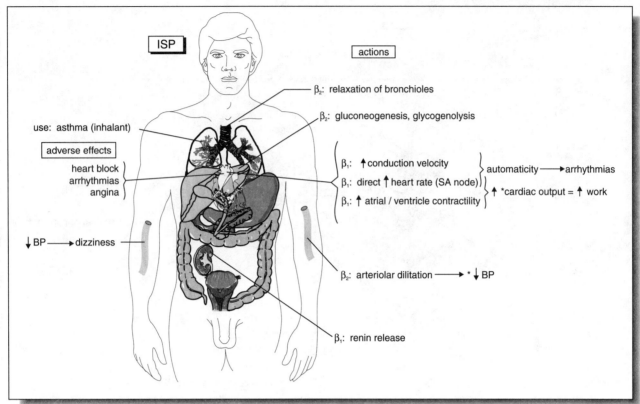

Figure 2.21

is short-lived, NE has no appreciable effects on nasal decongestion, salivation, etc. A summary of important effects expected after NE infusion is illustrated in **Figure 2.19**.

2. **Effects of Epi (Adrenalin) (α, β)**. Epi acts at all α and β adrenoceptors and is able to produce all of the effects of SNS stimulation. Clinically, epi is used i.v., as eye drops and as an inhalant. Actions and uses of epi are shown in **Figure 2.20**.

3. **Effects of Isoproterenol (Isuprel) (ISP) (β)**. Isp acts only at β adrenoceptors, not α receptors. Actions and uses of Isp are shown in **Figure 2.21**.

4. **Effects of Dopamine (Intropin) (DA)**. The predominate action of DA is an increase in cardiac contractility (β_1 site) **(Figure 2.22)**. At low doses DA may produce vasodilation in renal and splanchnic beds by inhibiting NE release (presynaptic DA D_2 site) and by a direct vasodilatory effect at the blood vessels (D_1 site). In high doses DA produces direct vasoconstriction (α_1 site). For treating shock, DA should be in infused at a level that produces an

increase in myocardial contractility and a decrease in afterload (vasodilation).

5. **Dobutamine (Dobutrex) (β_1, α_1)**
   This relatively selective β_1 agonist also acts at α_1 sites at high doses. The drug is used in congestive heart failure.

C. DIRECT-ACTING NONCATECHOLAMINE ADRENOCEPTOR AGONISTS

These drugs are effective orally, are not metabolized by COMT and are long acting.

1. **α_1-Selective Agonists**
   a. *Methoxamine* (Vasoxyl) has a profound pressor effect with reflex bradycardia and is used only occasionally for acute treatment of hypotensive states.
   b. *Phenylephrine* (Neo-synephrine) is sometimes used for hypotensive states, but more commonly as a mydriatic and decongestant.
   c. *Tetrahydrozoline* (Visine) is used to constricted dilated vessels that account for "red eyes."

2. **α_2-Selective Agonists.** These drugs reduce SNS actively by acting at α_2 sites (a) on sympathetic nerve endings and (b) in the CNS. They produce decreased BP and decreased heart rate and are used to treat hypertension.

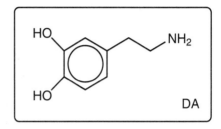

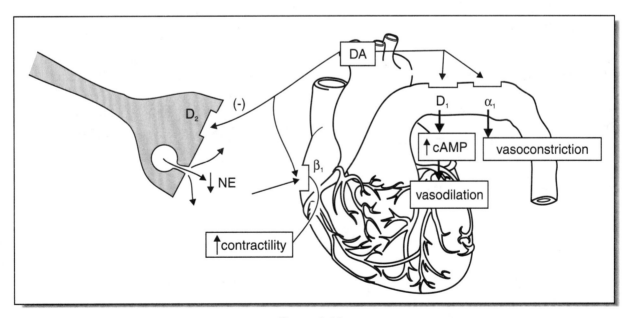

Figure 2.22

a. *Clonidine* (Catapres). This is also used as an adjunct in withdrawal from drugs of abuse.
b. *α-Methyldopa* (Aldomet). This drug is converted first to α-methyldopamine (α-Me-DA), then to α-methylnorepinephrine (α-Me-NE), a false transmitter that is stored in SNS granules. α-Me-NE has $α_2$-agonist activity. A parkinson side effect is sometimes seen, because α-Me-DA displaces DA in nigrostriatal dopaminergic nerves, depleting DA.

3. **$β_2$-Selective Agonists**. These drugs are used primarily for asthma, having a minimal direct effect on the heart.

   a. *Albuterol* (Proventil).
   b. *Terbutaline* (Brethine). This is sometimes used to suppress premature labor.
   c. *Ritodrine* (Bricanyl). This is used primarily to suppress premature labor.

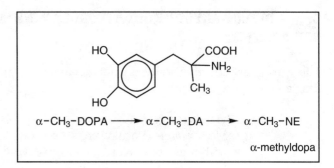

4. **Receptor Desensitization**. Repeated or prolonged stimulation of a receptor can result in a down-regulation of numbers of that receptor, and associated reduced responsiveness to the agonist. This phenomenon is known as receptor desensitization. It occurs with $α_1$ agonist decongestants and $β_2$ agonist bronchodilators. Receptor desensitization of DA receptors, muscarinic receptors, etc. will occur when their respective agonists are used for prolonged periods.

D. INDIRECT-ACTING NONCATECHOLAMINE ADRENOCEPTOR AGONISTS

These drugs lack the catechol nucleus and are therefore not metabolized by COMT (catechol-O-methyltransferase). These drugs are long-acting, orally-effective, and enter the CNS. They are accumulated in sympathetic nerves via uptake sites and displace NE from the cytoplasm. The NE exits the nerve and acts at adrenoceptors **(Figure 2.18)**.

A. Agents

   1. *Tyramine* is a dietary constituent of cheeses and fermented foods (smoked foods) and beverages (wines, beers). Normally, tyramine is metabolized by MAO in the gut.

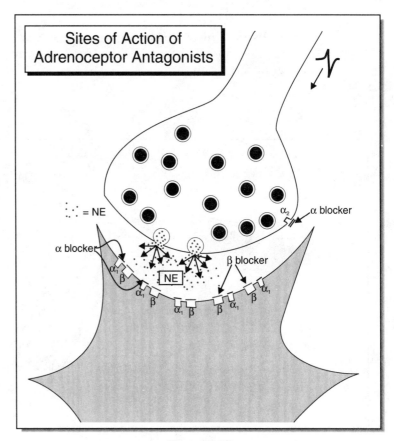

Figure 2.23

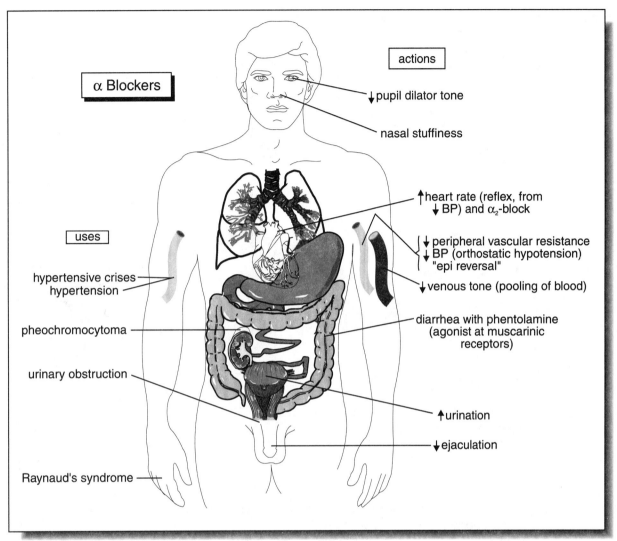

Figure 2.24

In people treated with an MAO-inhibitor, tyramine-containing foods present a medical emergency ("hypertensive crisis") because of the massive amounts of NE displaced at one time (causes increased BP, arrhythmias, stroke).

2. *Ephedrine* displaces NE, like tyramine, but has additional direct α and β agonist activity, so its spectrum is like epi, but long-acting. Ephedrine has an α-methyl group, making it resistant to MAO. Ephedrine also enters the brain and produces CNS stimulation (amphetamine-like). This drug is used as a nasal decongestant and pressor agent **(Figure 2.27)**.

3. *Amphetamine* (Benzedrine) has an α-methyl group, so it is not metabolized by MAO. It also has direct α and β agonist activity, so the spectrum of effects is like that of epi. However, unlike epi, amphetamine enters the brain, producing CNS excitation, euphoria and appetite suppression. It is a drug of abuse **(Figure 2.27)**.

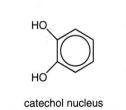

catechol nucleus

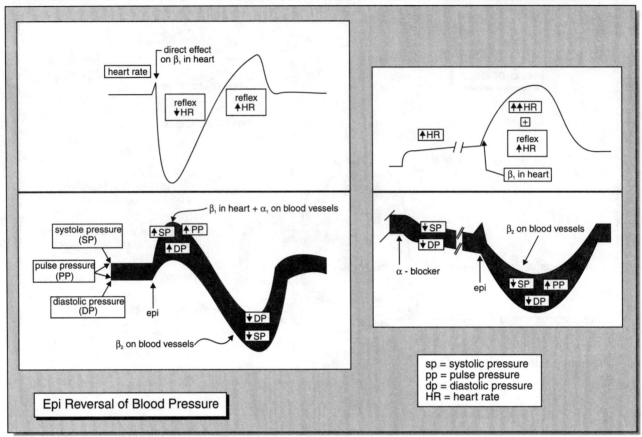

Figure 2.25

Other amphetamines include methamphetamine and hydroxyamphetamine. The former is also a drug of abuse, while the latter poorly enters the brain. Methylphenidate (Ritalin) is an amphetamine-like drug that is used (as is amphetamine) to treat attention deficit hyperactivity disorder (ADHD) **(Figure 2.27)**.

4. *Metaraminol* (Aramine) is indirect-acting like the other drugs in this category, but also releases epi and acts directly at α sites. Metaraminol is stored in granules as a false neurotransmitter **(Figure 2.27)**.

B. *Tachyphylaxis* is the phenomenon whereby there is a rapid diminution in response to a drug that is given repeatedly in a short time (minutes to less than 1 hour). This occurs with indirect-acting sympathomimetic amines (e.g., tyramine) and is related to depletion of NE stores that are released by these drugs (usually the cytoplasmic stores, not the granule stores).

E. ADRENOCEPTOR ANTAGONISTS

A. α-Blockers

These drugs block α_1 and/or α_2 receptors **(Figure 2.23)**, thereby attenuating many of the effects of NE and epi. The major effects are related to block of α receptors in smooth muscle, most notably in blood vessels. These drugs are used mainly in pheochromocytoma and hypertensive crises **(Figure 2.24)**.

Phenomenon of "epi-reversal" (Figure 2.25). Sympathetic nerve regulation of vasomotor tone (i.e., arteriole and venule pressure) is mainly related to actions of NE and epi at α and β receptors. Unlike NE which acts only at α receptors on the blood vessels, epi acts at α (vasoconstrictor effect) and β

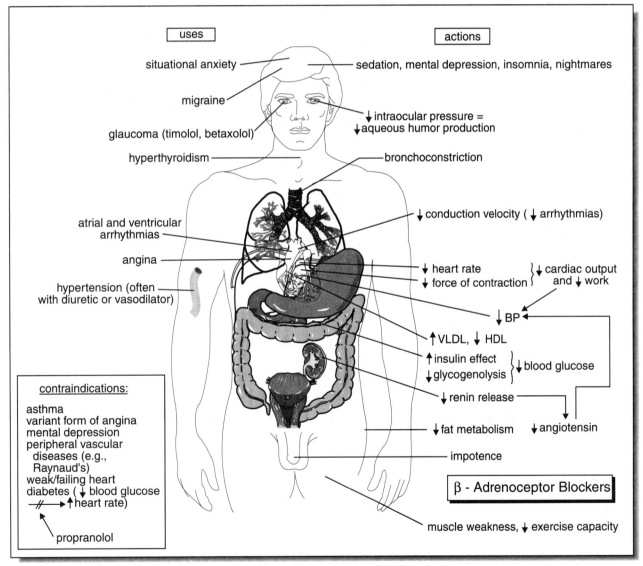

Figure 2.26

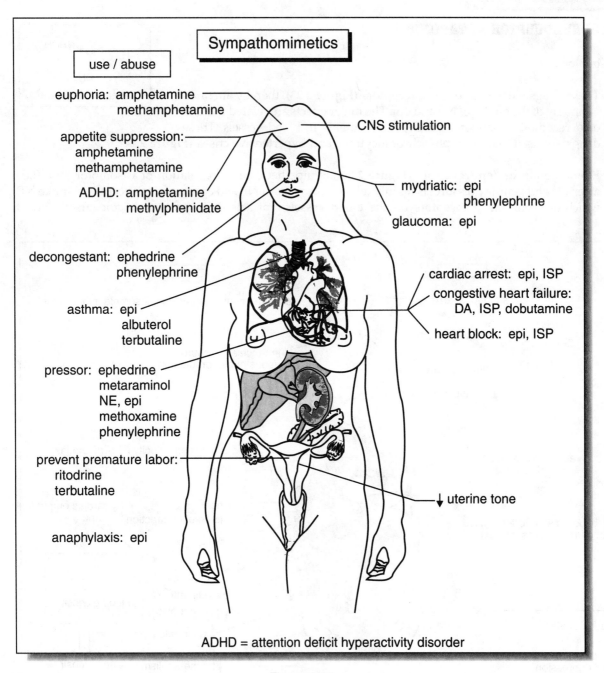

Figure 2.27

(vasodilator effect) receptors. α-sites are not as sensitive as β-sites but are more numerous. Therefore, after a moderate dose of epi, blood pressure increases because so many more α versus β-sites are stimulated. As epi is metabolized or taken-up into nerves, the low levels of epi excite only the sensitive β receptors. This effect produces a drop in blood pressure.

If an α-blocker is given beforehand, the moderate dose of epi is able to excite only β-receptors, so the observed effect is a reduction in blood pressure. This is known as "epi reversal." (The identical effect

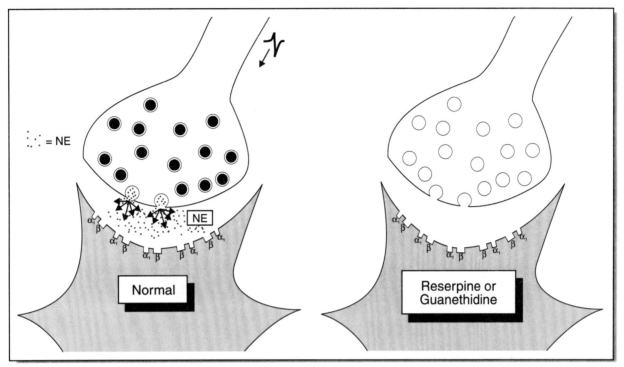

Figure 2.28

on blood pressure can be produced by a very low dose of epi, which is able to excite only the sensitive β receptors) **(Figure 2.27)**.

Specific Drugs.
1. **α_1-selective: prazosin**. *Prazosin* (Minipress) is selective for α_1 receptors, so it produces minimal tachycardia. This drug is useful in chronic hypertension.

2. **α_1, α_2: phentolamine (competitive) and phenoxybenzamine (non-competitive)**. *Phentolamine* (Regitine) is a potent α_1 and α_2 blocker that also is an agonist at muscarinic, H_1 and H_2 receptors. These actions contribute to its activity profile.

    Phenoxybenzamine (Dibenzyline) forms covalent bonds with a_1 and a_2 receptors, producing a non-competitive irreversible block of these receptors. This drug enters the brain and produces sedation (possibly due to its block of H_1, muscarinic and 5-HT receptors), as well as those effects common to α-blockers **(Figure 2.24)**. Because this drug blocks NE reuptake sites, direct acting sympathomimetic amines would produce enhanced responses.

3. **α_2 selective: yohimbine**. *Yohimbine* (Yohimex) is an aphrodisiac that is being tried as a treatment for impotence. However, it produces increased heart rate, increased BP, anxiety and tremor.

B. β-Blockers

These drugs block β_1 and/or β_2 receptors **(Figure 2.23)**, thereby attenuating many of the effects of (a) NE and epi on the heart (β_1) and (b) epi on smooth muscles (β_2). Major beneficial effects are related to β_1 block in the heart, while major adverse effects are related to β_2 block in smooth muscles. These drugs are used mainly to (a) reduce cardiac work (use in angina), (b) reduce heart rate (use in arrhythmias) and (c) reduce peripheral vascular resistance (use in hypertension) **(Figure 2.26)**.

1. **Nonselective β–Blockers (β_1, β_2)**
   a. *Propranolol* (*Inderal*) is the prototype β-blocker. It is well absorbed, extensively metabolized on the first pass through the liver (N.B., prolonged duration with liver disease) and has a half life of about 3-4 hours. It is highly bound to plasma proteins. Propranolol is used mainly for hypertension and cardiac disorders because of its ability to reduce blood pressure and reduce cardiac work and excitability. This and other actions of propranolol are shown in **Figure 2.26**.

   b. *Timolol* (*Timoptic*) is widely used for glaucoma because it is a full antagonist that does not have local anesthetic activity.

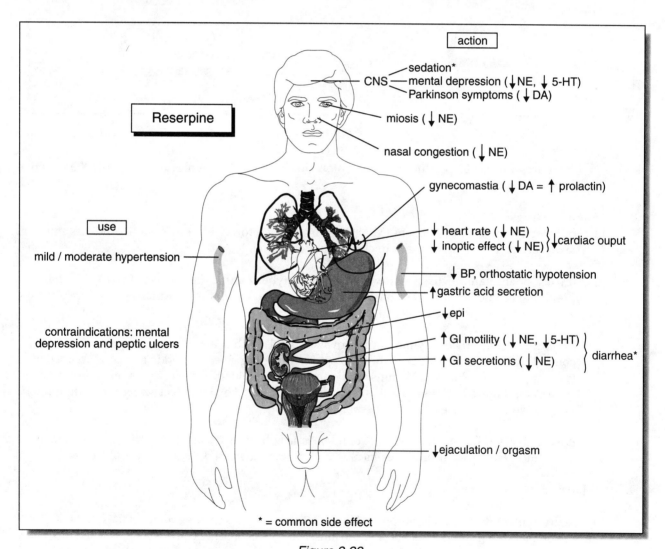

Figure 2.29

c. *Pindolol* (Visken) is a partial agonist at β sites, so it is less likely than propranolol to produce bradycardia, bronchoconstriction and metabolic effects. Pindolol also blocks 5-HT$_1$ receptors and this may contribute to pindolol's spectrum of effects.

d. *Nadolol* (Corgard) has a long t$_½$ (½-1 day).

e. *Labetalol* (Trandate) is non-selective β-blocker (RR isomer) with additional α$_1$-blocking activity (SR isomer). Labetolol is used in hypertension. Overall pharmacological actions reflect changes shown in **Figures 2.24 and 2.26**.

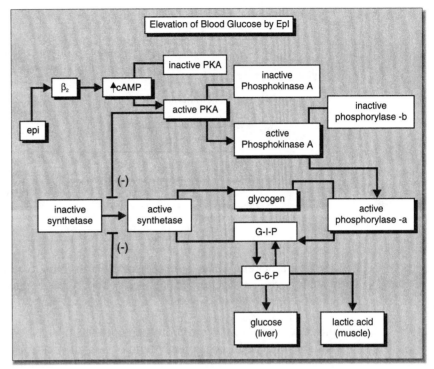

Figure 2.30

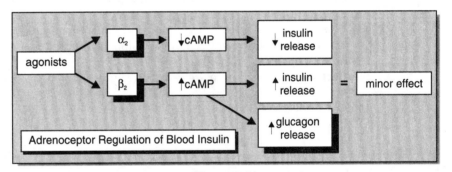

Figure 2.31

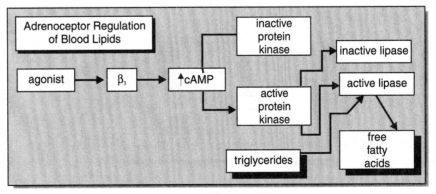

Figure 2.32

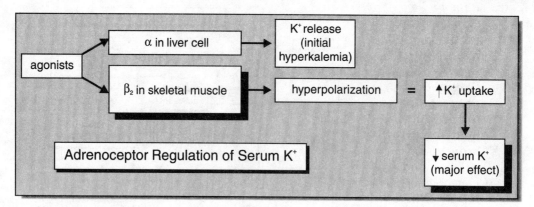

Figure 2.33

2. **β_1-selective Blockers**. These are used for their selective cardiac effects, and are thus less likely to produce bronchoconstriction, bronchosecretion, hypoglycemia, etc. Note that in high doses, these drugs can begin to act at β_2 sites.

   a. *Atenolol* (Tenormin) is the prototype of the β_1-selective antagonists. It does not enter the brain, so CNS effects do not occur. Atenolol is mainly used in hypertension and for stable angina pectoris.

   b. *Acebutolol* (Sectral) and metoprolol (Lopressor) have actions similar to atenolol.

   c. *Esmolol* (Brevibloc) is ultra-short ($t_{1/2}$ about 10 minutes) and is used for supraventricular arrhythmias, perioperative hypertension and acute relief of myocardial ischemia.

3. **β_2-selective blockers**. These are not clinically desirable (e.g., butoxamine).

4. **Uses of β-blockers (Figure 2.26)**
   a. β-blockers increases survival in patients with previous *myocardial infarcts*.

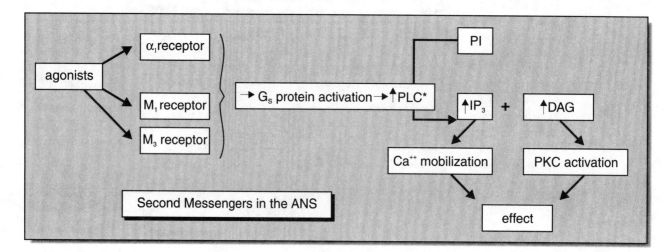

Figure 2.34

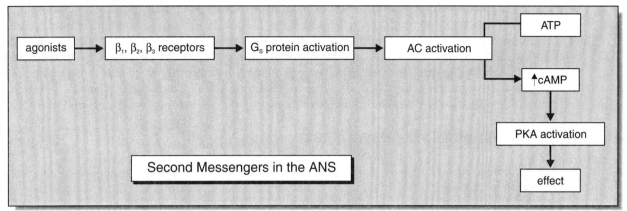

Figure 2.35

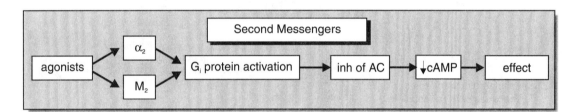

Figure 2.36

    b. β-blockers are useful in *hyperthyroidism* because they antagonize NE and epi effects on the heart.

5. **Discontinuing β-blockers**. Possibly because β-receptor proliferation (up-regulation) occurs with prolonged β-blockade, the β-blockers should not be abruptly discontinued in patients with ischemic heart disease.

F. ADRENOLYTICS

These drugs act within sympathetic nerves, altering (1) the NE storage pool and/or (2) excitation-secretion coupling **(Figures 2.28 and 2.29)**.

  A. **Reserpine**. Reserpine interferes with the ATPase which is essential for uptake and accumulation of NE by *storage granules* in sympathetic nerves. The NE then "leaks" from the granules into the cytosol. Cytoplasmic MAO metabolizes NE to inactive products before NE can diffuse through the cytosol into the synapse. Consequently, postsynaptic α and β sites are not activated as the nerves are becoming *NE depleted*. When sympathetic nerves are stimulated after reserpine treatment, there is an inadequate amount of NE to produce an effective response at α and β postsynaptic receptors.

Parasympathetic tone predominates when the SNS is thus inactivated and this imbalance accounts for most of the pharmacological effect of reserpine **(Figure 2.37)**.

Reserpine enters the brain and has the same amine-depleting effect in nerves containing NE, DA and serotonin (5-HT). Actions of reserpine are shown in **Figure 2.29**.

B. *Guanethidine* depletes NE storage granules, like reserpine, but additionally impairs the coupling of excitation (nerve action potentials) with NE secretion. (This is like a local anesthetic effect.) Acutely, guanethidine is accumulated in sympathetic nerves via the uptake sites used by NE. In effect, NE reuptake is inhibited and a sympathomimetic effect is initially seen with guanethine (in contrast to reserpine). Guanethidine produces all of the actions shown for reserpine, except for those in the brain. Guanethidine has a quaternary nitrogen and does not enter the brain. Guanethidine is more potent than reserpine. Side effects are greater and its range of therapy extends to severe hypertension **(Figure 2.37)**.

G. GANGLIONIC STIMULANTS AND BLOCKERS

Because of their wide spectrum of adverse side effects, these drugs are rarely used. The actions of ganglionic stimulants and blockers are predictable. By knowing the actions of the PSNS and SNS, and by knowing which

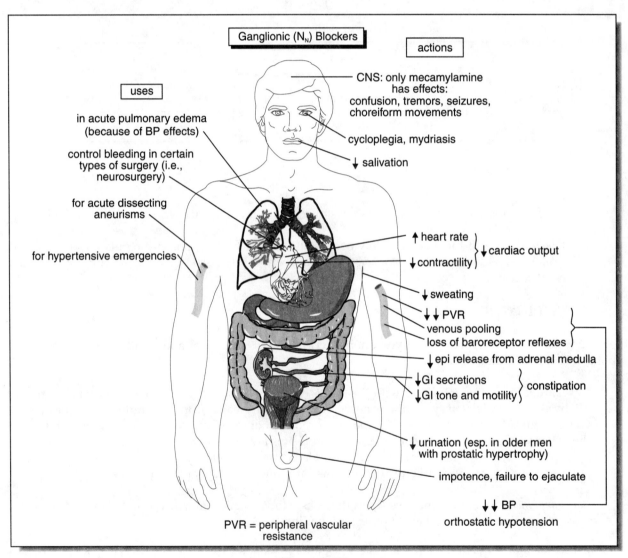

Figure 2.37

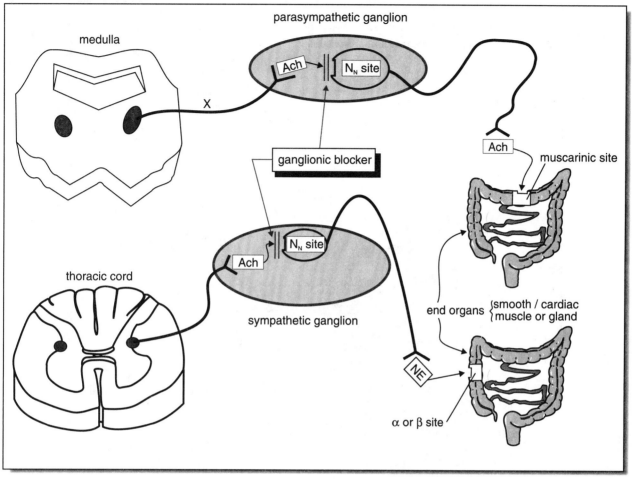

Figure 2.38

of these divisions of the ANS has predominate control over the different organs, one can reasonably predict the pharmacologic effects of ganglionic drugs **(Figure 2.37)**.

Because ganglionic blockers eliminate autonomic homeostatic (ganglionic) reflexes, exogenously administered agonists for muscarinic and adrenergic receptors tend to produce exaggerated responses **(Figure 2.38)**.

- A. **Specific Ganglionic Blockers**. All are competitive nondepolarizing nicotine (N) receptor antagonists (i.e., They do not initially excite N sites).

    1. **Hexamethonium** (C6). This blocks Na^+ channels associated with N_N receptors. The distance of 8 angstroms (Å) between the quaternary nitrogen atoms is the same distance separating critical anionic sites on the N_N receptor.

    2. **Trimethaphan (Arfonad)**. This blocks the N_N receptor *per se*, unlike hexamethonium. Because of its potent effect on

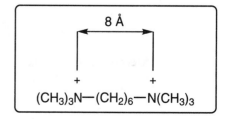

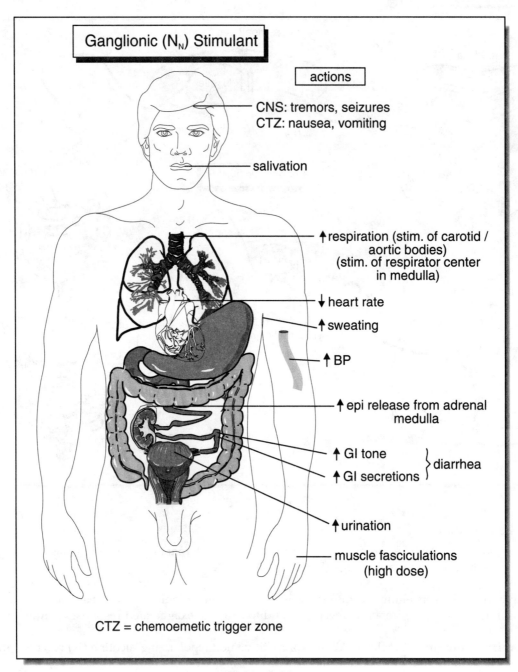

Figure 2.39

blood pressure and desired short duration of action (less than 10 minutes), this drug has some specialized uses **(Figure 2.37)**.

3. **Mecamylamine (Inversine).** This is a secondary amine, able to cross the blood-brain barrier and produce CNS effects, so it is almost never used today.

B. Ganglionic Stimulants (Nicotine)

The only common ganglionic stimulant is nicotine, a constituent of tobacco. In low amounts (as in smoking) nicotine stimulates N_N sites in the CNS and in SNS and PSNS ganglia **(see Figure 2.39)**. In higher amounts, nicotine also acts at N_M sites within the neuromuscular junction (NMJ), producing skeletal muscle fasciculations.

With poisoning, nicotine would ultimately cause depolarizing desensitization block of N_N sites in ganglia, and these actions theoretically would be the same as for ganglionic blockers. In reality, a person would have difficulty surviving the effects of initial massive N_N and N_M stimulation in ganglia and the NMJ, respectively.

CHAPTER 3
SOMATIC NERVES: SKELETAL MUSCLE

A. SKELETAL MUSCLE RELAXANTS

General. "Depolarizing" and "nondepolarizing" neuromuscular blocking agents produce skeletal muscle relaxation by virtue of their binding to nicotine (N_M) receptors on skeletal muscle and blocking of ion channels involved in depolarization of the skeletal muscle. The latter event is essential for muscle contraction **(Figure 3.1)**.

Uses
1. During surgery, to produce skeletal muscle relaxation and paralysis.
   Muscle susceptibility: first, small, rapidly contracting muscles; then, slow muscles; intercostals; and finally, the diaphragm (recovery is in reverse order).
2. For tracheal intubation.
3. For setting broken bones.
4. During epileptic seizures, drug-induced convulsions and electroshock therapy.
5. Diagnosis of myasthenia gravis.

NONDEPOLARIZING DRUGS ("PACHYCURARES")
These are bulky molecules that have two quaternary nitrogens separated by a distance that corresponds to the separation of two anionic sites in the N receptor ionophore complex (approximately 14Å). When bound to the anionic sites, these drugs block the ion channel in the receptor ionophone complex, so depolarization does not occur **(Figures 3.2, 3.5 and 3.6)**. Accordingly, miniature end plate potentials (mepps) are absent. The block can

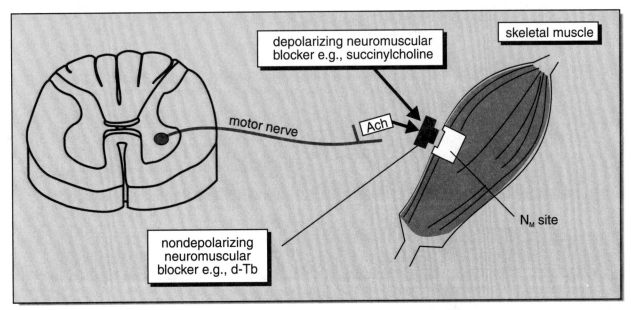

FIGURE 3.1

be overcome with high amounts of a nicotine agonist (e.g., synaptic acetylcholine [Ach] when a cholinesterase-inhibitor [ChE-I] is present).

A. History

These were used for centuries as arrow poisons for hunting game in South America. Muscle paralysis was produced in the wild game. Being quaternary amines, the poison is not absorbed appreciably from the gastrointestinal (GI) tract, so the game could be eaten by the hunters.

B. Specific Drugs

1. d-Tubocurarine (Tubarine)—the prototype.
2. Others: gallamine (Flaxedil), pancuronium (Pavlon), vecuronium (Norcuron).

C. Pharmacokinetics of d-tubocurarine

Absorption: This is poor by an oral route. Administer i.v.
Distribution: The charged molecules do not enter the CNS.
Redistribution, metabolism and excretion: These drugs are not appreciably metabolized, but they rapidly redistribute and are then excreted in urine. Duration of action is generally about 1 hour.

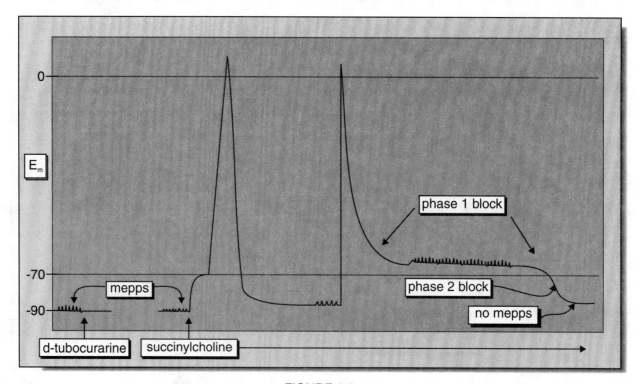

FIGURE 3.2

D. Adverse Effects

1. d-Tubocurarine is a potent releaser of histamine from mast cells. Hypotension occurs. Histamine effects are largely prevented when an H_1-blocker is administered prophylactically.
2. Some anesthetics (halothane, isoflurane, enflurane) and antibiotics (aminoglycosides such as gentamicin) potentiate the effects of d-tubocurarine, probably by inhibiting Ca^{++} uptake by presynaptic cholinergic fibers and thus inhibiting Ach release.
3. Tachycardia may occur because of a vagolytic effect (e.g., *ganglionic block*).

E. Other N_M Pachycures Blockers Versus d-tubocurarine

Some have little effect on histamine release and some are metabolized by the liver and esterases.

DEPOLARIZING DRUGS ("LEPTOCURARES")

These are "skinny" linear chain molecules that also have two quaternary nitrogens separated by about 14Å. When bound to the anionic sites on the nicotine receptor, these drugs initially permit Na^+ to pass through the ion channel to produce depolarization **(Figures 3.5 and 3.7)**. Muscle contractions, sometimes fasciculations occur (pain) **(Figure 3.2)**. These actions correspond to phase 1 block during which nicotine agonists would add to the effect while nicotine antagonists (e.g., d-tubocurarine) would attenuate the effect. Mepps still occur.

Eventually there is a sustained partial depolarization of the skeletal muscle membrane, to a level above that of the resting membrane potential (E_m), so that the Na^+ ion channel is inactivated and action potentials cannot occur.

In time, the skeletal muscle membrane repolarizes and Na^+ ion channels become operative. However, the nicotine receptors become desensitized and unable to respond to agonists like Ach, so mepps are absent **(Figure 3.2)**. ChE-Is will not antagonize the blocker in this phase and d-tubocurarine adds to the effect. This corresponds to phase 2 block.

A. Specific Drugs

Succinylcholine (Amectine) is the protype and only important agent. Chemically, it is two molecules of Ach joined end to end.

Decamethonium with a 10 carbon chain separating the N^+-groups is another agonist at the N_M receptor. A distance of 14 Å separates anionic sites on the N_M receptor, versus the 8 Å of the N_N receptor. Recall that hexamethonium (with only 6 carbons between N-groups) is the preferred agonist at N_N receptors **(Figure 3.3)**.

B. Pharmacokinetics of Succinylcholine

Administration is i.v.
Metabolism occurs rapidly by pseudocholinesterase (nonspecific ChE) so the duration of action is less than 10 minutes.

C. Adverse Effects of Succinylcholine

1. About 1 in 10,000 people have a genetic variant of nonspecific ChE, whereby the enzyme has low affinity for succinylcho-

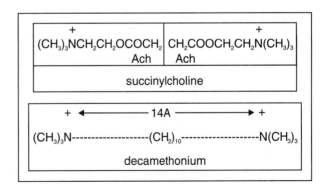

Figure 3.3

line. A reduced rate of metabolism of succinylcholine extends its duration excessively, producing apnea and prolonged paralysis.
2. Low levels of nonspecific ChE can occur with liver disease.
3. The elderly are more susceptible to succinylcholine.
4. Succinylcholine may release large amounts of K^+, especially in patients with existing muscle paralysis (and associated nicotine receptor proliferation), producing hyperkalemia and cardiac arrhythmias.
5. Succinylcholine is an agonist at autonomic ganglia and at muscarinic receptors, resulting in a variety of adverse effects.
6. Malignant hyperthermia occurs rarely with succinylcholine.

B. SPASMOLYTICS

General. These drugs are used to reduce muscle spasms associated with several neurological disorders, such as stroke, multiple sclerosis and cerebral palsy.

Figure 3.4

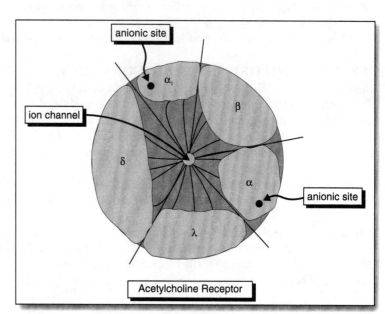

FIGURE 3.5

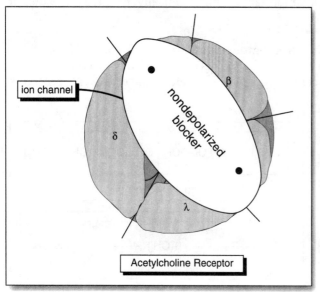

FIGURE 3.6

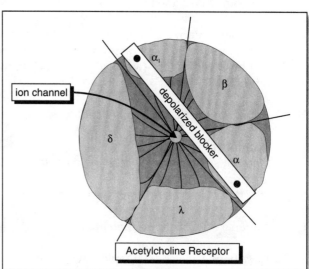

FIGURE 3.7

Specific drugs (Figures 3.8 and 3.9)

1. Diazepam (Valium) **(see Chapter 9)**. Diazepam enhances γ-aminobutyric acid (GABA) actions at pre- and postsynaptic sites in the CNS. In the spinal cord diazepam enhances presynaptic inhibition of sensory fibers by internuncial neurons **(Figure 3.9)**.

2. Baclofen (Lioresal). This chemical analog of GABA blocks postsynaptic GABA receptors in the CNS **(Figures 3.4 and 3.9)**. In the spinal cord, baclofen enhances presynaptic inhibition of sensory fibers by internuncial neurons—like diazepam—and inhibits release of excitatory neurotransmitters. Motor neuron firing rate is thus diminished **(Figure 3.9)**. Baclofen, like diazepam, is taken orally but is less sedating than diazepam.

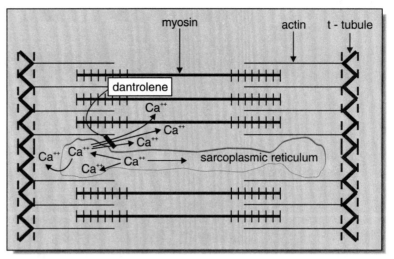

Figure 3.8

3. Dantrolene (Dantrium). Dantrolene impedes release of Ca^{++} from the sarcoplasmic reticulum (SR) of skeletal muscle, thereby diminishing actin:myosin interaction and excitation-contraction coupling **(Figure 3.8)**. Muscle strength and tone is reduced. Rapidly contracting muscles are most affected.

A special use of dantrolene is in the treatment of malignant hyperthermia, a hereditary disorder characterized by an impaired sequestration of Ca^{++} by the sarcoplasmic reticulum. In this case, muscles become spastic and produce much lactic acid and heat. A crisis is sometimes induced by general anesthetics and neuromuscular blocking drugs. Dantrolene is used prophylactically and for overt treatment of this condition. Dantrolene dramatically reduces mortality in malignant hyperthermia. Adverse effects of dantrolene include hepatic toxicity and seizures.

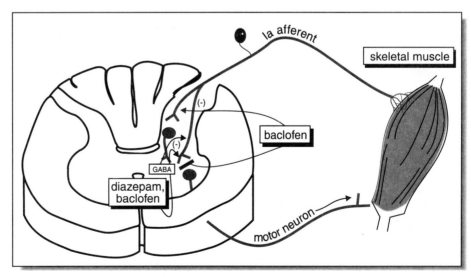

Figure 3.9

CHAPTER 4
DIURETICS

Diuretics are drugs that increase urine volume ("di" and "uretic") by facilitating water (H_2O) excretion and more importantly, sodium (Na^+) excretion in urine. This reduces extracellular fluid volume, including blood volume. Blood vessel walls become less distended and blood pressure is often lowered. Also the back-pressure against which the heart pumps (LVEDP, left ventricular end-diastolic pressure) is thus reduced, so the amount of cardiac work is diminished. Consequently, diuretics have become first-line drugs for treating hypertension, congestive heart failure and edematous states. Some diuretics are used for glaucoma (i.e., to reduce aqueous/vitreous humor volume) and cerebral edema.

Most diuretics are first secreted into the proximal tubule of the nephron and then act on the luminal surface of the nephron. If there is not adequate renal perfusion (e.g., renal failure), most diuretics cannot reach their site of action and therefore cannot produce diuresis.

A. Renal Physiology

The two kidneys normally receive 25% of the cardiac output (5000 ml/min x 0.25 = 1250 ml/min). About 125 ml/min of plasma is then filtered there by first passing through glomeruli. Some substances are actively secreted into the tubules of the nephron, like organic acids (e.g., uric acid). Most solutes within the nephron are reabsorbed, while undesired metabolites are usually excreted by passage through the nephron into what becomes urine. It is critical that almost all of the salt and water be reabsorbed or else electrolyte depletion and dehydration would rapidly occur. Less than 1 ml/min of urine is formed and less than 1 mg/min of Na^+ and other ions is lost in urine.

Some important aspects of plasma filtering by the nephron are illustrated (**Figure 4.1**). The indicated volume of glomerular ultrafiltrate at different sites within the nephron is for the whole kidney, not a single nephron. The approximate osmolality at different positions outside the nephron is also shown. Major sites of action of the major diuretics are indicated.

About 65% of water and electrolytes are reabsorbed in the proximal tubule. The loop of Henle, the distal tubule and collecting duct are major sites for concentrating fluid that becomes urine. Antidiuretic hormone (ADH) regulates the permeability of the distal tubule and collecting duct. Aldosterone acts primarily at the distal tubule and collecting duct to conserve Na^+ at the expense of K^+.

B. Diuretics

1. *Carbonic anhydrase inhibitors* (CAIs) are older diuretics that work primarily at the proximal tubule. Since the glomerular filtrate must still pass through those elements of the nephron that concentrate urine, and since those sites can easily handle a greater filter load, the CAIs are not very potent.

Mechanism of CAI action (Figure 4.2). CAIs act primarily in the proximal tubule (PT), by noncompetitive inhibition of carbonic anhydrase (CA), an enzyme that catalyzes formation of carbonic acid (H_2CO_3) **(Figure 4.2)**. The net effect is reduced H^+-Na^+ exchange with more bicarbonate (HCO_3^-) excretion and less Cl^- excretion. Eventual mild hyperchloremia (elevated plasma Cl^- concentration) and metabolic acidosis (reduced plasma HCO_3^- concentration) overides the CA inhibition, causing refractoriness in less than 48 hours and only mild diuresis.

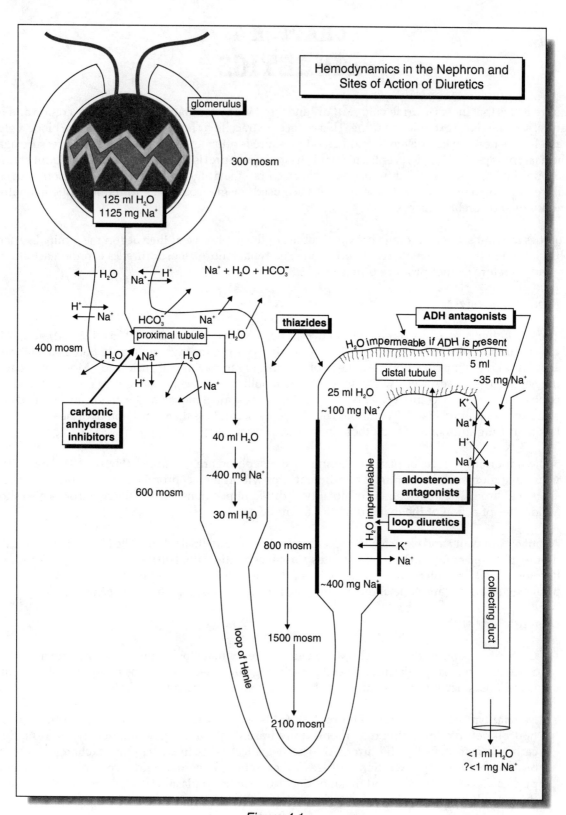

Figure 4.1

Chemistry and Pharmacokinetics of CAIs. These drugs are sulfonamides (RSO_2NH_2). Acetazolamide (Diamox) is rapidly absorbed, exerts peak effects in approximately 2 hours and is excreted unmetabolized in less than 12 hours. Other CAIs are metabolized.

Uses
a. CAIs are too short-lived to be practical as diuretics, except for brief use in conditions like cor pulmonale in which there is an elevated P_{CO_2}. Most often, CAIs are given when there is metabolic alkalosis, as in CHF, or to offset the alkalosis of other classes of diuretics.
b. Because CAIs alkalinize urine, they are used to enhance excretion of acidic drugs and toxins, or to increase uric acid excretion in gout.
c. CAIs are used as adjuncts in epilepsy, occasionally in acute altitude sickness (to reduce CSF formation) and in glaucoma (to reduce aqueous humor formation).

Adverse effects. These include hypokalemia and a tendency towards nephrolithiasis (elevated pH of urine and hypercalcuria). Taste is sometimes affected. Hypersensitivity occurs rarely. CAIs are contraindicated in cirrhosis, since the excess ammonia formed in cirrhosis has to be excreted as NH_4^+ in acidic urine; otherwise, hepatic encephalopathy occurs.

2. *Thiazides* are also sulfonamide analogs that still retain some CAI activity. Thiazides produce moderate diuresis, but their action requires an adequate glomerular filtration rate (GFR).

Mechanism of Thiazides. Thiazides are actively secreted into the renal tubule by the organic acid transport system. Thiazides

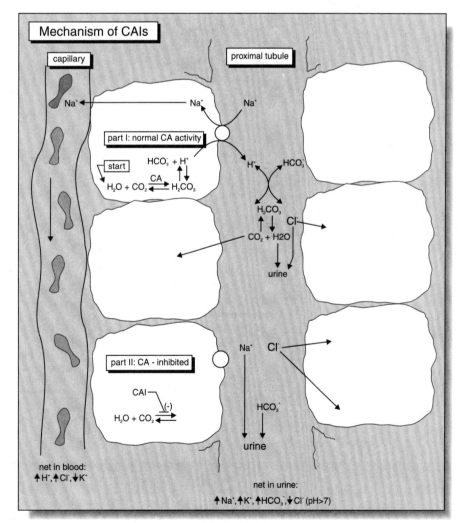

Figure 4.2

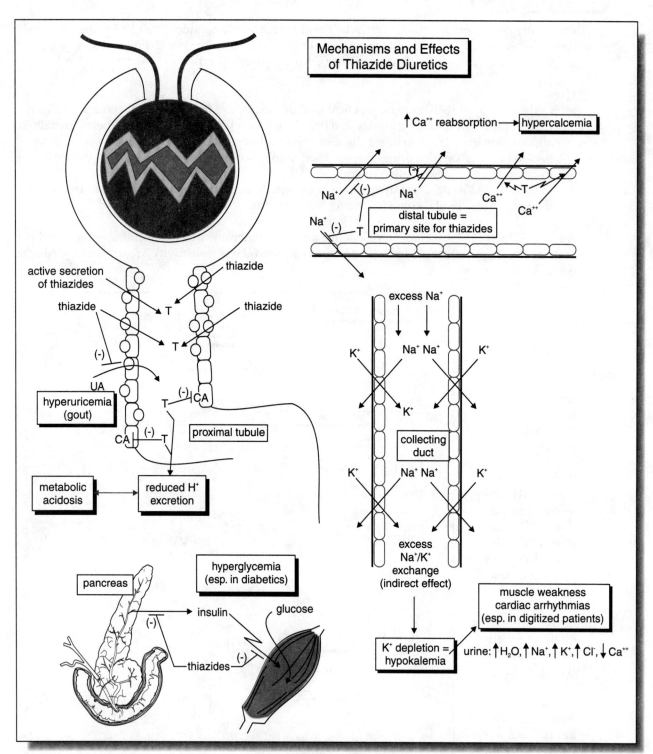

Figure 4.3

compete for these sites with acidic drugs and metabolites, and each interferes with excretion of the other (e.g., NSAIDs, uric acid). Thiazides then travel within the nephron to their primary site of action, the distal convoluted tubule, and interfere with reabsorption of Na^+ and Cl^-. The main effects of thiazides are shown in **Figure 4.3**.

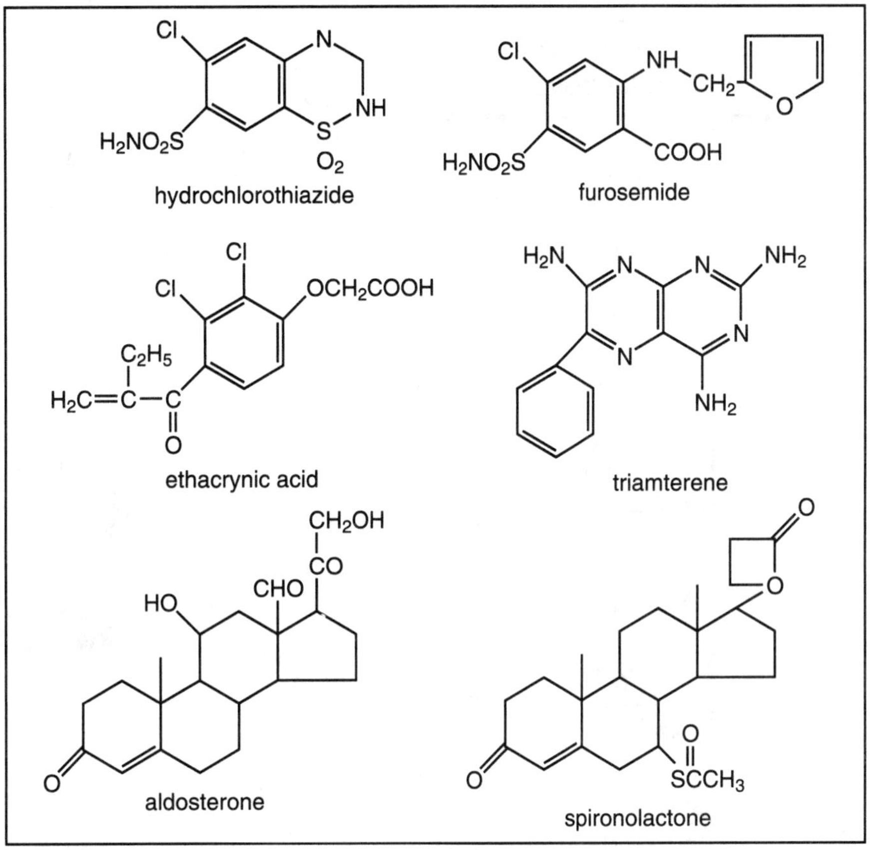

Figure 4.4

Pharmacokinetics. Hydrochlorothiazide (Hydrodiuril) is the prototypic short-lived thiazide, while chlorthalidone (Hygroton) is the prototypic long-lived thiazide-like drug **(Figure 4.4)**. Onset of action is in less than 2 hours, metabolism is minimal and elimination is in less than 6 hours. There is cross resistance among the thiazides.

Thiazides are used to treat
  a. Hypertension. A reduction in peripheral vascular resistance (PVR) is not due to diuresis alone.
  b. Congestive heart failure (CHF: right side failure).
  c. Cirrhotic edema. However, thiazides should be used cautiously with chronic hepatic or renal disease.
  d. Nephrogenic diabetes insipidus. Less urine with higher osmolality is formed.
  e. Hypercalciuric renal stones.

Adverse effects of thiazides. These are outlined in **Figure 4.3**. An anti-insulin hyperglycemic effect is related to reduced insulin release or action. Increased VLDL-cholesterol and triglyceride levels occur. Hypersensitivity is rare (rash, agranulocytosis).

2B. *Thiazide-like diuretics* include chlorthalidone (Hygroton), metolazone (Diulo) and indapamide (Lozol). These are long-lived thiazide analogs that act at virtually the same site in the nephron as thiazides. Metolazone and indapamide are effective even when there is renal impairment, unlike thiazides. Indapamide also directly reduces PVR.

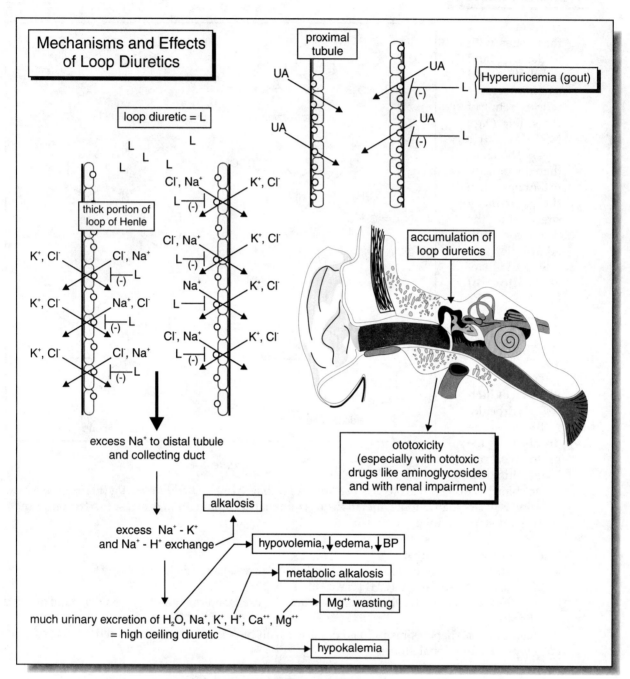

Figure 4.5

3. *Loop diuretics* act on the thick portion of the loop of Henle. These are "high ceiling" diuretics, because of their potential to eliminate a high percentage of the ultrafiltrate—far more than for other diuretics. Loop diuretics act even if the GFR is low, and often in patients who are refractory to other diuretics. These diuretics should be given in low doses initially. Plasma electrolytes, especially K^+, should be monitored.

Mechanism and actions of loop diuretics. Loop diuretics inhibit a cotransporter system for 1 Na^+ + 1 K^+ + 2 Cl^-, thereby inhibiting Na^+ and Cl^- reabsorption. The distal tubule is unable to handle this increased ionic load. Actions are shown in **Figure 4.5**. Also, since less Na^+ and Cl^- are reabsorbed, the peritubular osmotic gradient partially degrades. Loop diuretics also increase venous capacitance. The nephrotoxicity of cephalosporins is increased by furosemide.

Pharmacokinetics. Furosemide (Lasix), bumetanide (Bumex) and ethacrynic acid (Edecrin) are rapid-acting diuretics that can be administered orally or parenterally. Furosemide is not metabolized; the others, partially. All are highly bound to plasma proteins, so they interact with other highly bound drugs like NSAIDs. Unlike furosemide and bumetanide, ethacrinic acid is not a sulfonamide **(Figure 4.4)**. Furosemide and bumetanide produce cross-sensitivity reactions in patients allergic to other sulfonamides (e.g., thiazides, antimicrobial sulfonamides).

Loop diuretics are used to treat
   a. Hypertension, CHF and edema (including pulmonary edema). There is synergism with thiazide diuretics.
   b. Hypercalcemic crisis.

4. *Potassium-sparing diuretics* interfere with Na^+-K^+ exchange in the distal tubule and collecting duct. Some are aldosterone receptor antagonists **(Figure 4.6)**. All are weak and are commonly used with other diuretics.

4A. *Aldosterone antagonist.* Spironolactone (Aldactone) is a chemical analog of aldosterone **(Figure 4.4)** that prevents aldosterone binding to cytoplasmic receptors. If aldosterone is not present in plasma (e.g., adrenalectomy), the drug has no action. Conversely, if aldosterone levels are excessively high (e.g., high salt diet), spironolactone

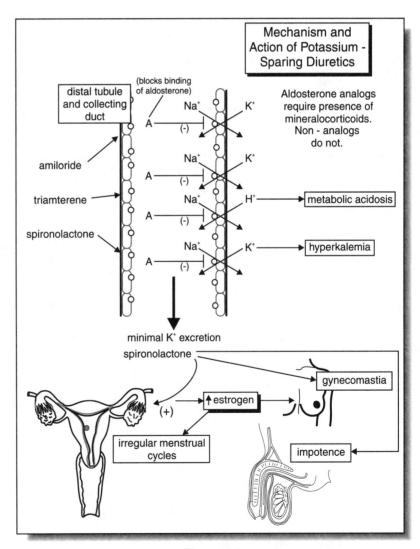

Figure 4.6

cannot effectively compete with aldosterone binding, and therefore is ineffective.

Pharmacokinetics of spironolactone. Spironolactone **(Figure 4.4)** is orally absorbed and metabolized mainly in the liver to canrenone and canrenoate, active long-lived ($t_{1/2}$ 20 hours) metabolites that are excreted

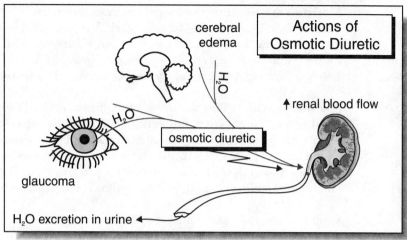

Figure 4.7

mainly in the urine. Effects are delayed for several days, until previously produced aldosterone-induced Na^+-transporter proteins degrade. This drug should not be used if the liver or kidney is impaired.

4B. *Non-aldosterone antagonists*. Triamterene (Dyrenium) and amiloride (Midamor) "spare K^+" by irreversibly inhibiting Na^+-K^+ exchange in renal tubule cells **(Figure 4.6)**.

Pharmacokinetics of Triamterene and Amiloride. Triamterene and amiloride are orally effective, producing peak effects in several hours. Triamterene is converted to active metabolites which produce most of the effect. Triamterene is eliminated in urine, mainly as active metabolites. Amiloride is a base that is actively secreted into the proximal tubule. (Any other basic drug would compete for this site, affecting elimination of that drug as well as amiloride.) Amiloride is excreted largely unmetabolized in urine. Postassium supplements should be discontinued when these drugs are taken. To avoid hyperkalemia, the K^+-sparing diuretics should not be used together.

Potassium sparing diuretics are used
a. to counteract the K^+- and H^+-depletion of other diuretics. Since Na^+-K^+ and H^+-K^+ exchange in the collecting tubule account for only 2-3% of Na^+ reabsorption, these drugs are too weak to be used alone.
b. Spironolactone is used to treat primary aldosteronism (Conn's syndrome, ectopic ACTH tumors) and secondary aldosteronism (cirrhosis, CHF).

5. Osmotic diuretics are metabolically inert small molecules that do not enter cells. They are filtered at the glomerulus, but are not reabsorbed in renal tubules.

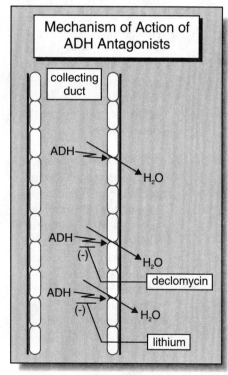

Figure 4.8

Their retention in extracellular fluid compartments, like blood and the lumen of the nephron, leads to increased osmotic retention of water and increased urine flow with water elimination—as per definition of a diuretic **(Figure 4.7)**.

Uses. Mannitol (Osmitrol), and less commonly urea (Ureaphil), are administered i.v. to
a. maintain urine flow and thereby prevent anuria during surgical procedures in which excess bleeding is anticipated;
b. prevent precipitation of a drug or toxin in the kidneys, by "flushing" the kidneys (e.g., in drug overdose) and
c. osmotically remove water from edematous cells, as when there is increased intraocular pressure (e.g., glaucoma) or increased intracranial pressure (e.g., head trauma).

Glycerin (Glyrol) and isosorbide (ismotic) are administered orally for ophthalmic surgical procedures. Topical glycerin is used to reduce corneal edema.

Adverse effects. An expansion of extracellular fluid volume is the major adverse effect, because of the consequent increased work load on the heart and increased potential for pulmonary edema. CHF is a contraindication for osmotic diuretics. A test dose of mannitol is recommended when renal insufficiency is present, since urine flow would not necessarily occur in this condition. Blood volume expansion could be fatal.

6. *ADH antagonists* attenuate ADH effects on water reabsorption in the collecting duct. These are used for the rare "syndrome of inappropriate ADH release." The tetracycline analog demeclocycline (Declomycin) and the anti-manic lithium (Li^+) each have this effect on ADH **(Figure 4.8)**. In excess, nephrogenic diabetes insipidus occurs (see respective chapters).

CHAPTER 5
CARDIOVASCULAR PHARMACOLOGY

A. CARDIOVASCULAR PHARMACOLOGY

Heart disease continues to be the number one killer in the United States. As might be expected, cardiovascular drugs are among the most widely used drugs today. These are used to
    a. decrease high BP (hypertension),
    b. decrease blood volume,
    c. stabilize cardiac rhythm,
    d. increase cardiac work efficiency and
    e. decrease the tendency for formation of emboli or
    f. atheromas.

The following sections describe drugs for each of these entities.

B. TREATMENT OF HYPERTENSION

Hypertension is defined according to the extent of blood pressure (BP) elevation:
    Borderline (mild): 140-160/90-105
    Moderate : -180/ -120
    Severe : >180/ >120

These values are approximate. Different upper limit values are sometimes used. Hypertension is associated with increased incidence of stroke, coronary artery disease, congestive heart failure (CHF), aneurysm, renal failure, retinopathy, and other disorders. In about 90% of cases, there is no diagnosis of an underlying basis for the elevated BP. This is termed essential hypertension. Secondary hypertension represents the instances where the cause of hypertension can be defined (e.g., renal artery stenosis).

When hypertension is mild, non-drug approaches towards lowering blood pressure include (a) reducing salt intake, (b) reducing body weight and reducing fat consumption, (c) eliminating specific risk factors like smoking and (d) instituting an exercise program. When these measures are inadequate, drugs must also be instituted. This section describes the most commonly used antihypertensive drugs.

ANTIHYPERTENSIVES

1. **Diuretics used for hypertension**. Diuretics enhance the renal excretion of H_2O. The net effect is a reduction in blood volume, accompanied by expected increases in renin and aldosterone levels. (Consequently, these would not be used for hypertensives with elevated renin levels.) Reduced blood volume is associated with reduced pre-load, reduced afterload and reduced cardiac output. Diuretics are used alone for mild hypertension or in combination with other antihypertensives for more severe cases.

    a. *Thiazides* like chlorothiazide (Diuril) are well-tolerated but may produce
       i. hypokalemia (skeletal muscle weakness, cardiac arrhythmias and death) and hypercalcemia, (Potassium supplements are often included as part of thiazide therapy.);
       ii. hyperglycemia;
       iii. increased VLDL and LDL and
       iv. hyperuricemia (gout).

b. *Thiazide-like* diuretics (e.g., chlorthalidone [Hygroton]). These diuretics are longer-acting than thiazides and may produce fewer adverse effects. Metolazone (Zaroxolyn) and indapamide (Lozol) may be effective in patients with impaired renal function.

c. *Loop diuretics* like furosemide (Lasix), bumetanide (Bumex) and ethacrynic acid (Edecrin) are very potent in promoting H_2O excretion. These diuretics are shorter acting than thiazides and are not as effective for routine treatment of hypertension. Loop diuretics are most effective for patients with renal insufficiency. Loop diuretics are usually combined with potassium-sparing diuretics. Loop diuretics are often used in combination with antihypertensives that tend to increase fluid volume (e.g., guanethidine and vasodilators).

d. *Potassium-sparing diuretics* like amiloride (Midamor), triamterene (Dyrenium) and spironolactone (Aldactone) are only used for hypertension with other diuretics, in order to prevent hypokalemia. Caution must be used with these drugs
   i. in patients with renal impairment or
   ii. in patients also receiving ACE-inhibitors which reduce aldosterone levels and enhance the K^+-sparing.

All of these diuretics are described more completely in **Chapter 4, Diuretics**.

2. **Calcium channel blockers** are used as monotherapy in mild to moderate hypertension and as part of polytherapy (with β-blockers, ACE-inhibitors, α-methyldopa) in moderate to severe hypertension. In contrast to α-blockers, Ca^{++}-channel blockers can be used safely in asthmatics and diabetics. The Ca^{++}-channel blockers are most effective in hypertensives with low renin levels (e.g. most blacks and geriatric patients). These drugs are more fully described in **Section C, Anti-anginal Drugs and in Chapter 2**.

Advantages of Ca-channel blockers. These drugs do not cause fluid retention and can be used safely even if there is renal impairment. Verapamil and diltiazem, because of direct actions on the SA node, have little net effect on heart rate (direct inhibition of Ca channels of SA node counteracts the baroreceptor-mediated reflex increase in sympathetic tone to SA node).

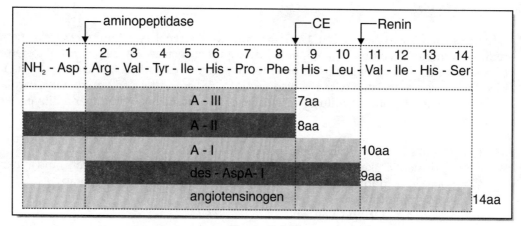

Figure 5.1

Adverse effects
a. Dihydropyridines produce headache and flushing with peripheral edema.
b. Diltiazem produces bradycardia.
c. Verapamil is associated with constipation.

3. **Angiotensin converting enzyme (ACE) inhibitors**

   The renin-angiotensin system. Angiotensinogen is a peptide of 14 amino acids (aa) that is normally hydrolyzed to the decapeptide angiotensin I (A-I) by renin, a hormone produced by cells in the juxtaglomerular apparatus in afferent arterioles that supply blood to the glomerulus. Plasma converting enzymes further hydrolyze A-I to the octapeptide, A-II, an active species. A-II, in turn, can be hydroxyzed by aminopeptidases to a heptapeptide, A-III, which is also active. Further metabolism of this peptide by peptidases results in loss of biological activity. In an alternate pathway, des-Asp^1-peptides are formed, to active products which are illustrated in **Figure 5.1**.

   Physiological effects and regulation of angiotensin (Figure 5.2). When there is either decreased BP or blood volume in renal afferent arterioles, juxtaglomerular (JG) cells release the plasma hormone, renin. Increased sympathetic activity, which releases norepinephrine (NE) onto β_1 receptors on the JG cells also releases renin, as does a reduced concentration of Na$^+$ in distal tubules (DT) of the nephron.

   Renin, in turn, acts on angiotensinogen to ultimately form active A-II and A-III, potent vasoconstrictors (causing increased BP). Also, by acting on the adrenal zona glomerulosa, aldosterone is released. This hormone (a) promotes release of antidiuretic hormone (ADH), which enhances water absorption in the DT and collecting duct (CD) of the nephron (causing increased blood volume). Aldosterone also enhances Na$^+$-K$^+$ exchange in the DT, enhancing Na$^+$ absorption (causing increased BP plasma volume). These actions tend to maintain BP.

   ACE inhibitors are drugs that inhibit converting enzyme and thereby prevent conversion of A-I to A-II. This action reduces plasma levels of A-II and A-III—effects that tend to decrease BP.

   a. *Captopril* (Capoten) is rapidly absorbed, has a bioavailability of about 65%, has peak activity at 1 hour and is excreted in urine as both active species and metabolites. Duration of action after a single dose is about 12 hours.

   b. *Enalapril* (Vasotec) is a prodrug that is activated by serum esterases. The onset of action is slower versus captopril, but duration of action is longer (approximately 24 hours).

   c. *Lisinopril* (Zestril) is an active analog of enalapril that is excreted largely unmetabolized in urine. Duration of action is similar to that of enalapril.

   Adverse effects of ACE inhibitors
   a. Occasional profound decreased BP, especially if plasma renin level is high.
   b. Skin rash and ageusia (loss of taste) (captopril more than enalapril).
   c. Cough in greater than 10% of patients, unresponsive to drug treatment.
   d. Enhanced retention of K$^+$, when used in combination with K-sparing diuretics.
   e. Proteinuria and neutropenia (rarely).
   f. Inhibition of many types of CEs, including one that inactivates bradykinin (BK), an algesic that produces vasodilation and decreased BP.

   Advantages of ACE inhibitors. These drugs do not alter sympathetic activity, so reflexes are intact. Also, bronchoconstriction, anti-insulin effects and inhibition of sexual climax are not produced, versus adrenoceptor blockers.

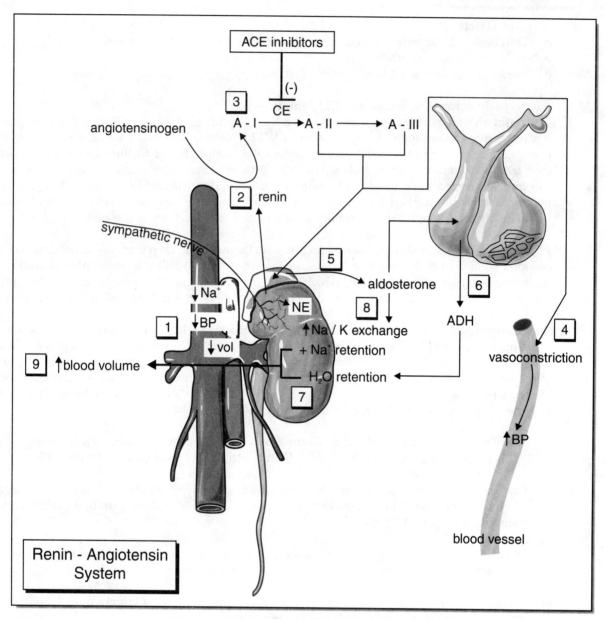

Figure 5.2

Use in hypertension. Because of their favorable profile of activity, ACE inhibitors are useful for treating hypertension. Patients tend to be compliant with these drugs. Blacks are partly resistant to the actions of ACE inhibitors, but concurrent use of diuretics seem to restore activity.

4. **β-blockers** are sometimes used as monotherapy in mild to moderate hypertension. For moderate to severe hypertension, β-blockers are often one of the drugs in polytherapy. β-blockers are generally less-effective in blacks and elderly patients. Actions of β-blockers are summarized in **Figure 5.3** and described more completely in **Section C, Anti-anginal Drugs and Chapter 2**.

a. Propranolol (Inderal) (β_1, β_2) produces an antihypertensive effect in the absence of postural hypotension. A hypertensive crisis with arrhythmias may occur if propranolol is abruptly withdrawn.
b. Nadolol (Corgard) (β_1, β_2).
c. Atenolol (Tenormin (β_1 much greater than β_2). CNS effects are not seen with nadol or atenolol.
d. Metoprolol (Lopressor) (β_1 much greater than β_2). Metoprolol and atenolol are safer than propranolol in asthmatics and diabetics.
e. Pindolol (Visken) and acebutolol (Sectral) (β_1, β_2) have actions much like propranolol, but with additional intrinsic sympathomimetic action (so β-receptors are partially activated). Consequently, heart rate and cardiac output are less impaired. There is also less risk in asthmatics and diabetics. Unlike other β-blockers, pindolol and acebutolol do not elevate triglyceride levels.

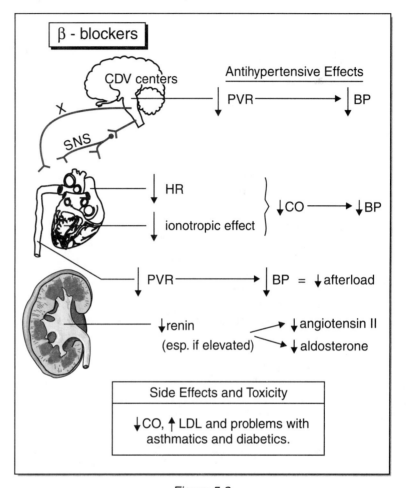

Figure 5.3

f. Labetalol (Trandate) (β_1 and β_2 more than α) is a partial β_2 agonist, so there is less risk in asthmatics and diabetics, versus propranolol. Heart rate and cardiac output are minimally affected. Because BP is rapidly reduced, labetalol can be used for hypertensive crisis.

5. **α-adrenoceptor antagonists**. These drugs cause increased Na^+ and H_2O retention, which tends to offset the α-blocking effect. Consequently, α-blockers are usually used with other antihypertensives (diuretic, β-blocker). Postural hypotension is a common side effect of α-blockers. Actions are summarized in **Figure 5.4**.

   a. Prazosin (Minipress) (α_1) does not block α_2 sites, so the sympathetic system is only partially inhibited. Accordingly, there is less reflex tachycardia and less orthostatic hypotension. These actions, plus a reduced preload and afterload, make prazosin a good antihypertensive in patients with CHF.
   b. Phentolamine (Regitine) (α_1, α_2).
   c. Phenoxybenzamine (Dibenzyline) (α_1, α_2).
   Phentolamine and phenoxybenzamine are useful for the hypertensive crisis in pheochromocytoma.

6. **α₂-agonists** are used for all stages of hypertension, often with a diuretic. Actions of α₂-agonists are summarized in **Figure 5.5**. There is a more complete discussion of α₂-agonists in **Chapter 2**.

   a. *Clonidine* (Catapres) (α₂) seems to predominately reduce sympathetic activity to the heart, but has no consistent effect on peripheral vascular resistance. Prolonged use may lead to depression. Antidepressants (α-block) reverse the antihypertensive action of clonidine. Abrupt removal of clonidine precipitates withdrawal symptoms (agitation, tachycardia) and increased BP (hypertensive crisis).

   b. *Methyldopa* (Aldomet) (α₂) seems to predominately reduce sympathetic activity to blood vessels. There are fewer cardiac effects versus clonidine.

   Methyldopa (α-Me-DOPA) is a prodrug that owes its activity to the metabolites α-Me-DA and α-Me-NE, which act as false neurotransmitters in DA and NE nerves, respectively. α-Me-NE acts as an agonist at α₂ receptors.

   Since α-Me-DA lacks the intrinsic activity of DA, side effects related to low DA activity are seen. These include:
   i. **Hyperprolactinemia** (cuases gynecomastia, galactorrhea). DA is the physiologic regulator (inhibitor) of prolactin release from the pituitary.

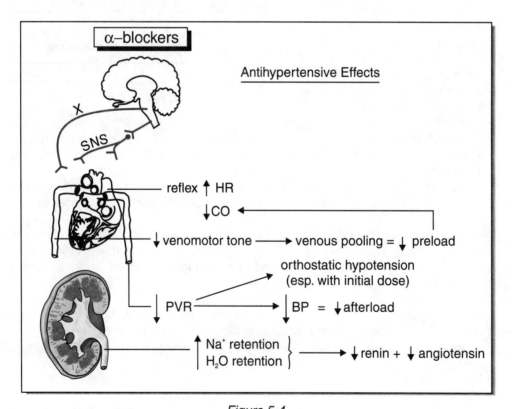

Figure 5.4

ii. **Drug-induced parkinsonism.** DA in nigrostriatal nerves regulates extrapyramidal motor activity.

7. **Adrenolytics.** These drugs tend to inactivate excitation-secretion coupling or to deplete sympathetic nerves of the neurotransmitter, NE. The net effect is inactivation of sympathetic nerves. These drugs are more-fully described in **Chapter 2, Adrenolytics.**

   a. *Reserpine* (Serpasil) is used in mild to moderate hypertension, often with a diuretic. Reserpine depletes NE

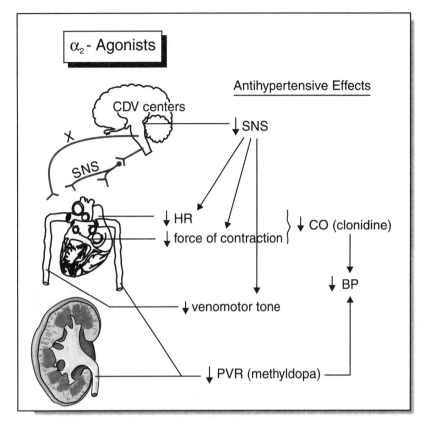

Figure 5.5

from sympathetic nerves, releasing NE as inactive metabolites, so that acute sympathomimetic effects are generally not observed. Eventual reduced sympathetic tone produces decreased vasoconstriction, decreased PVR, decreased BP, decreased heart rate and decreased inotropy with a reduced cardiac output that further decreases BP.

   b. *Guanethidine* (Ismelin) is used only in severe hypertension. Guanethidine depletes NE from sympathetic nerves like reserpine and also interferes with excitation-secretion coupling. These actions inactivate sympathetic nerves causing decreased PVR and decreased cardiac output, which reduces BP. Unlike reserpine, guanethidine does not get into the brain.

   Initially, guanethidine releases NE intact, producing a sympathomimetic action, increased BP etc. This could be fatal if a pheochromocytoma is present.

   Drugs that act on the NE transporter (uptake site) (e.g., cocaine, amphetamine, tricyclic antidepressants) will prevent guanethidine from entering nerves. These drugs would prevent guanethidine actions and could precipitate a hypertensive crisis.

8. **Musculotropic vasodilators.** These drugs directly relax smooth muscles of arteries or veins, or act on cells that release substances having this effect. These drugs are most frequently used for severe hypertension.

   Because compensatory changes are likely to offset the beneficial effect (e.g., decreased renal perfusion causes fluid retention, increased blood volume; or decreased BP causes increased renin causing increased A-II; or decreased BP causes increased cardiac output and increased cardiac work), these drugs are usually combined with antihypertensive drugs that inactivate the compensatory processes. The musculotropic vasodilators do not have side effects in common with sympatholytic drugs that produce orthostatic hypotension, inhibition of orgasm, etc.). Their actions are outlined in **Figure 5.6**.

a. **Arteriolar vasodilators**. All of these produce decreased BP, which is accompanied by activation of baroreceptor reflexes that increases sympathetic tone and heart rate.

  i. *Hydralazine* (Apresoline) seems to stimulate production of NO and interfere with mobilization of Ca^{++} in vascular smooth muscle.

  Toxic effects include a reversible lupus-like syndrome accompanied by arthralgia and myalgia.

  ii. *Minoxidil* (Loniten) is a prodrug, metabolized to the active species, minoxidil sulfate, which opens K^+-channels in arteriolar smooth muscle causing hyperpolarization, relaxation and vasodilation. Minoxidil is often reserved for patients that do not respond well to hydralazine. Side effects and toxic effects include severe fluid retention and hirsutism. (Minoxidil is used to treat balding.)

  iii. *Diazoxide* (Hyperstat) i.v. is reserved for hypertensive emergencies. Its action on K^+ channels is analogous to that of minoxidil. Diazoxide is a thiazide analog that increases Na^+ and H_2O retention, although this is not of importance since diazoxide is only used for brief periods in controlling severe hypertension. Side effects are related to its severe hypotensive effect, which may cause stroke and MI.

  Because it inhibits insulin release, diazoxide is used to treat insulinomas.

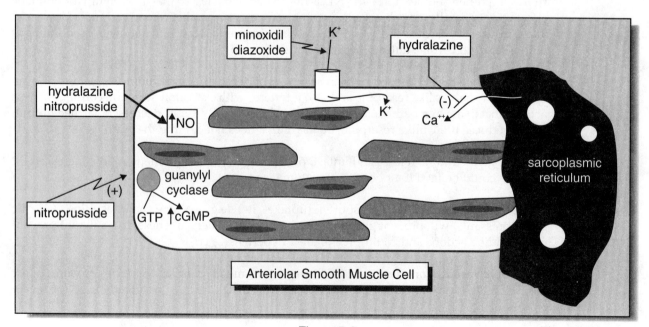

Figure 5.6

b. **Arteriolar plus venule vasodilators**. These also tend to activate compensatory reflexes that increase heart rate, etc.

Nitroprusside (Nipride[$Fe(CN)_5NO$]), administered by i.v. infusion, dilates both arterioles and venules by activating guanylyl cyclase (causing increased cGMP) and increasing NO production. The net effect is decreased preload and decreased afterload. Cardiac output and cardiac work are reduced. This drug can be used for emergency relief of angina and heart failure. It is often the drug of choice for hypertensive crises.

In excess, thiocyanate (SCN^-) toxicity (causing convulsions, psychoses) could be produced:

$$Fe(CN)_5NO \xrightarrow{RBC} CN^- \xrightarrow{S} SCN^-$$

Because CN^- is so toxic, sodium thiosulfate (Na_2SO_3) is sometimes given as a sulfur donor:

$$CN^- + S_2O_3^{-2} \longrightarrow SCN^-$$

9. **Ganglionic blockers**. Such a wide spectrum of adverse actions are produced, that most patients are unable to tolerate these drugs. Consequently, they are reserved for hypertensive emergencies. These drugs are more-completely described in **Chapter 2**.

   a. *Trimethaphan* (Arfonad), i.v., owes its therapeutic action to blockade of sympathetic ganglia. This decreases sympathetic activity to arteries and veins, producing vasodilation, decreased BP and venous pooling. There is reduced pre- and after-load. Side effects are due to block of both sympathetic and parasympathetic ganglia.

SUMMARY OF ANTIHYPERTENSIVE DRUGS
General use for each class

1. Mild-moderate hypertension	2. Severe hypertension	3. Hypertensive crisis
Diuretics	Diuretic and ACE-inhibitor	Nitroprusside
Ca-channel blockers	β-blocker	Labetalol
ACE-inhibitors	α-blocker	Guanethidine
β-blockers	Ca-channel blocker	
———	reserpine	
α-blockers (and diuretic or β-blocker)	β-blocker and α-blocker	
Reserpine	Musculotropic vasodilator and β-blocker	
	Guanethidine	
	Reserpine	

Other drug combinations are also used to treat hypertension. There are many drug interactions involving antihypertensives. In particular, nonsteroidal anti-inflammatory drugs (NSAIDs) tend to reduce the effectiveness of most antihypertensives.

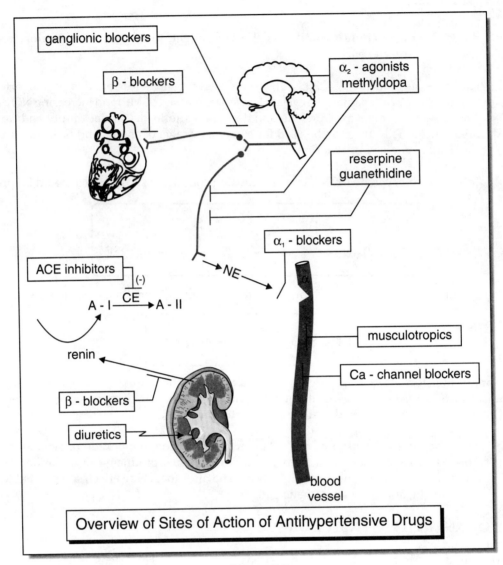

Figure 5.7

SUMMARY ON USES OF ANTIHYPERTENSIVE DRUGS
1. Only diuretics and β-blockers have been shown, so far, to reduce the mortality in patients with hypertension.
2. Blacks are relatively resistant to β-blockers and ACE-inhibitors. Diuretics and Ca-channel blockers are effective.
3. Diabetics respond well to ACE-inhibitors, which do not adversely affect glucose metabolism.
4. Patients with hyperlipidemia are adversely affected by most β-blockers. Nonetheless, β-blockers reduce mortality of MI patients. ACE-inhibitors and Ca-channel blockers are effective.
5. Patients with asthma should not be given non-selective β-blockers.
6. Patients with congestive heart failure should not be given Ca-channel blockers or β-blockers.

C. ANTI-ANGINAL DRUGS

Angina (pain) pectoris (chest—cardiac muscle) represents the pain that occurs when oxygen demand outstrips oxygen supply to the heart. This occurs when (a) coronary blood flow is insufficient because of an atheromatous plaque and/or coronary vasospasm (angiospastic or variant angina) and (b) cardiac oxygen demand increases, as during exercise (classic angina). These problems are treated respectively with vasodilators which increase blood flow to the heart and with β-blockers to reduce sympathetic impulses which would increase work of the heart. Unstable angina represents angina at rest, probably due to irregular coronary vasospasms and/or small clots in the coronary vessels.

Vasodilators for angina

A. *Organic nitrates* (RNO_3) and *nitrites* (RNO_2) are polyalcohol esters that are hydrolyzed to NO_3^- and NO_2^- which decay to nitric oxide (NO), a potent vasodilator that is normally produced by vascular endothelial cells **(Figure 5.8)**.

Pharmacokinetics of RNO_3 and RNO_2 (Figure 5.8). RNO_3s and RNO_2s are rapidly cleaved to dinitro (active) and mononitro (active or inactive: examples below) compounds by hepatic nitrate reductase. Because of first pass metabolism, only 10% of the parent drug is bioavailable after oral administration. Sublingual, nasal and transdermal routes of administration are used to bypass the liver. Ultimately, glucuronide metabolites are formed and excreted in urine.

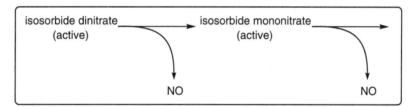

Mechanisms of Action of Anti-angina Drugs. When nitrates (RNO_3s) and nitrites (RNO_2s) are metabolized in vascular smooth muscle, (NO) is formed. The NO then reacts with sulfhydryl (-SH) groups on guanylyl cyclase, thereby activating the enzyme. This increases cGMP levels which inhibit contraction (inhibit vasoconstriction) by preventing phosphorylation of myosin-light chain (M-LC) **(Figure 5.6)**. Under usual physiological conditions, activated M-LC kinase and cAMP would stimulate M-LC phosphorylation and increase vascular and/or cardiac contraction.

Actions of RNO_3s and RNO_2s
1. Vasodilation causes venous pooling of blood and decreased preload to heart. This reduces cardiac work (decreased CO), and O_2 demand, in addition to decreased BP or hypotension (syncope, throbbing headache especially because of temporal artery dilation) and decreased afterload. Large arteries and veins are more sensitive than arterioles and venules.

2. When hemoglobin (Hb-Fe^{+2}) combines with NO, methemoglobinemia (Hb-Fe^{+3}) occurs and thus there is reduced O_2 carrying capacity of Hb. This effect is negligible for anti-angina drugs, since the relative amount of Fe^{+3} formation is low.

3. The most common adverse effects are syncope and headache.

Uses of RNO_2s and RNO_3s.
1. Angina pectoris. Anti-anginal drugs cause decreased cardiac O_2 demand.

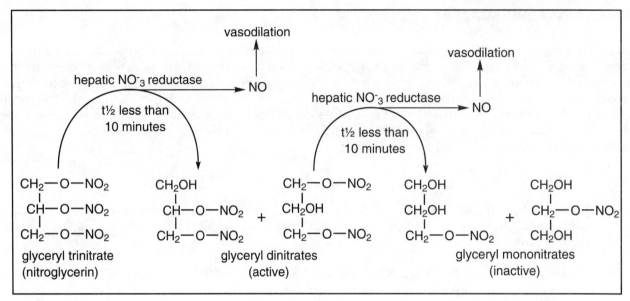

Figure 5.8

2. Variant angina. Anti-anginal drugs cause decreased coronary vasospasm.
3. Unstable angina. Anti-anginal drugs produce both effects.

$NaNO_2$ poisoning may inadvertently occur in infants whose intestinal flora converts NO_3^- to NO_2^-, which is absorbed and then produces methemoglobinemia (above). NO_3^-'s may be in well water and in foods cured with $NaNO_2$.

NO_2^- tolerance develops in industrial workers exposed to NO_2^-. Variant angina occurs after an absence of several days (e.g., over the weekend); headaches and dizziness are common at the start of the work week, being less frequent on subsequent days.

NO_3^- and NO_2^- combine with organic amines (RNH_2), forming nitrosamines (RN-NO) in the gut—suspected carcinogens.

B. Calcium Channel Blockers

Chemistry. Three of the Ca-channel blockers are analogs of dihydropyridine, as illustrated. The other two are each in different families **(Figure 5.10)**.

Pharmacokinetic data are outlined below (Figure 5.11). These drugs have a bioavailability of less than 50%, are extensively bound to plasma proteins and have a $t_{1/2}$ of about 4 hours. However, active metabolites may have a longer $t_{1/2}$. Almost all of the drug is metabolized, so dosage adjustments must be made when there is impaired liver function. Virtually none of the active drug is excreted in urine. Consequently, when there is renal impairment, it is safe to administer these drugs and no dosage adjustment is needed.

Mechanisms and actions. Calcium channel blockers bind to subunits of voltage-dependent L-type Ca-channels and inhibit Ca flux across the channel. Verapamil has the greatest effect on SA and AV nodes and in producing negative inotropy. Dihydropyridines are most potent in producing arteriolar, but not venule, relaxation or vasodilation, possibly owing to an additional effect of inhibiting phosphodi-

CARDIOVASCULAR PHARMACOLOGY 105

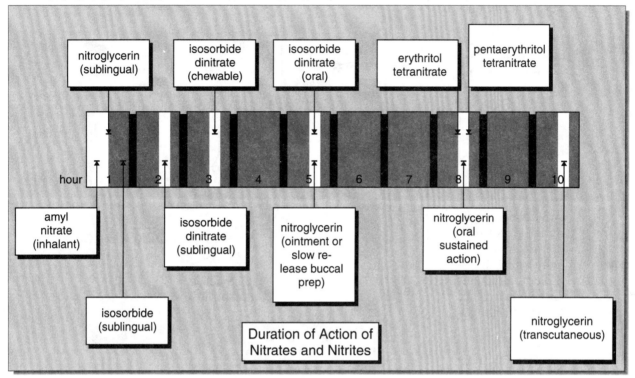

Figure 5.9

esterase (PDE) activity, which would result in prolonged intracellular cAMP effects. Afterload but not preload would be reduced. Diltiazem is intermediate in actions between verapamil and dihydropyridines.

Vascular smooth muscle is affected more by the Ca-channel blockers than bronchiolar, GI or uterine smooth muscle. Skeletal muscles are not affected by Ca-channel blockers.

Effects specifically associated with anti-angina activity

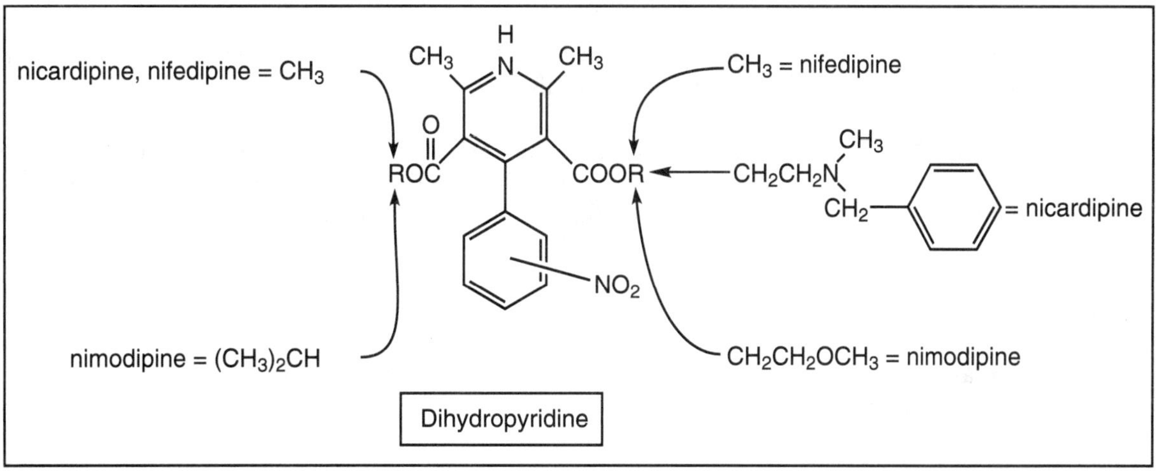

Figure 5.10

1. Most smooth muscles like vascular smooth muscle are dependent on transmembrane Ca^{++} influx for contractile responses and maintenance of normal resting tone. A Ca-channel blocker, accordingly, produces relaxation. This is the basis for use of nifedipine (Procardia), diltiazem (Cardizem) and amlodipine (Norvasc) as anti-angina drugs. Order of sensitivity:

   Arterioles are more sensitive than venules, so orthostatic hypotension is not a major problem.

2. Verapamil and diltiazem produce reduced chronotropy and inotropy, thereby reducing cardiac work and O_2 demand. These actions are important for use of these drugs in treating angina.

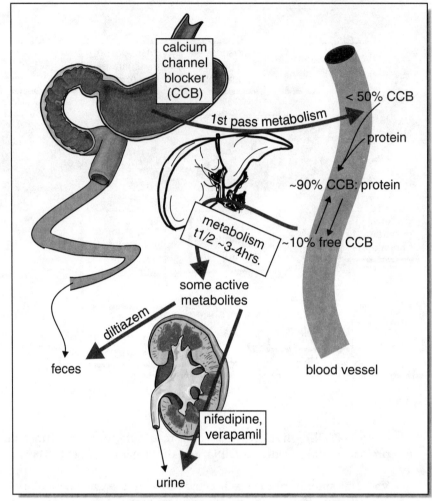

Figure 5.11

Diltiazem, nifedipine and verapamil are each used to treat angina. Nifedipine is preferred in angina patients with hypertension or cardiac nodal disorder.

Adverse effects. Ca-channel blockers should not be used with β-blockers because of the increased potential for producing A-V block or severe cardiac depression. Because verapamil displaces digoxin from its binding sites and enhances digoxin actions (causing toxicity), verapamil cannot be used to treat digoxin toxicity.

C. β-Adrenoceptor Antagonists

Mechanism and actions of β-blockers. These are describe in **Chapter 2**. Only those actions related to effectiveness in angina are outlined here:

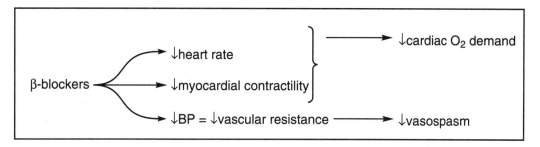

These actions compliment the other in production of an anti-angina effect. However, β-blockers also produce an undesired increase in EDV (end-diastolic volume) because of their negative inotropic effect.

D. Summary

1. For variant angina, RNO_3/RNO_2 and Ca-channel blockers are more effective than β-blockers.
2. RNO_3/RNO_2 are often used in combination with Ca-channel blockers or β-blockers. Combined therapy is more efficacious than monotherapy. In unstable angina all these classes of anti-anginal drugs are sometimes supplemented with the antithrombotics, aspirin or dipyridamole.

D. ANTIARRHYTHMICS

Cardiac arrhythmias are caused either by abnormal discharge of cardiac pacemaker cells or by conduction defects.

Electrophysiological basis of pacemaker activity and impulse conduction
1. **SA and AV Nodes.** Cells in these pacemakers are driven primarily by Ca current. The resting membrane potential is unstable, primarily reflecting leakage of Ca^{++}. During the rise of the action potential there is a large increase in both Na and Ca conductance. At the peak of the action potential, Na current is turned off, Ca current fluctuates and potassium current increases, becoming greatest during rapid repolarization **(Figure 5.12)**.

2. **Atrial Muscle, Purkinje Cells and Ventricle Muscle.** These cells are primarily governed by fluxes in Na or K current. The rapid rise in the action potential (phase 0) is due to a large increase in Na current. At the peak of the action potential there is a decline in Na current, a fluctuation in Ca current and an increase in K current. During rapid repolarization, K current is highest and Na current is lowest **(Figure 5.13)**.

Innervation of heart
1. *Parasympathetic* (vagal) fibers innervate the SA node, AV node and atrial muscle. The greatest density of these fibers is in the SA node. Ventricles are not innervated by these nerves. Resting heart rate (SA node activity) is governed by these nerves **(Figure 5.14)**.

2. *Sympathetic fibers* innervate all regions in the heart. Acceleration of heart rate is due primarily to increased activity by these nerves **(Figure 5.14)**.

    Although the autonomic nervous systems regulates heart rate and force of contraction, a totally denervated heart (i.e., transplanted heart) will beat and contract, but at a slower rate and with less force.

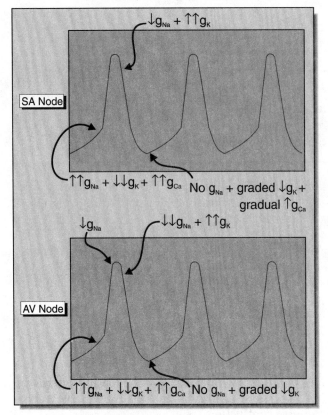

Figure 5.12

Firing of pacemaker cells
There are four main factors that determine pacemaker activity (heart rate) **(Figure 5.15)**:
  a. The membrane potential (E_m). As E_m increases (i.e., becomes more negative), heart rate slows.
  b. The rate of diastolic depolarization (phase 4). As this rate of depolarization slows, heart rate slows.
  c. The threshold potential (TP). As threshold potential decreases (i.e., becomes more positive), heart rate slows.
  d. Duration of the action potential (effective refractory period—ERP). The longer the action potential, the less frequently the cells can fire.

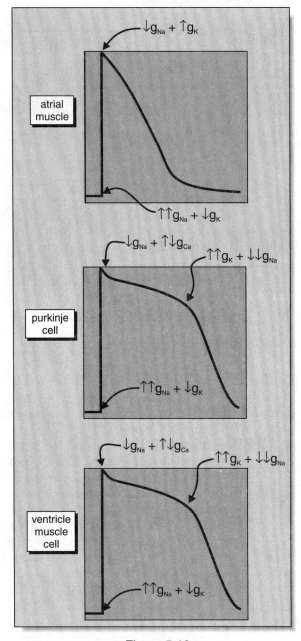

Figure 5.13

Impulse conduction in the heart. Impulses from the SA node are conducted via atrial muscle cells to the AV node. AV node cells fire and then transmit the signal through Purkinje fibers to the ventricle myocardium. According to the theory of circus movements, impulses do not keep traveling in the myocardium, because waves of impulses traveling around muscle bands in opposite directions, cancel each other out.

Arrhythmias can be caused by ectopic foci (spontaneously firing cells), by propagation defects in the myocardium that do not allow impulse waves to cancel out each other or by after depolarizations (extrasystoles), caused by membrane instability or defects in sequestering of Ca in cells.

ANTIARRHYTHMIC DRUGS

A. Class IA drugs: Na Channel Blockers

These drugs (a) decrease the firing rate of ectopic foci versus SA node and/or (b) reduce conduction velocity in depolarized versus normally polarized cells.

Since the Na channel blockers have greatest access to the Na channel when the channel is open (activated Na channels), these drugs tend to localize in those cells with highest activity, the ectopic foci **(Figure 5.16)**. These actions are identical to those of local anesthetics, and indeed all of these class I drugs are local anesthetics.

Mechanism (Figure 5.17). By blocking Na channels these drugs
    a. decrease the rates of phase 4 and phase 0 depolarization,
    b. decrease the rate of phase 3 repolarization,
    c. reduce excitability and

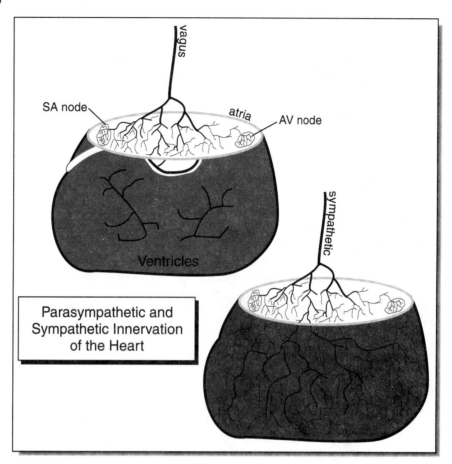

Figure 5.14

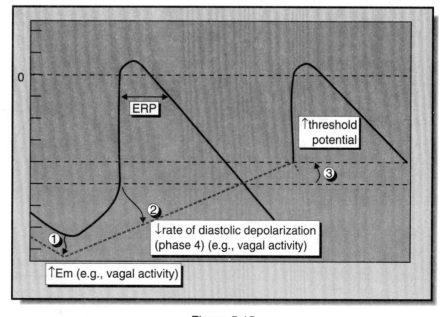

Figure 5.15

d. reduce conduction.

A prolongation in the effective refractory period (Q-T interval) and a reduction in conduction leads to a reduction in impulse generation in ectopic foci **(Figure 5.17)**. (In toxic amount, asystole can occur.)

1. **Quinidine** (Cin-quin)

    Cardiac effects. In subthreshold amount quinidine can increase firing and cause ventricular tachcardia because of a *vagolytic action*. For this reason A-V block is usually produced (e.g., with propranolol, verapamil, or digitalis) prior to quinidine treatment.

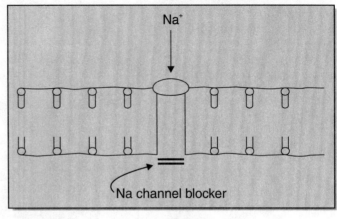

Figure 5.16

Other effects of quinidine are outlined in **Figure 5.18**. In toxic amount quinidine can produce asystole or precipitate arrhythmias, especially when there is hyperkalemia.

Use. Supraventricular and ventricular arrhythmias.

2. **Procainamide** (Pronestyl)

    Cardiac effect. Like quinidine, but with less anticholinergic activity.

    Other effects. A reversible lupus-like syndrome develops in a high percentage of patients, especially slow-acetylators, who cannot as easily form the N-acetylated metabolite. This product is active like the parent drug, necessitating dosage reduction in renal-impaired patients.

    Ganglionic blocking actions of procainamide are associated with overdose hypotension.

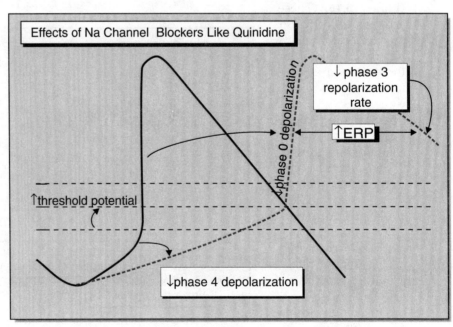

Figure 5.17

Use. Supraventricular and ventricular arrhythmias.

3. **Disopyramide** (Norpace)

   Cardiac effect. Like quinidine, but even greater antimuscarinic activity. Use an A-V blocker when starting disopyramide treatment, as per quinidine.

   Other effects. Prominent anticholinergic activity results in anhydrosis, urinary retention, etc.

   Use. Ventricular arrhythmias that are refractory to quinidine and procainamide.

B. Class IB Drugs: also Na channel blockers, but no prolongation of ERP.

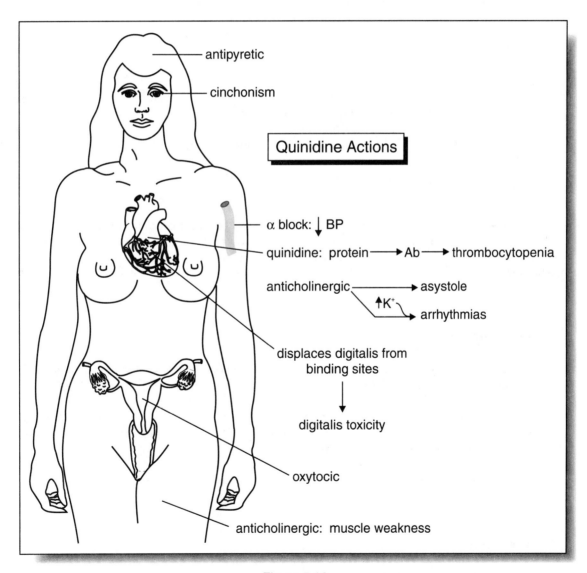

Figure 5.18

1. **Lidocaine** (Xylocaine)

   Cardiac effect. Lidocaine blocks both activated and inactivated Na channels. Consequently, a large percentage of sodium channels become blocked in cardiac cells with long plateaus in phase 2 (Purkinje fibers and ventricle muscle). During diastolic depolarization (phase 4) lidocaine dissociates from normally polarized cells. Consequently, the cells that are rapidly firing are mostly affected, while normal cells are unaffected by lidocaine. Cells without plateaus in phase 2 (atrial, SA node, AV node) are largely unaltered by lidocaine.

   Other effects. In toxic amount, lidocaine produces tremor and convulsions (predominate block of Na channels in inhibitory neurons).

   Note on pharmacokinetics. Since greater than 95% of lidocaine would be inactivated by first-pass metabolism, an i.v. route is used.

   Use. Lidocaine is the drug of choice for ventricular arrhythmias. It is used as prophylaxis against ventricular fibrillation following a myocardial infarct (MI). Lidocaine has a low incidence of side effects and is highly effective.

2. *Mexiletine* (Mexitil) is an analog of lidocaine, orally effective because of resistance to first-pass metabolism. Actions and uses are similar to lidocaine.

3. *Phenytoin* (Dilantin) acts like lidocaine and is used mainly for digitalis-induced arrhythmias that are resistant to lidocaine.

C. Class II Drugs: β-adrenoceptor Antagonists

These drugs block sympathetic inotropic, chronotropic and dromotropic effects in the heart. Propranolol (Inderal), a $β_1$- and $β_2$-blocker, increases survival rate of MI patients. Several other drugs act like propranolol. Propranolol can produce crises in asthmatics and diabetics. This drug is discussed in **Chapter 2**.

D. Class III Drugs: K Channel Blockers

1. **Bretylium (Bretylol)**

   Cardiac effects. By blocking K channels, bretylium delays phase 3 repolarization and thereby prolongs the duration of the action potential. The effect is to slow the rate of firing.

   Adverse effects. Acutely, bretylium releases norepinephrine (NE) stores from sympathetic nerves, producing a hypertensive effect. Chronically, bretylium uncouples excitation-secretion coupling in sympathetic nerves (decreased NE release).

   Use. Bretylium is used for lidocaine-resistant ventricular arrhythmia.

E. Class IV Drugs: Ca Channel Blockers

1. **Verapamil**

   Cardiac effects. Verapamil blocks both activated and inactivated Ca channels. Consequently, cells with high firing rates are largely affected; also, those cells reliant on Ca currents for activation (SA node and AV node). These effects suppress the rate of rise of phase 0 depolarization and prolong the plateau phase 2, prolonging the ERP. A-V block may occur.

Other effects. The pharmacology of verapamil is discussed with anti-angina drugs in this chapter.

Use. Supraventricular arrhythmias, particularly paroxysmal supraventricular tachycardia.

E. TREATMENT OF CONGESTIVE HEART FAILURE

The failing heart (Figure 5.19). In chronic congestive heart failure the heart is not adequately pumping enough blood to comfortably meet the O_2 demands of the body (heart failure). Also, the associated strains on this heart (outlined below) further compromise cardiac reserve. Added work-load (e.g., exercise) can precipitate acute myocardial infarction (MI).

1. The failing heart dilates in an attempt to work more efficiently. This is the Starling mechanism **(Figure 5.19)**.

2. Over a long period of time the thickness of the left ventricle increases.

3. Sympathetic reflexes also are activated, to produce
   a. inotropic effects (increased cardiac output) and
   b. arterial constriction (increased BP which increases "pressure head" to more-adequately perfuse body tissues). The increased BP causes increased back pressure on the left ventricle (afterload) and increased LVEDP (left ventricle end-diastolic pressure).

4. A compromised heart overdistends, pumps less efficiently, and blood in the pulmonary system backs-up (increased pulmonary pressure) causing pulmonary edema.

5. Usually, secondary to these changes, there is increased preload, increased RAP (right atrial pressure) and increased CVP (central venous pressure).

6. Renal circulation is compromised in heart failure, so that Na^+ and H_2O are retained and blood volume increases, further increasing workload on the heart and producing generalized edema.

7. Inadequate renal perfusion leads to an increase in renin release, increased angiotensin II and aldosterone levels and increased sympathetic tone, leading to further increased BP, which adds to the afterload on the heart.

Drugs discussed in this section are those that reverse the downward spiral of effects on the failing heart, by helping the heart to pump more efficiently without increasing O_2 demands on the heart. Although digitalis glycosides increase O_2 consumption in individual myocardiocytes when contractility is enhanced, there is greater efficiency in pumping (from reduced heart size and reduced overstretching of myocardiocytes). Consequently, net O_2 demand is reduced or unchanged.

Digitalis glycosides, isolated from the leaf of the foxglove and other plants, are a mainstay in the treatment of heart failure. The therapeutic index is very low, however, so that side effects (particularly arrhythmias) occur frequently.

Other useful drugs for heart failure are listed at the end of this chapter. Each belongs to a drug class that is discussed elsewhere, so their pharmacological properties are only outlined here.

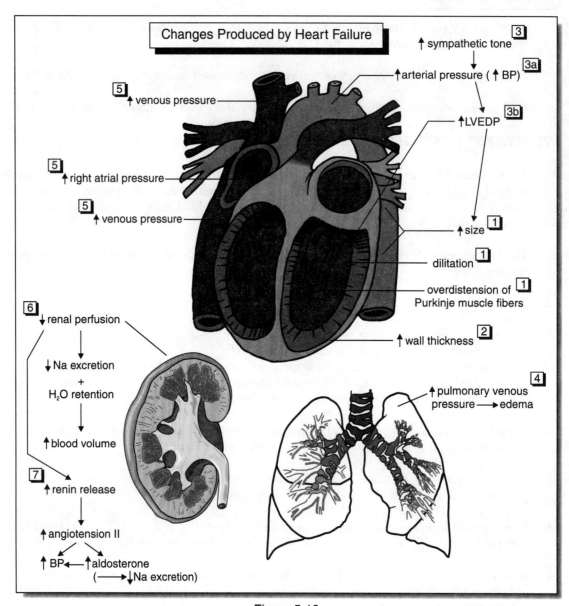

Figure 5.19

DIGITAL GLYCOSIDES

Chemistry. The structures of the two main digitalis glycosides, digoxin and digitoxin, are shown **(Figure 5.20)**. The 12-hydroxydroxy group of digoxin imparts added water solubility, which vastly influences pharmacokinetic properties to make this the superior agent.

Mechanism (Figure 5.21). Digitalis glycosides inhibit Na-K-ATPase in myocardial cells, thereby reducing Na^+-K^+ exchange. This effect increases intracellular levels of Na^+, which then (by mass action) inhibit Na^+-Ca^{++} exchange. The resulting elevation of intracellular Ca^{++} concentration leads to greater interaction of actin and myosin, which is associated with greater contractility and force of contraction. Greater intracellular levels of Na^+

and Ca^{++} additionally reduce the resting membrane potential (E_m) which augments Ca^{++} influx and Ca^{++} release from the sarcoplasmic reticulum. These effects would also increase contractility and inotropy.

The increased concentration of Ca^{++} in myocardial cells is not without danger. In nodal tissue this increases automaticity. In ventricle muscles excess Ca^{++} leads to generation of afterpotentials which can generate extra action potentials (extrasystoles). It is noted that digitalis glycosides have a low therapeutic index and that life-threatening caridac effects are produced, in addition to adverse effects on other systems.

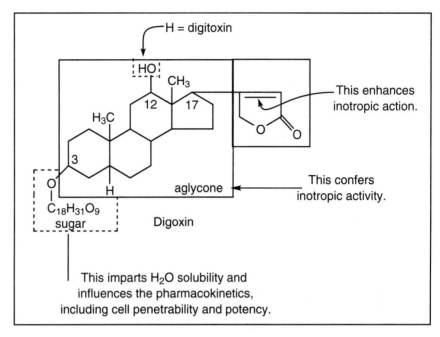

Figure 5.20

Actions of digitalis glycosides. Digitalis glycosides produce their major beneficial effects on the heart (1) by the direct actions on myocardial cells that were just described and (2) by indirect actions mediated by parasympathetic vagal nerves. In the heart the actions on nodal tissue (SA and AV nodes) and ventricle muscle are most important.

1. **Actions of digitalis on nodal tissue (Figure 5.22)**
   Basically, therapeutic effects of digitalis glycosides on SA and AV nodes are related to enhanced vagal tone. When digitalis moves into the toxic range, its direct actions on SA and AV nodes override the indirect vagal effects. The direct actions are generally opposite to indirect vagal effects.

   In the therapeutic range, digitalis stimulates the vagal center, sensitizes baroreceptors and sensitizes cardiac nodal tissue to acetylcholine (ACh). Simultaneously, sympathetic activity is reduced. These indirect effects tend to override the direct effects of digitalis on the atrium. Consequently, heart rate is reduced (hyperpolarization and reduced phase 4 depolarization) and conduction velocity through SA node, atrial muscle, and AV node is reduced. This tends to produce A-V block. However, the refractory period is reduced, so atrial rate may increase. As digitalis moves towards a toxic range, sympathetic activity increases and vagal activity decreases.

   Digitalis glycosides directly and indirectly also decrease conduction velocity. In the therapeutic range heart rate is decreased. As digitalis moves into the toxic range, toxic depolarization of nodal tissue can totally inhibit generation of action potentials. Also, conduction velocity through the AV node becomes so inhibited (coupled with increased refractory period) that AV nodal block is produced.

2. **Actions of digitalis on ventricle muscle.** These are illustrated in **Figure 5.23**.
   Direct effects of digitalis prevail on ventricle muscle. (Recall that parasympathetic vagal nerves do not innervate the ventricle.)

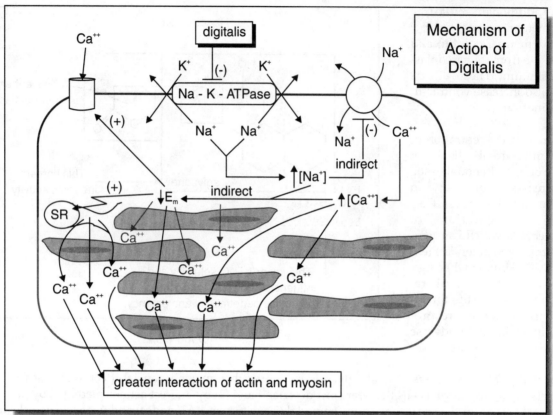

Figure 5.21

In therapeutic amounts digitalis slightly shortens the duration of the ventricle action potential. This effect is associated with the reduced QT interval of the EKG.

Digitalis—in non-nodal cardiac tissue
a. decreases the E_m (becomes less negative).
b. decreases the rate of phase 0 depolarization, which reflects the decreased conduction velocity. There is a consequent voltage-dependent closing of Na^+ channels.
c. decreases the duration of the action potential, thereby reducing the effective refractory period (ERP). There is an increased responsiveness to stimuli and a reduced QT interval of the EKG.
d. decreases the rate of phase 2 and phase 3 repolarization, which leads to inversion of the T wave of the EKG.
e. increases the rate of phase 4 depolarization (Purkinje fibers) which increases automaticity and heart rate.
f. The appearance of delayed afterdepolarizations coincides with extrasystolies. This is related to oscillations of high concentrations of free intracellular levels of Ca^{++}.

In toxic amounts, digitalis produces
a. decreased E_m;
b. decreased conduction velocity and decreased rate of depolarization of the ventricle action potential;
c. decreased phase 0 depolarization of ventricle;
d. decreased amplitude and duration of the AP and

e. decreased rate of repolarization **(Figure 5.23)**.

In the therapeutic range digitalis increases automaticity (e.g., reduced refractory period and production of after depolarization). In the toxic range extrasystoles are produced and this can lead to ventricular tachycardia and fibrillation.

This sequence of effects explains how digitalis often slows a failing heart, while producing AV block and automaticity of the ventricle. The indirect effects of digitalis on intracellular Ca^{++} levels explain the inotropic actions and the associated increase in stroke volume, cardiac output and reduction in heart size.

The most life-threatening effect of digitalis is ventricular fibrillation. Atrial arrhythmias, A-V block and ventricular extrasystoles are common. However, other cardiac dysrhythmias occur.

3. **Effects of digitalis on heart size**. Increased contractility of the myocardium is synonymous with greater efficiency of the heart. The heart size is reduced, preload (venous filling pressure) is reduced and LVEDP is reduced.

4. **Non-cardiac effects of digitalis**. Secondary effects related to digitalis action on the heart include (a) improved renal perfusion, which in turn produces elimination of water (and edema and blood volume), and (b) reduction of renin (reducing angiotensin), which helps to reduce elevated blood pressure, and thereby reduce afterload.

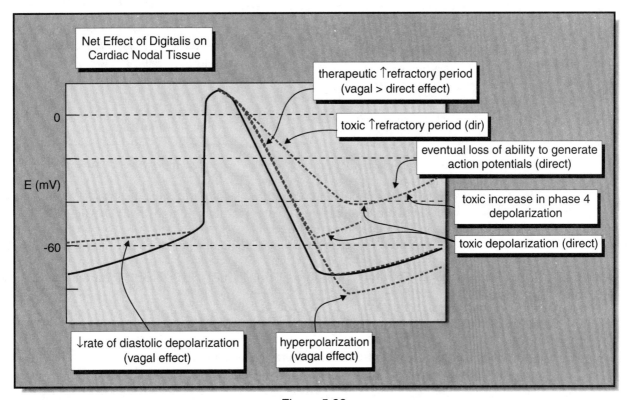

Figure 5.22

5. **Other effects of digitalis**
   a. Arterioles and venules: Digitalis directly constricts vessels, while a reduced sympathetic tone indirectly relaxes vessels. Usually the improvement in cardiac output overrides these actions so that blood pressure is not changed.
   b. GI: Vomiting (stimulation of chemoemetic trigger zone) and diarrhea (direct action and vagal stimulation) occur.
   c. Eyes: Blurred vision, halos around dark objects and altered color perception occurs.
   d. CNS: Anxiety, disorientation, hallucinations and convulsions can occur.
   e. Gynecomastia is rare.
   f. Hypokalemia causes enhanced digitalis binding with enhanced effect and toxicity.

6. **Pharmacokinetics**
   a. Lipophilicity of digitoxin is much greater than that of digoxin.
   b. Absorption of digitoxin (95%) is greater than that of digoxin (67%), but digoxin has a quicker onset of action (2 versus 8 hour peak).
   c. Protein binding of digitoxin (90%) is greater than that of digoxin (25%). Digitoxin is displaced by many drugs.
   d. Duration of action of digitoxin (6 days) is greater than that of digoxin (1.5 days). Enterohepatic circulation occurs for each (approximately 30%).
   e. Extent of metabolism of digitoxin is much greater than that of digoxin (25%). Digitoxin dosage should be reduced if there is hepatic dysfunction.
   f. Renal excretion: Digoxin is excreted unchanged, so its dose must be reduced if there is renal insufficiency.
   g. Deposition in tissues: Both are concentrated 25:1 in kidneys, liver, heart, GI, and skeletal muscle versus plasma.

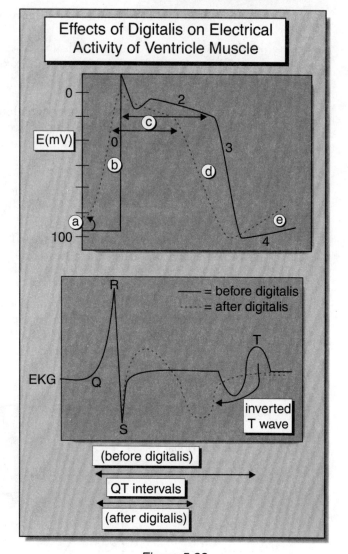

Figure 5.23

7. **Main use for digoxin (Lanoxin).** Congestive heart failure.

8. **Main use for digitoxin (Crystodigin).** Supraventricular arrhythmias.
   a. Atrial fibrillation and flutter.
   b. SA and A-V nodal paroxysmal tachycardia.

TREATMENT OF DIGITALIS TOXICITY

1. Do not administer next dose(s).

2. Administer K^+ (if hyperkalemia and AV block do not exist). K^+ displaces digitalis from binding sites on ATPase, reducing activity. Note, excess K^+ induces arrhythmias and cardiac arrest.

3. Antiarrhythmics are given.
   a. Ventricular arrhythmia: Phenytoin and lidocaine (no effect on AV node);
   b. atrial or ventricular ectopic foci: Procainamide and quinidine (depress AV node); quinidine, which also displaces digoxin from binding sites;
   c. supraventricular arrhythmias: Ca-channel blockers and
   d. excess vagal effect: Atropine.

4. Administer digibind, a specific antibody for digitalis, for severe toxicity.

GENERAL APPROACH TO TREATMENT OF HEART FAILURE

Success is contingent on aggressive treatment coupled with patient education and patient compliance. Usually, patients must restrict Na^+, H_2O and fat intake. Diuretics are often the first drugs to be administered. An exercise program may be slowly begun with monitoring of cardiac parameters as required for each patient. The first drug of choice may be selected from among digoxin, antihypertensives, vasodilators, sympathomimetics and others—according to the status of the patient.

OTHER DRUGS FOR TREATING CONGESTIVE HEART FAILURE

1. *Amrinone* (Inocor), *milrinone* (Corotrope) and *flosequinan* (Manoplax) are inhibitors of the type of phosphodiesterase found in the heart (PDE III). This action elevates cAMP levels, thereby increasing contractility.

   PDE in vascular smooth muscle would also be inhibited, producing relaxation of the muscle and vasodilation. These are often used for short-term treatment and are usually held in reserve for those patients unresponsive to digoxin and other drugs.

   Amrinone and milrinone produce adverse GI effects (pain, diarrhea), hepatotoxicity and blood dyscrasias. Fluosequinan has anticoagulant activity, but more needs to be learned about this newer drug.

2. β_1-agonists: dobutamine
3. β_2-agonists: albuterol
4. Dopamine: direct (dopamine receptor) and indirect (β) actions on heart
5. α-blockers: prazosin
6. ACE-Inhibitors: enalapril, captopril
7. Ca-Channel blockers
8. Vasodilators: hydralazine, nitrates

F. ANTIHYPERLIPIDEMICS

Fat Transport and Disposition. Fats in the diet are emulsified by bile and digested by pancreatic lipase to fatty acids (FAs) and glycerides. These are absorbed and reconstituted into large globules called chylomicrons which contain triglycerides (TGs) and cholesterol esters (ChEs) **(Figure 5.24)**. As chylomicrons pass through the capillaries of adipose tissue and skeletal muscle (Sk mus), endothelial

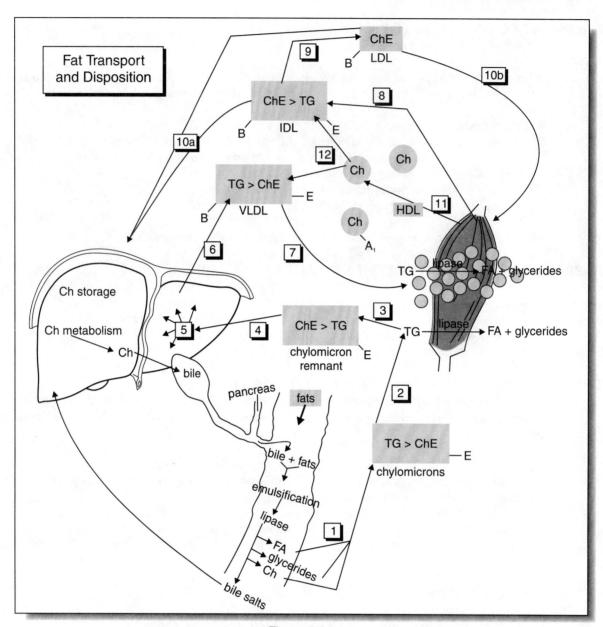

Figure 5.24

cells digest the TGs, leaving mainly ChE in the smaller chylomicron remnant, which is endocytosed by hepatocytes and digested there into bile, VLDL (very low density lipoproteins) and cholesterol.

The liver serves as an energy sink, releasing VLDLs when needed by different tissues. VLDLs are digested by adipose tissue and skeletal muscle, in a manner analogous to that of chylomicrons: TGs are cleaved, leaving mainly ChE in the smaller IDL (intermediate density lipoprotein). Some IDLs are digested by hepatocytes, others are transformed into smaller LDL particles which are composed mainly of ChEs. Adipose tissue and skeletal muscle digest the IDLs. Ch is continually released into the blood as cells die, forming small HDLs which consist mainly of free Ch. HDLs carry 2/3 of plasma Ch. HDLs, in turn, are converted into IDL and VLDL particles.

These are the main biological cycles for digestion, transport and disposition of fats in the body. The action of some of the hypocholesterolemic agents is indicated in **Figure 5.24**.

Hyperlipidemia represents an excessively high level of plasma lipids. The primary forms are single-gene (e.g., familial hypercholesterolemia) or multi-gene related (e.g., hypertriglyceridemia). Secondary forms are associated with various pathological disorders (e.g., diabetes, hypothyroidism), dietary excesses (e.g., alcoholic hyperlipidemia) or drug treatment (e.g., oral contraceptives).

Hypercholesterolemia has been associated with increased frequency of myocardial infarct and atherosclerosis; hypertriglyceridemia, with increased incidence of pancreatitis. Different apolipoproteins are associated with the assorted lipoproteins. Elevated levels of apo-B (on VLDL and LDL; **Figure 5.24**) are associated with increased risk of coronary artery disease.

NON-DRUG MEASURES FOR CONTROLLING HYPERLIPIDEMIA
Dietary changes constitute the first measure for controlling elevated plasma lipid levels. This includes reducing total caloric intake, reducing fats to less than 30% of total calories and reducing saturated fats to less than 10% of calories.

In addition, life-style changes may be needed. This includes an exercise program, elimination of risk factors like smoking and instituting good control of predisposing disorders such as diabetes.

DRUGS USED FOR HYPERLIPIDEMIA
When all non-drug approaches fail to adequately reduce plasma lipid levels, a drug approach must be instituted.

A. Niacin or Nicotinic Acid (Nicolar)

Pharmacokinetics. Niacin must be taken in large amounts (much greater than vitamin dose). It is rapidly absorbed from the GI tract and is excreted as a liver metabolite.

Mechanism of niacin (Figure 5.25). Niacin
1. decreases Ch synthesis and esterification in the liver.
2. decreases VLDL formation, which secondarily causes reduced formation of IDLs and LDLs.
3. decreases lipolysis in adipose tissue.
4. increases lipase activity in adipose tissue. Lipase enhances metabolism of TGs.
5. increases LDL uptake by the liver.
6. decreases HDL degradation, thereby increasing the HDL/LDL ratio.

Net effect. There is a reduction in plasma TG level and LDL-Ch. This is delayed for several days until already-present VLDLs are metabolized.

Adverse effects of niacin include
1. cutaneous flush (prevented by aspirin, 30 minutes before niacin), pruritus, hyperpigmentation.
2. GI vomiting, diarrhea.
3. hepatic jaundice, increased SGOT and SGPT.
4. competition with uric acid excretion in the nephron, resulting in hyperuricemia and gout.
5. hyperglycemia.

Major uses of niacin. Hypertriglyceridemias.

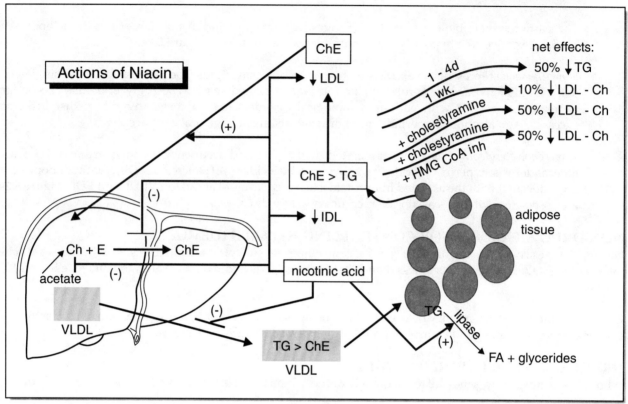

Figure 5.25

B. Bile-acid Sequestration: Cholestyramine (Questran) and Cholestipol (Colestid)

Pharmacokinetics. These large MW polymers are taken orally but are not absorbed. Their direct action is exerted exclusively in the lumen of the intestine. The main effect is adsorbtion (sequestration) of bile acids, which are essential for (a) emulsification of fats and (b) absorption of fats and cholesterol. Consequently, more cholesterol and fats (and fat-soluble vitamins) are excreted in feces (steatorrhea).

Mechanism (Figure 5.26)
1. The increased fecal excretion of bile acids means that more hepatic cholesterol must be utilized to synthesize bile acids. Since the latter product is largely being excreted, there is less of the normal feedback inhibition of the 7-α-hydroxyation of cholesterol.

2. The increased hepatic utilization of cholesterol induces proliferation of LDL receptors and greater endocytosis of LDL, with metabolism of the cholesterol. This leads to decreased plasma LDL and decreased plasma Ch. This effect can be enhanced by HMG CoA reductase inhibitors or niacin.

Use. Cholestyramine and cholestipol are most effective for hypercholesterolemia. The incidence of coronary heart disease is reduced by the resins.

If there is a genetic deficiency for LDL receptor formation in the liver (homozygous familial hypercholesterolemia), these resins do not decrease plasma Ch.

Adverse effects of cholestyramine and cholestipol.
1. Plasma TG levels increase in the first few weeks.
2. GI effects occur, including nausea, indigestion and constipation (irritation of hemorrhoids). Treat with high fiber foods which promote defecation.
3. Drugs may be adsorbed to the resins in the GI tract, impairing drug absorption. Drugs should be taken 1 or more hours before or over 4 after the resin.

C. HMG CoA Reductase Inhibitors: Lovastatin (Mevacor), Pravastatin (Pravachol) and Simvastatin (Zocor)

Pharmacokinetics. All of these are administered orally. Lovastatin and simvastatin are prodrugs that are hydrolyzed to the active species in the liver.

Mechanism. These drugs reduce cholesterol formation by inhibiting hepatic HMG CoA reductase (3-hydroxy-3-methylglutaryl coenzyme A reductase), the rate-limiting enzyme in cholesterol synthesis. Although there is an eventual compensatory increase in HMG CoA reductase which tends to offset the inhibition, these drugs still reduce plasma cholesterol levels. That effect appears to be related to an increased number of hepatic LDL receptors, which leads to enhanced binding and increased engulfment of LDLs at these sites **(Figure 5.27)**.

The net effect is a dose-dependent reduction in plasma levels of LDL, Ch and TG with an accompanying elevation of plasma HDL (HDL is good cholesterol). Lovastatin specifically has been shown to promote regression of coronary atheromas.

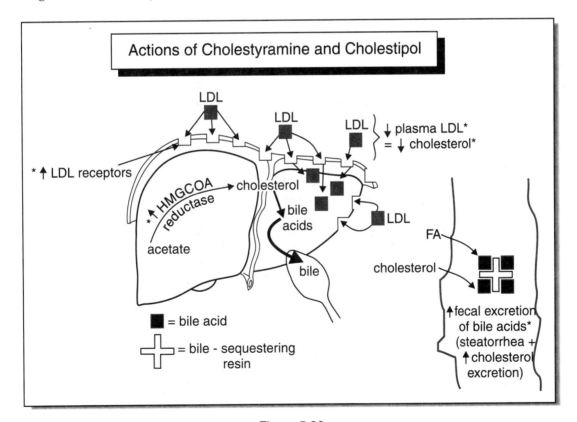

Figure 5.26

Use of HMG CoA reductase inhibitors. These are used to treat hypercholesterolemia, even if there is a deficiency of LDL receptors (i.e., familial heterozygous hypercholesterolemia). Bile-acid binding resins (e.g., cholestyramine) enhance the effect.

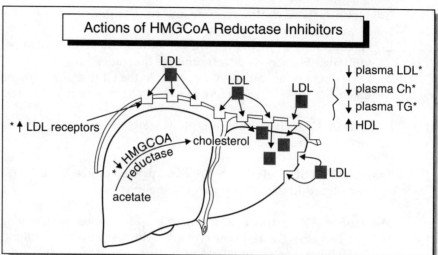

Figure 5.27

These drugs are better tolerated than the other antihyperlipidemics. Adverse effects of lovastatin are uncommon, but include
    a. elevated serum transaminase. Jaundice is rare.
    b. elevated serum muscle-creatine phosphokinase. Myalgia occurs, but myopathy is rare.
    c. cataracts in test animals.
    d. lupus-like syndrome with lovastatin and simvastatin.

D.   Fibric Acid Analogs: Gemfibrozil (Lopid) and Clofibrate (Atromid-S)

Pharmacokinetics. These are administered orally and are bound to plasma proteins. These drugs displace warfarin from its binding sites and may alter warfarin binding to its receptor, thereby activating warfarin. Prothrombin time must be determined regularly if warfarin is co-administered.

Mechanism. These drugs produce a large reduction in serum TG, a moderate reduction in serum VLDL and LDL, and sometimes a reduction in HDL-Ch. Mechanisms involve
    a. enhancing lipoprotein lipase activity, which metabolizes TG;
    b. inhibiting hepatic synthesis and release of VLDL and
    c. inhibiting synthesis of apo-B **(Figure 5.28)**.

Use. Gemfibrozil is the drug of choice for hypertriglyceridemia. Both fibric acids are particularly effective in reducing both TG and Ch in familial type III hyperlipoproteinemia (abnormal apo E on VLDL and other lipoproteins). Gemfibrozil, not clofibrate, reduces the incidence of coronary artery disease. Niacin is sometimes given simultaneous with gemfibrozil.

Adverse effects
1. GI distress, including pain and diarrhea.
2. Increased incidence of gallstones, by increasing secretion of Ch into bile, while reducing conversion of Ch into bile acids. This reduces solubility of Ch in bile **(Figure 5.28)**.
3. Atrial and ventricular arrhythmias (Clofibrate only).
4. Muscle cramps, muscle weakness and increased plasma creatine phosphokinase.
5. Alopecia, rash, leukopenia, anemia and impotence may occur (DNA effects?).

E. Probucol (Lorelco) is a synthetic antioxidant.

Pharmacokinetics. Less than 10% of probucol is absorbed. The drug is deposited in fat, persisting there for months after the last dose.

Mechanism. Unknown.

Effect: Probucol reduces seurm HDL-Ch, more so than serum LDL-Ch.

Use. Probucol is less effective in lowering Ch than niacin, bile-acid sequestrants and HMG CoA reductase inhibitors. Probucol is useful only if used with one or more of the other drugs.

Adverse effects
1. GI: pain, diarrhea.
2. Infrequent increase in the Q-T interval.
3. Eosinophilia, angioneurotic edema.

G. DRUGS THAT ALTER BLOOD COAGULATION/HEMOSTASIS/CLOT DISSOLUTION

A variety of drugs influence different components of either blood coagulation or blood clot dissolution. The site of action of these drugs is outlined in **Figure 5.29**.

A. Anticoagulants

Blood coagulation is a multistep process involving thirteen identified blood-clotting factors and several cofactors. Blood coagulation represents a rapid means for closing breaks in blood vessels before too much blood has been lost. The lack of one or more of the reactions in the cascade would disrupt the process. This ensures that there are adequate safeguards to prevent inappropriate coagulation of blood. Nevertheless, inadvertent coagulation produces intravascular blood clots (thrombi) that can dislodge from vessel walls and migrate (emboli). Most emboli become lodged in the lung (pulmonary embolism), although most any tissue could be involved. Thrombus formation is associated with myocardial infarct (MI) and stroke, and is most likely to occur after surgery.

1. *Heparin* is a sulfated polysaccharide consisting of 8 to 15 repeats (in varying ratios) of the 2 disaccharides, iduronic acid to glucosamine and glucuronic acid to glucosamine **(Figure 5.30)**. According to the length of the saccharide chain heparin can be low or high molecular weight (MW). Heparin is naturally produced by various tissues in the body and is obtained as an extract of animal tissue.

   Pharmacokinetics. Since heparin is a saccharide that would be cleaved by amylases in salivary and pancreatic secretions, a parenteral route is necessary. Heparin is metabolized in liver and excreted in urine.

   Mechanism. After initially binding to endothelial cells in blood vessels, heparin forms a complex with the plasma protease inhibitor, antithrombin III, that rapidly binds to clotting factor proteases. Heparin then dissociates from the inactive antithrombin III: protease complex and catalyzes formation of more of these complexes **(Figure 5.31)**.

   Standardization. Each commercial preparation of heparin has different ratios of the disaccharides, different lengths of the saccharide chain and different MW. High molecular weight heparin has the

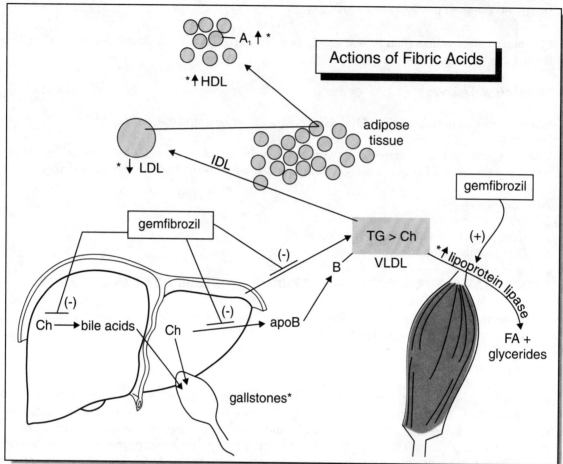

Figure 5.28

activity described above. Low molecular weight heparin is an inhibitor of activated clotting factor X. This heterogeneity of heparin requires a standardization on the basis of biological activity.

Uses. Heparin is used to prevent pulmonary embolism in patients with thrombi and to prevent surgical venous thrombosis.

Adverse effects
a. Excess bleeding.
b. Allergy to animal protein co-extracted with heparin.
c. Thrombocytopenia, resulting from heparin-induced anti-platelet antibodies.
d. Osteoporosis (long-term use).

Fast reversal of heparin action
Protamine sulfate is a basic peptide that binds with heparin, inactivating it (1 mg protamine sulfate for every 100 units of heparin that remains).

2. **Coumarin analogs (Warfarin) are vitamin K antagonists.**

Pharmacokinetics. These orally effective drugs are 99% bound to plasma proteins and have a long $t_{1/2}$ (more than 1 day). Many drugs displace warfarin from its binding sites, potentially producing

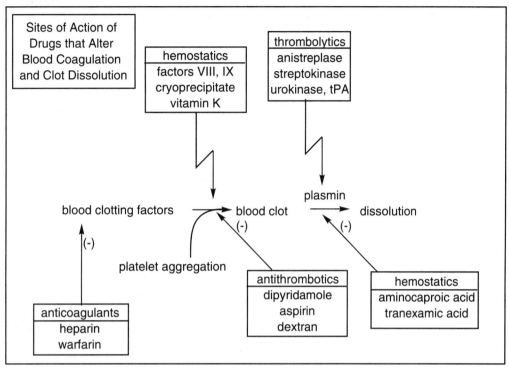

Figure 5.29

bleeding episodes. Also, drugs that alter metabolic rate (e.g., phenobarbital, cimetidine) markedly influence activity of warfarin. The levo form of warfarin has 4 times the potency of the dextro form.

Mechanism. Coumarin analogs inhibit conversion of inactive vitamin K epoxide to active vitamin K hydroquinone, which is necessary for γ-carboxylation of glutamate moieties in coagulation factors II (prothrombin), VII, IX and X **(Figure 5.32)**. Since circulating levels of these factors (especially VII) are not effectively biodegraded for 8-12 hours, there is a delay in the action of coumarins.

Use. As per heparin.

Adverse effects. Warfarin inhibits formation of protein C and may produce cutaneous necrosis. Since warfarin crosses the placenta and inhibits γ-carboxylation of glutamate

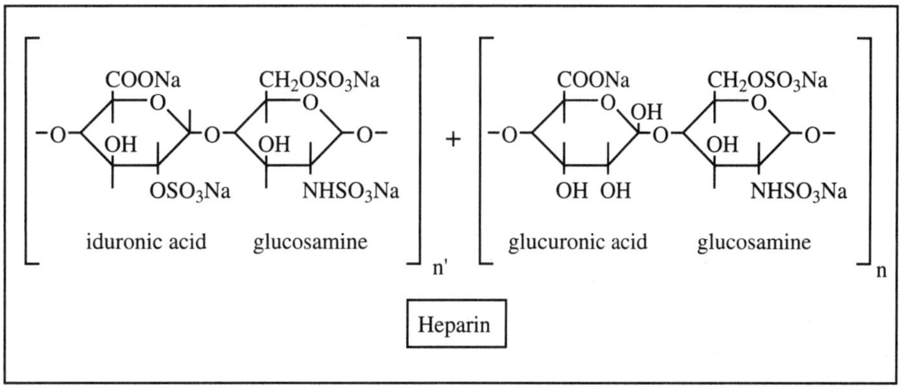

Figure 5.30

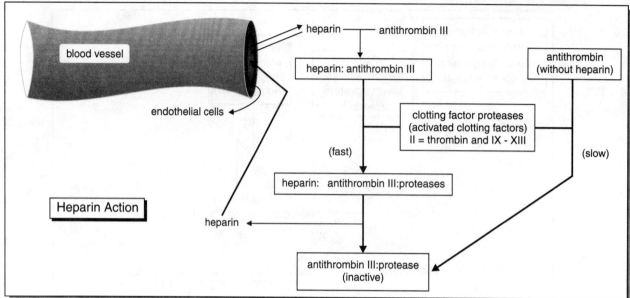

Figure 5.31

moieties in developing bone, bone deformities may occur in the fetus—as well as bleeding tendency.

Drug interactions are serious **(see Pharmacokinetics above)**, especially for those that tend to increase bleeding (e.g., NSAIDs, glucocorticoids). Third generation cephalosporins inhibit vitamin K activation (as per warfarin) and kill colonic bacteria that produce vitamin K. Oral contraceptives reduce warfarin effects by reducing antithrombin III and increasing levels of clotting factors.

Fast reversal of warfarin action. A large dose of vitamin K_1, fresh-frozen plasma or factor IX concentrate (which contains prothrombin) can be given.

B. Antithrombotics are drugs that inhibit platelet participation in blood clotting. Normally a disruption in the endothelial lining of blood vessels induces thromboxane synthesis and platelet aggregation.

1. Dipyridamole (Persantine) inhibits platelet phosphodiesterase, thereby increasing platelet cAMP levels and preventing platelet adhesion.
2. Aspirin acetylates thromboxane synthetase which permanently inactivates the enzyme and thereby prevents thromboxane synthesis. Aspirin is effective in preventing a second MI or second stroke.
3. Dextran 40, dextran 70 and dextran 75 are large MW polysaccharides (40-, 70- or 75-thousand) that inhibit platelet activation and fibrin polymerization.
4. Timolol is useful in preventing MIs, but not necessarily because of effects on blood coagulation.

C. Thrombolytics are enzymes that "dissolve" blood clots by activating plasminogen **(Figure 5.33)**.

1. Streptokinase is naturally produced by streptococci. Anistreplase is acylated streptokinase-plasminogen complex, which is more potent and selective for blood clots.
2. Urokinase is naturally produced by the kidneys.
3. Tissue plasminogen activators (tPA) are produced in the clot. This selectively cleaves plasminogen that is bound to fibrin.

Mechanism. These drugs activate conversion of plasminogen to plasmin, which degrades fibrin and fibrinogen in blood clots **(Figure 5.33)**.

These thrombolytics are able to act, since antiplasmins in the plasma do not easily penetrate into blood clots. However, these antiplasmins do preclude use of plasmin, which would immediately be inactivated by plasma.

Use. Thrombolytics are used to dissolve pulmonary emboli and deep venous or coronary thrombi. Effectiveness is increased by β-blockers and aspirin.

Adverse effects. Patients with antibodies to streptococcal protein have allergic reactions or resistance to streptokinase.

D. Hemostatics are substances that promote blood coagulation.

1. Specific blood fractions are used for people with genetic absence of clotting factors:
   a. Hemophilia A is the absence of VIII: administer lyophilized VIII.
   b. Hemophilia B is the Christmas disease or absence of IX: administer lyophilized IX.
   c. Cryoprecipitate is a plasma protein fraction of whole blood that can also be used for factor VIII or IX deficiencies.

   To inactivate viruses (e.g., HIV, hepatitis), pasteurization, UV exposure or detergent/solvent extraction is used.

2. Vitamin K_1 (phytonadione, from food) and K_2 (menaquinone, from bacteria) activate blood factors II, VII, IX and X.

   Use. Vitamin K is routinely administered to newborns since they are commonly deficient in vitamin K, particularly premature infants. Vitamin K is also given to people with vitamin K deficiency produced by malnutrition or antibiotic-impaired synthesis of vitamin K by colonic bacteria.

3. Aminocaproic acid and tranexamic acid are analogs of lysine

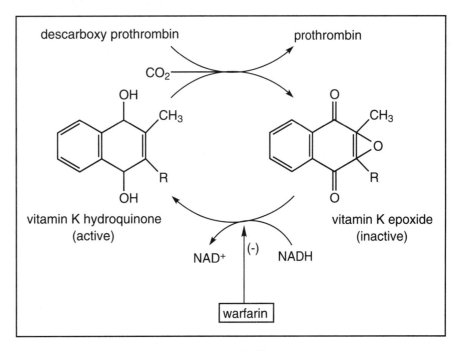

Figure 5.32

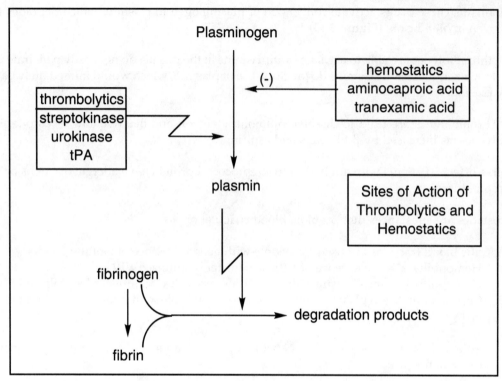

Figure 5.33

that inhibit plasminogen activation and thereby inhibit dissolution of blood clots **(Figure 5.33)**. These are used for post-surgical bleeding; in hemophilia; or bleeding from overuse of fibrinolytic drugs. Thrombus formation is the major adverse effect.

CHAPTER 6
TREATMENT OF ALLERGY

A. THE ALLERGIC RESPONSE

Allergens are antigens (e.g., proteins or haptens) that initiate a cascade of reactions known as the allergic response. In immunology this is *immediate hypersensitivity*. IgEs are the major antibodies reacting with allergens. On initial exposure of an allergic person to an allergen, the body produces IgEs, with the magnitude of this response being in proportion to the antigenicity of the allergen. Because IgE production occurs gradually over a period of hours to days, the first exposure to the allergen usually is not associated with an allergic reaction (unless there has been previous exposure to a very similar substance—cross sensitivity).

Mast cell membranes and basophil cell membranes contain tens of thousands of receptors for IgE. Consequently, after IgEs are produced they reside largely (and permanently) on the surface of mast cells and basophils. No unpleasant effect occurs as IgEs become localized on these cells. On subsequent exposure, this allergen forms a complex with adjacent IgEs, initiating Ca^{++} influx and inositol phosphate hydrolysis, which ultimately lead to fusion of intracellular granules with the outer cell membrane **(Figure 6.1)**.

The granules of mast cells and basophils contain large amounts of histamine complexed to heparin. Thus, allergens release histamine from mast cells and basophils **(Figure 6.1)**. Actions of histamine are shown in **Figure 6.2**. Histamine H$_1$-blockers are effective in preventing many of the unpleasant allergic reactions. These drugs are described later this chapter. Many other autacoids, local hormones, are released with histamine during an IgE: allergen reaction. These include prostaglandins (PGs), leukotrienes (LTs), platelet aggregating factor (PAF), kinins, proteolytic enzymes and chemotactic factors. The chemokinesis of leukocytes results in formation of complement and a generalized inflammatory reaction.

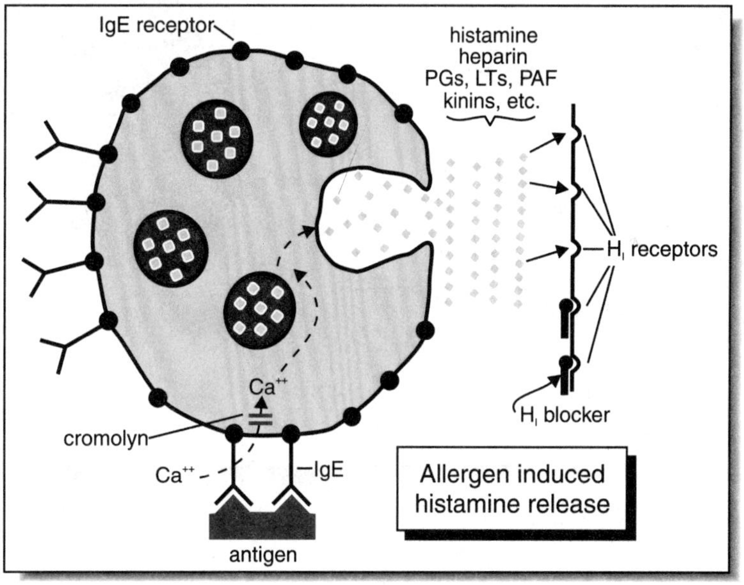

Figure 6.1

Mast cells are located largely at those sites where allergic reactions occur:
- **Lungs**: hayfever, bronchial asthma.
- **Nasal mucosa**: hayfever, allergic rhinitis.
- **Skin**: dermal allergies such as hives, eczema, contact dermatitis.
- **GI tract**: food allergies.
- **Gastric and duodenal mucosa**: local histamine stimulates release of HCl from parietal cells.

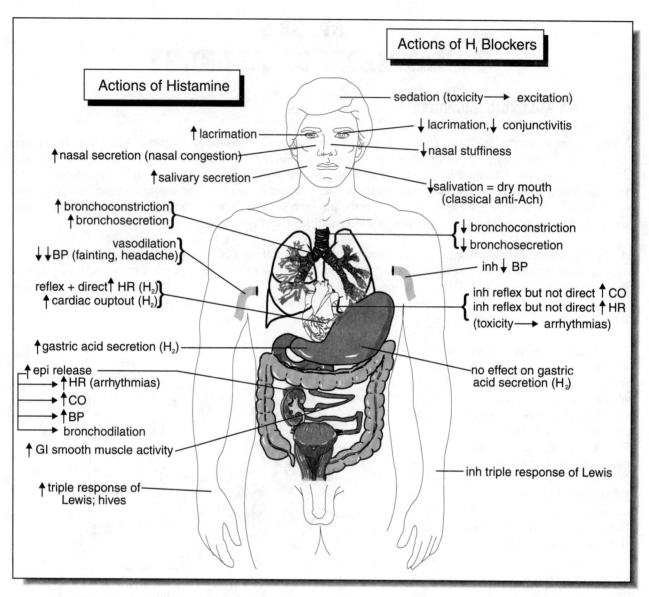

Figure 6.2

An *anaphylactic reaction* to drugs and other substances like bee stings is caused by massive release of histamine and other autocoids from mast cells and basophils, resulting in concentrations of these substances that are high enough to produces effects on cardiac muscle, smooth muscle and exocrine glands throughout the body.

A. Histamine Action

Histamine ("histos" amine means tissue amine) is produced in virtually all cells but is present in large amounts primarily in tissue mast cells and blood basophils. Dietary histamine is inactivated in the gut by an N-acetyltransferase (NAT). Histamine in food, therefore, does not constitute a source of body histamine. Body stores of histamine are made primarily from the essential amino acid, histidine, according to the scheme in **Figure 6.4**. Histamine is inactivated by diamine oxidases (DAO) in tissues and blood, and by methyltransferases in tissues. (These enzymes are analogous to the MAO and COMT involved in NE metabolism.)

Triple response of Lewis. A series of reactions occurs when a small amount of allergen is introduced just below the surface of the skin (e.g., mosquito bite). This allergen releases histamine locally. Histamine (a) dilates arterioles (redness is the first response), (b) increases capillary permeability (local edema is the second response) and (c) stimulates nerve endings, with antidromic impulses leading to dilatation of arterioles 1-2 cm distant (more vasodilation and a greater area of redness is the third response) **(Figure 6.3)**. This scheme highlights the cardiovascular effects that are typical of histamine. This sequence of reactions illustrates how hives occurs during allergy, and illustrates how profound hypotension occurs during anaphylaxis.

B. Cellular Stores of Histamine

1. **Exocytolic and non-exocytolic release**. Histamine is present in cells primarily in granules, but a small amount is in the cytoplasm. Physiological and allergic reactions promote release of histamine from granules. This is *exocytotic release*, analogous to neurotransmission.

    Non-exocytotic release of histamine, from cytoplasm, occurs from:
    a. *Cell lysis* (e.g., hitting your finger with a hammer).
    b. Cell membrane effects caused by *physical processes* such as heat/cold (e.g., rubbing an ice cube on your arm) or touch (e.g., scratching).
    c. *Chemical displacement* of histamine from its binding sites. Drugs with basic moieties (e.g., amino group) have this effect, notably *morphine* (e.g., constant nose dripping in heroin addicts), *d-tubocurarine* (e.g., severe hypotensive reaction in some people), radiocontrast media.

2. **Histamine receptors**
    H_1: microcirculation, large blood vessels, respiratory and GI tract
    H_2: parietal cells, lamina propria of stomach, macrophages, immune cells, large blood vessels
    H_3: autoreceptors on nerves

3. **Pathologies involving excess histamine release or effect**
    Dermal: urticarial reactions (e.g., urticaria pigmentosa)
    Intestinal: carcinoid tumors
    Systemic: mastocytosis, basophilic leukemia, angioneurotic edema, cirrhosis (impaired inactivation of circulating histamine).

C. Medical Uses of Histamine Analogs

Histamine exerts prominent effects throughout the body, so it is not so useful as a drug. To

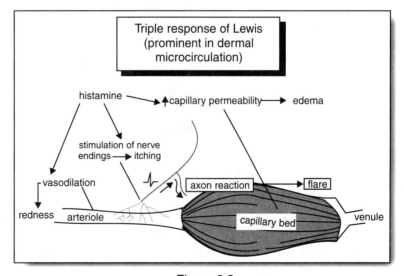

Figure 6.3

test for bronchial hypersecretory activity, as in chronic asthmatics, histamine can be inhaled. This test is uncommon. At one time histamine was used to test for pheochromocytoma, evoking large and dangerous amounts of epi/NE via an action at histamine receptors on these cells.

Betazole (Histalog) is a selective H_2 agonist, used for diagnosis of gastric function. With Zollinger-Ellison syndrome and ulcers, large amounts of gastric acid are released. With pernicious anemia and gastric carcinoma, little gastric acid is released. *Pentagastrin*, a pentapeptide analog of endogenous gastrin, has largely replaced betazole for these diagnoses, since pentagastrin does not produce systemic effects.

D. Approaches to reducing histamine actions in the body

1. Inhibit histamine synthesis. This is impractical and not desirable.
2. Block histamine receptors with H_1- or H_2-blockers.
3. Inhibit histamine release with mast cell Ca^{++}-channel blockers.
4. Give a pharmacological antagonist to counteract histamine actions. For example, epinephrine produces responses in smooth muscle and glands that are generally the opposite of histamine action.
5. Acute desensitization involves repeated administration of increasingly greater amounts of the allergen over a period of hours, to deplete histamine in graded amounts from the granules. When this is achieved, a therapeutic amount of that substance can be given (e.g., a vaccine).
6. Prolonged desensitization involves repeated administration of increasingly greater amounts of the allergen over a period of months, to induce increasingly greater production of IgG antibodies to the allergen, thereby suppressing the allergic reaction. This approach is used to desensitize from bee venom, etc. and is practical only because IgE antibody production is not simultaneously induced.

B. HISTAMINE H_1-BLOCKERS

Histamine H_1-blockers are commonly used to treat allergy. When taken before exposure to allergens, H_1-blockers are often effective in attenuating allergic responses. When taken after exposure, H_1-blockers are not so effective. Because of (a) the prevalence of allergy, (b) disparity in patient response to individual drugs and (c) varying degrees of tolerance to side effects of individual drugs, there are many prescription and over-the-counter (OTC) preparations available.

1. **Limitations of H_1-blockers.** Because histamine is only one of many autacoids released during allergic reactions, H_1-blockers are only partially effective in treating allergy. In asthma, LTs especially, and PGs secondarily, play the major role in airway obstruction. Consequently, H_1-blockers are not effective in treating asthma.

2. **Pharmacokinetics of H_1-blockers.** The OTC H_1-blockers are generally well absorbed after oral administration, attain peak levels in ½-2 hours, distribute to all sites including brain, have a duration of 3-4 hours and are extensively metabolized. Excretion occurs in about 1 day. The prescription H_1-blockers are distinguished by a duration of action of several days.

3. **Actions of H_1-blockers.** The OTC H_1-blockers are competitive antagonists of histamine at the H_1 receptor. The prescription drugs, terfenadine and astemizole, are non-competitive antagonists. H_1-blockers do not inhibit histamine release. In fact, since they have amino groups, the H_1-blockers tend to release histamine from mast cells and basophils. However, this effect is generally a small one and has little significance since H_1 receptors would be blocked by these same drugs. Pharmacological effects of H_1-blockers in treating allergy are shown in **Figure 6.2**.

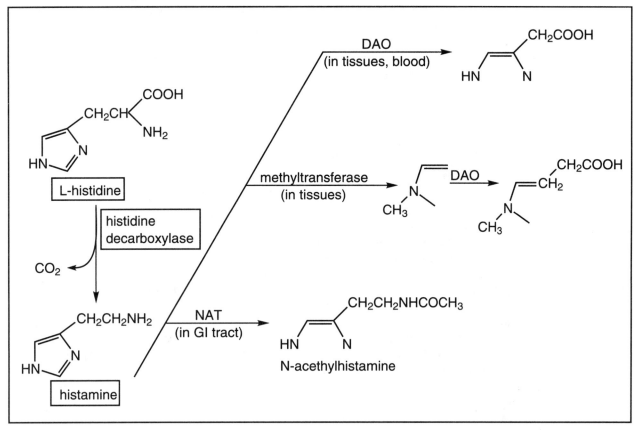

Figure 6.4

Especially for the OTC drugs, H_1 blockers often block muscarinic and serotoninergic receptors, and have local anesthetic actions. These activities contribute to the profile of undesirable actions of these drugs.

4. **Toxicity of H_1-blockers is especially prominent in children. Effects include:**
   a. *CNS excitation* including confusion, hallucinations, convulsions; ultimately, coma and death.
   b. *Hyperthermia* with fixed dilated pupils and flushed face.
   This resembles the toxicity of atropine, a belladonna ("beautiful woman:" warm skin, rosy cheeks, big eyes—dilated pupils). In adults there is less fever and less excitation, but arrhythmias occur. Treatment is supportive—maintain respiration and give an antiseizure drug.

5. **Clinical uses of H_1-blockers**
   a. Treatment of *allergy* (hives, rhinitis, hayfever, conjunctivitis), but not so effective against GI allergy.
   b. Prevention of *motion sickness*, especially cyclizine, marezine, promethazine. Promethazine finds special use as an anti-emetic, because of its simultaneous block of dopamine D_2 receptors.
   c. In *sunburn preps*, because of local anesthetic action (e.g., promethazine).
   d. Production of *sedation* (e.g., diphenhydramine, promethazine).

e. Diphenhydramine and a couple other H_1-blockers are used to treat *Parkinsonism*, probably because of simultaneous block of muscarinic receptors.

6. **Chemistry and properties of competitive H_1-blockers.** Most H_1-blockers have the general formula shown, where X determines the drug class.

$$-X-CH_2-CH_2-N\begin{matrix}R_1\\R_2\end{matrix}$$
H_1-blocker

a. **Ethanolamines (X = O)** -O-C-C-N=
Generally these produce prominent sedation and antimuscarinic activity, but GI effects are mild (e.g., *diphenhydramine* [Benadryl]).
b. **Ethylenediamines (X = N)** =N-C-C-N=
Generally these have fewer sedating and anti muscarinic actions, but more GI effects that the ethanolamines (e.g., *pyrilamine* [Neo-Antergan]).
c. **Alkylamines (X = C)** -C-C-C-N=
Generally these produce CNS excitation and are good for daytime use (e.g., *chlorpheniramine* [Chlor-Trimeton]).
d. **Piperazines**
Generally these produce little sedation or antimuscarinic effect, and are especially effective for motion sickness (e.g., cyclizine [Marezine]). Because of teratogenicity in animals, they should be avoided in pregnancy.
e. **Phenothiazines**
Generally these drugs owe their effect to block of dopamine D_2 receptors, although they do also block H_1, muscarinic and serotoninergic receptors. These are used as anti-emetics and for motion sickness (e.g., *Promethazine* [Phenergan]).
f. **Miscellaneous drugs**
These drugs don't fall into the above chemical classes, but are widely used.
   a. *Hydroxyzine* (Atarax) is commonly used in children.
   b. *Cyproheptadine* (Periactin) has prominent anti-serotonin activity. A frequent side effect is increased appetite with weight gain.

7. **Chemistry and Properties of Non-Competitive H_1-Blockers.** These newer drugs don't fall into the above chemical classes. They do not easily enter the brain, so they do not produce much sedation. Side effects are generally less than with the above drugs. They rarely induce ventricular arrhythmias known as torsade de pointes (increased QT interval).

Terfenadine (Seldane) has a $t_{1/2}$ of approximately 1 day; *Astemizole* (Hismanal) has a $t_{1/2}$ of about 10 days; loratadine (Claritin) has a half life of about 12 hours and its active metabolite, descarboethoxyloratadine, has a $t_{1/2}$ of approximately 20 hours. There is delayed onset with astemizole, so it is less preferred for acute allergies. The long $t_{1/2}$ makes these drugs particularly well-suited for treatment of chronic allergies. Side effects include increased appetite and weight gain.

C. MAST CELL CA++-CHANNEL BLOCKERS

Cromolyn (*Intal*) and nedocromil (Tilade) stabilize mast cell membranes and inhibit IgE-mediated release of histamine and other autacoids. These drugs, however, are not acutely effective.

Pharmacokinetics. These drugs are poorly absorbed orally. They are taken as inhalants or topically as ophthalmic preparations. Serum $t_{1/2}$ is only about 80 minutes, but these drugs continue to exert effects on the mast cell for a much longer time. They poorly penetrate the blood-brain barrier. They are excreted in bile in urine, mostly unmetabolized.

Mechanism. When taken several times daily for a month or more, cromolyn and nedocromil effectively block Ca^{++} channels in mast cells, preventing IgE-mediated degranulation and associated histamine release.

Uses. If taken for a long enough time, prophylactically, cromolyn and nedocromil are useful for respiratory allergies and food allergies. Also, they may be used for ulcerative colitis and systemic mastocytosis. Since basophils and dermal mast cells are resistant, these drugs are not used for dermal allergies. They are not useful for acute asthma.

Adverse effects. These include throat irritation from the powder and overt hypersensitivity reactions (urticaria, anaphylaxis).

D. PHARMACOLOGICAL ANTAGONISTS TO HISTAMINE AND AUTACOIDS

These drugs, by actions at receptors other than H_1, produce physiological responses that are generally the opposite to that of histamine and other autacoids such as LTs and PGs. These pharmacological antagonists thus improve breathing and restore blood pressure. In addition, these drugs inhibit further mast cell release of histamine and other autacoids. Except for methylxanthines, these drugs are useful when even the actions of histamine are already prominent (e.g., in asthma and anaphylactic shock). Substances include:

1. Sympathomimetic amines (e.g., epinephrine, isoproterenol)
2. Selective β_2-agonists (e.g., albuterol)
3. Methylxanthines (e.g., theophylline)
4. Corticosteroids (e.g., cortisol)

E. ANTIASTHMATICS

Asthma is a disorder characterized by excessive bronchoconstriction tone, excessive bronchosecretion of mucus and inflammation—edema of the respiratory mucosa. All three factors contribute to airway obstruction and difficulty in breathing. Physical (cold air, exercise), chemical (ozone, allergens) and infection (viral) processes can precipitate asthma.

A. Classification of Asthma

1. Extrinsic asthma is related to allergen sensitivity. Allergens react with IgEs on mast cells, resulting in mast cell release of histamine and other autacoids, which then adversely affect the bronchi. Leukotrienes B_4, C_4 and D_4 seem to be the most important mediators.

2. Intrinsic asthma is not attributable to antigens.

B. Drugs Used to Treat Asthma

1. **Mast cell Ca^{++}-channel blockers.** Cromolyn sodium (Intal) and nedocromil (Tilade) have already been described. For prophylaxis of asthma, cromolyn and nedocromil treatment for 4 weeks reduces allergen-induced, as well as cold and exercise induced asthma.

2. **Nonspecific β-adrenoceptor agonists.** These drugs are discussed more thoroughly in **Chapter 2**. They are effective in acute asthma, but have prominent $β_1$ effects (increased HR, inotropic action, tremor).

   a. Epinephrine (epi)
   b. Isoproterenol (Medihaler)(ISP)
   c. Metaproterenol (Metaprel)
   d. Ephedrine

   Mechanism. By stimulating adenylyl cyclase these drugs increase levels of cAMP in bronchial smooth muscle and mast cells, respectively, relaxing bronchioles and inhibiting mast cell release of histamine and other autacoids.

3. **Specific $β_2$-adrenoceptor agonists**
   These are the most effective drugs for acute bronchospasm and exercise-induced asthma. Their pharmacology is discussed more thoroughly in **Chapter 2**. For asthma, inhalation is as effective as s.c. administration, while adverse effects are less frequent.

   a. Albuterol (Proventil)
   b. Terbutaline (Brethaire)
   c,d. Pirbuterol (Maxair) and Bitolterol (Tomalate) are newer.

   These drugs have a longer duration than epi and ISP. The $β_2$-agonists are less likely to produce $β_1$-mediated effects (increased HR, tremor, hypokalemia) vs. epi and ISP. The mechanism of action in asthma is the same as for epi.

4. **Corticosteroids.** Asthma is an allergen-induced inflammatory disorder of the airway. Corticosteroids inhibit inflammation and prevent production of phospholipase A_2 and the subsequent cascade that leads to generation of PGs and LTs.

   a. Beclomethasone (Beclovent)(inhalant)
   b. Triamcinolone (Azmacort)(inhalant)
   c. Prednisone or prednisolone (oral)
   d. Funisolide (AeroBid)(inhalant)

Corticosteroids are effective even in patients not responsive to bronchodilators.

Inhaled corticosteroids seems to be more effective in prophylaxis than $β_2$-agonists. Inhaled corticosteroids tend to produce oral candidiasis but little serious toxicity. However, the spectrum of side effects may be similar to that discussed in **Chapter 10.V**: suppression of the pituitary-adrenal axis and reduced bone formation (increased bone resorption and slowed growth in children).

Systemic corticosteroids are very effective in decreasing the frequency and severity of asthma attacks. Side effects are discussed in **Chapter 10.V**.

5. **Methylxanthines**
   a. Theophylline (oral).
   b. Aminophylline (theophylline ethylenediamine).

These drugs reduce bronchiole tone by a mechanism that does not seem to be correlated with changes in cAMP levels. The xanthines reduce the incidence of acute asthma and also enhance the effects of β-agonists and corticosteroids, so that the dose of corticosteroids can be reduced.

Serum levels should be monitored since side effects are related to serum concentration:

<10 mg/ml, few adverse effects.
  10-20 mg/ml, anxiety and nausea.
>20 mg/ml, vomiting, arrhythmias, hypotension, anxiety, tremor and seizures, hypokalemia and hyperglycemia.

Methylxanthines interact with many drugs.

6. **Muscarine receptor antagonist.** Ipratropium (Atrovent) is a quaternary antimuscarinic drug that produces bronchodilitation without affecting bronchosecretion. This is used primarily as a supplement for methylxanthines or β-agonists, not for acute asthma.

F. TREATMENT APPROACHES

$β_2$-agonists, as needed, may be adequate when asthma occurs infrequently. b_2-agonists, when taken before exercise, also decrease the incidence of exercise-induced asthma. Cromolyn may eliminate the need for corticosteroids in more severe forms of asthma. Theophylline is sometimes held in reserve.

CHAPTER 7
GASTROINTESTINAL (GI) DRUGS

Drugs affecting GI function are among the most-widely used drugs in medicine. Many of these are over-the-counter (OTC) and are taken without a physician's advice. These drugs are used to treat stomach ulcer, constipation, gas and diarrhea, as well as some pathologies involving the GI tract.

A. TREATMENT OF ULCERS

There is an old adage, "No acid, no ulcer." The most effective anti-ulcer drugs owe their effectiveness to reduction in gastric acid secretion (H_2-blockers, antacids, proton pump inhibitor). Useful, often less effective agents, owe their effect to promotion of a cytoprotective barrier in the stomach (sucralfate, bismuth). The histamine H_2-blockers are the most important anti-ulcer drugs, having rendered surgical removal of ulcers mostly a thing of the past. A comparison of the healing rate for these drugs is made in **Figure 7.1**. It is noted that duodenal ulcers are healed more quickly than gastric ulcers.

1. **Histamine H_2-Blockers**. The different H_2-blockers are about equally effective in healing duodenal ulcers (75% healed at 4 weeks, 90% at 8 weeks) and gastric ulcers (60% healed at 4 weeks, 85% at 8 weeks). Generally a treatment dose is administered for 4 to 8 weeks, followed by a lower maintenance dose for several months. Relapse rates are high when patients come off these and other anti-ulcer drugs.

    Pharmacokinetics. All are well absorbed, attain serum peak levels in about an hour and have a duration of action of 4 to 24 hours. A high percentage of the H_2-blockers are excreted unmetabolized.

    Mechanism. These drugs block the action of histamine at the H_2 receptor, decreasing both basal and food-stimulated acid secretion. It appears that the site of action is on macrophages and other immune cells in lamina propria of the stomach, or on parietal cells. Local inflammation, possibly due to unsuspected infections, may be the cause of GI ulcers.

    | **Specific drugs:** | Cimetidine (Tagamet) |
    | | Ranitidine (Zantac) |
    | | Famotidine (Pepcid) |
    | | Nizatidine (Axid) |

    Adverse effects. Severe side effects are uncommon. Rarely, hepatitis or hematologic toxicity develops. High dose cimetidine, which is no longer recommended, causes an increased incidence of hyperprolactinemia, gynecomastia (increased prolactin) and an anti-androgen action. Cimetidine, more than ranitidine, competes with other drugs for cytochrome P_{450} (e.g., benzodiazepines, phenytoin, warfarin) and delays their metabolism. Doses should be adjusted. Famotidine and nizatidine do not interfere with cytochrome P_{450}.

    Other uses. Reflux esophagitis and hypersecretory states (Zollinger-Ellison syndrome, mastocytosis, basophilia and cirrhosis) are effectively treated by H_2-blockers.

2. **Antacids**. These are weakly basic substances that chemically neutralize stomach acid. This class of drugs is sometimes used as monotherapy of ulcer, often without physician intervention. Most often, antacids are used as an adjunct to other anti-ulcer drugs, since the antacids often produce immediate

relief of ulcer pain. All antacids, especially Ca^{++}-antacids, cause rebound (increased) secretion of stomach acid. This may not be a problem if the antacid is taken in adequate amounts. The objective of antacid therapy is partial neutralization of stomach acid and elevation of intragastric pH to greater than 3.5. With less irritation of the ulcer surface by acid, the ulcer is able to heal faster. Antacids are used for reflux esophagitis, as well as stomach ulcer.

Specific Antacids

a. *Sodium bicarbonate* is the only *systemic antacid* (i.e., largely absorbed), making it the least desirable antacid. It is preferred that the action of antacids be solely within the stomach, and that the neutralized products be retained within the lumen of the GI tract for excretion in feces. Otherwise, systemic effects can be considerable, as for sodium bicarbonate **(Figure 7.2)**. Since gram amounts of bicarbonate are taken at one time, the risks indicated are substantial.

b. *Magnesium antacids* are only slightly absorbed, so systemic effects are insignificant if absorbed Mg^{++} can be eliminated in urine **(Figure 7.3)**. Mg-antacids should not be used if there is renal insufficiency. A common prep is "milk of magnesia." Mg-antacids typically produce laxation or diarrhea.

c. *Aluminum antacids* are not appreciably absorbed. Since Al forms a precipitate (ppt) with phosphate (PO_4^{-3}), PO_4^{-3} absorption is impaired and this could be a problem with prolonged use of Al preps. Al forms complexes with some drugs, like tetracycline (TTC) and digoxin, so that their absorption is also impaired. Al preps typically cause constipation **(Figure 7.4)**.

d. *Calcium antacids* are partially absorbed, more so if ulcer is present. Ca forms a ppt with PO_4^{-3}, so PO_4^{-3} absorption can be impaired. Prolonged use of Ca-antacids could adversely affect systemic PO_4^{-3} levels and/or predispose to nephrolithiasis. Ca preps typically cause constipation **(Figure 7.5)**. The Ca in Ca-antacids could be an advantage to aged and pregnant women who generally need more Ca. However, absorption of Ca is a complex process, as discussed in the section on parathyroid hormone, and one should not assume that extra ingested Ca will increase body stores of Ca.

Common formulations are combinations of Mg with either Al or Ca preps, so that the laxative effect of Mg is counterbalanced by the constipating effect of the Al or Ca prep.

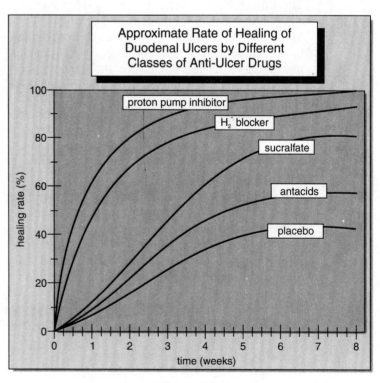

Figure 7.1

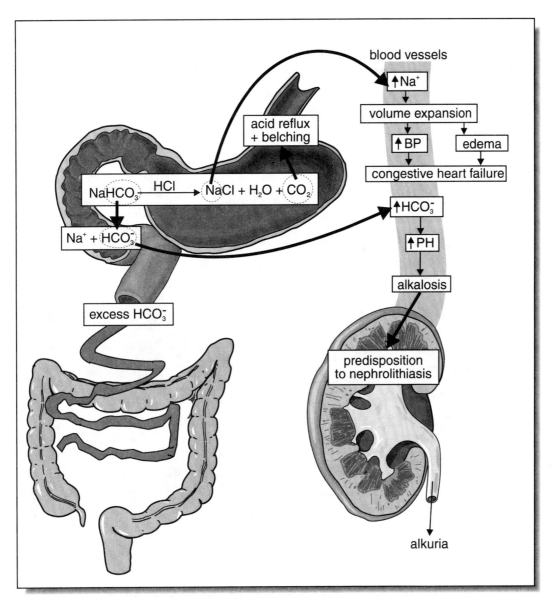

Figure 7.2

Although Na+ may not be the active ingredient, it is often present in the binder for Al, Mg and Ca preps.

Ideally, antacids should be taken at timed intervals after meals, for a total of about 7 times daily. Few people can comply with this regimen.

3. **The proton pump inhibitor**, omeprazole (Losec), is the most potent inhibitor of stomach acid production. A single 20-30 mg dose inhibits more than 90% of 24 hour stomach acid secretion (versus 50-80% after a single dose of an H_2-blocker).

Mechanism. The prodrug, omeprazole, enters the parietal cell from the blood and is metabolized to a sulfide, which irreversibly inhibits the parietal cell HK-ATPase. This effect renders the

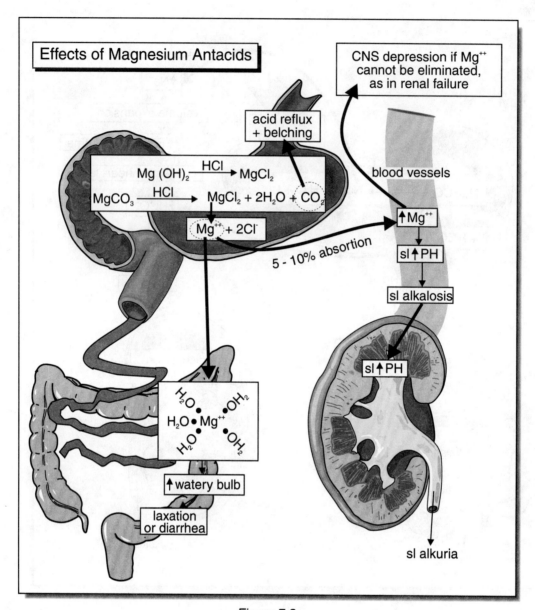

Figure 7.3

parietal cell incapable of secreting acid **(Figure 7.6)**. The parietal cell does not recover until new ATPase is made.

Pharmacokinetics. Omeprazole is inactivated by stomach acid, so it must be administered in an enteric-coated form or as a buffered suspension. Since absorption is from the small intestine, the time required is 3-4 hours. Synthesis of new HK-ATPase, to replace irreversibly inactivated HK-ATPase, requires about 3 days.

Use. Duodenal ulcers are healed more quickly and pain is relieved more rapidly with omeprazole than with other anti-ulcer drugs **(Figure 7.1)**. Although not approved for treatment of gastric ulcers, these are also healed more rapidly with omeprazole. Omeprazole is highly effective in healing the approximate 5% of peptic ulcers which are refractory to H_2-blockers. This is a primary use of omeprazole (approximately 40 mg per day for 8 weeks). For hypersecretory states, like Zollinger-Ellison syndrome, omeprazole is likewise more effective than H_2-blockers.

Adverse effects. Increased serum gastrin levels occur after omeprazole, a potential for induction of carcinoid tumors (which are produced in lab animals). Effects on cytochrome P_{450} are like that of cimetidine. Gynecomastia and painful erections can occur after omeprazole.

4. **Cytoprotective agents** include sucralfate (Carafate) and colloidal bismuth.

   Sucralfate is an aluminum hydroxide sulfated sucrose complex that forms a viscous coating on the surface of the ulcer and stimulates gastric secretion of mucus, PGs and bicarbonate---all of which are protective from stomach acid. Sucralfate is not appreciably absorbed, acting almost totally

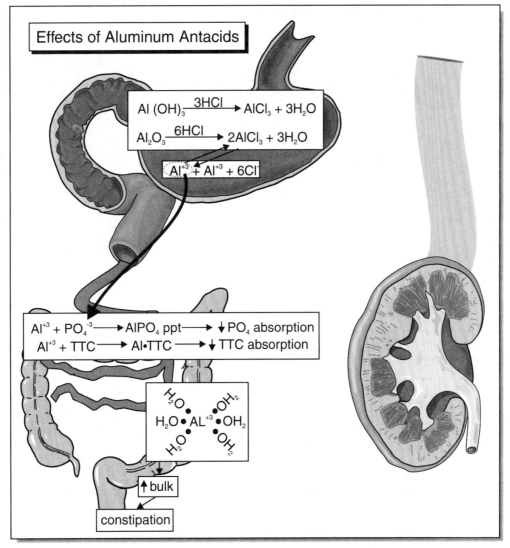

Figure 7.4

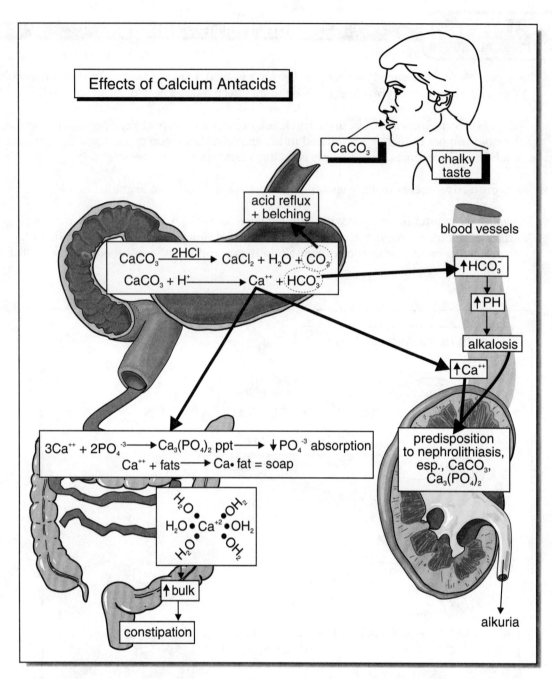

Figure 7.5

within the lumen of the GI tract, like an aluminum antacid. Side effects of sucralfate are also like those of aluminum antacids: constipation and reduced PO_4^{-3} absorption. Duodenal ulcer healing rate for sucralfate is shown in **Figure 7.1**.

Colloidal bismuth (Pepto-Bismol) forms a coating on the surface of the ulcer, analogous to sucralfate. Since bismuth is toxic to the kidney, long-term use presents a potential for renal tubule injury; encephalopathy may occur as well. Black tarry stools can occur with colloidal bismuth.

5. **Miscellaneous Agents**
   a. **PGE Analogs**
   Misoprostol (Cytotec), a prodrug that is converted to a *PGE_1 analog*, is used specifically to prevent the induction of ulcers by nonsteroidal antiinflammatory drugs (NSAIDs). In Europe misoprostol is used as an antiulcer drug.

   Enprostil is a PGE_2 analog, with the same spectrum as misoprostol. It is not yet available in the United States.

   Both PGE analogs have a spectrum of action like PGE, including an abortion potential.

   b. **Mucus stimulant**
   Carbenoxalone stimulates production of mucus, a cytoprotectant on the ulcer base. Its effect relies on an *aldosterone action*, which also means that this drug has a potential for increasing K^+ loss in the kidney, Na^+ retention, fluid retention and hypertension. Carbenxoalone is used for duodenal and gastric ulcers in Europe, but has not been approved in the United States.

   c. **Antibiotics** specific for Helicobacter pylori may become a more common approach to ulcer treatment. H. pylori has been suspected as a causative factor in ulcer formation. Recent localization of H_2 receptors on inflammatory cells in the stomach lining, gives credence to this possibility.

B. TREATMENT OF CONSTIPATION: LAXATIVES AND CATHARTICS

Laxatives and stool softeners are used to treat constipation, which is defined as the infrequent passage of a formed stool. Cathartics and enemas are used to promote more thorough evacuation of the bowel, through passage of a watery unformed stool. Fluids must be consumed with these agents to avoid elimination of excess body water which can cause dehydration. Generally, a simple high fiber diet eliminates the need for laxatives.

LAXATIVES

1. **Substances high in fiber** include cereals (bran, psyllium, cellulose, methyl-, carboxy-cellulose), fruits and vegetables. These poorly absorbed polysaccharides and cellulose derivatives adsorb H_2O and increase fecal bulk. A soft watery stool is formed. Latency is about 2-3 days. High fiber diets are beneficial for ulcerative colitis and diverticulitis.

2. **Osmotic Laxatives**
   a. *Saline laxatives* are ionic salts and complexes which are not appreciably absorbed, but which adsorb H_2O. This hydrated ionic complex is retained in the lumen of

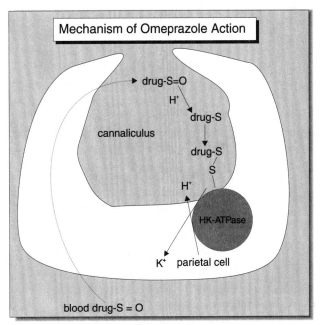

Figure 7.6

the GI tract, increasing bulk and forming a watery stool. With large amounts of the salt complex, a catharsis occurs. Since adsorption begins in the upper intestine, latency for laxation is about 3-6 hours. Common preps are MgO, $Mg(OH)_2$, $MgSO_4$, Mg citrate, KNa tartrate and Na_2HPO_4/NaH_2PO_4. Colyte is an osmotic salt solution that does not alter body electrolyte balance.

    b. *Lactulose* (Cephulac) is a synthetic disaccharide that is hydrolyzed only by colonic bacteria, to small organic acids (formic, acetic, lactic, etc.), which then produce an increase in osmotic strength and consequent greater retention of H_2O in the lumen. This results in laxation. The acidic byproducts of lactulose metabolism neutralize ammonia (NH_3), thereby reducing NH_3 absorption, and thus making lactulose especially valuable in treating portal-systemic encephalopathy. Because galactose is one of the saccharides in this sugar, it should be avoided in people with lactose intolerance. Latency for laxation is 1-3 days.

    c. *Castor oil*, a triglyceride of ricinoleic acid, must first be emulsified by bile. Do not use castor oil if there is a bile obstruction. Castor oil, once emulsified, is digested by pancreatic lipase to 1 glycerol and 3 ricinoleic acid molecules. The latter is a surfactant that impairs H_2O and salt absorption, increasing osmolality and thereby promoting H_2O adsorption and excretion. Glycerol per se can penetrate the stool and act as a stool softener. Latency for laxation is less than 3 hours.

3. **Contact agents**

By acting on the colonic mucosa, contact agents interfere with H_2O absorption and produce laxation in 6-8 hours. Because they are partially absorbed and excreted in bile, this enterohepatic circulation prolongs the duration of these agents to several days.

    a. *Cascara, aloe, senna and rhubarb* are glycosides that are cleaved to anthraquinones, which (a) stimulate the myenteric plexus, (b) adsorb H_2O, and (c) interfere with H_2O absorption. Cascara and senna may produce a red urine and reversible pigmentation of the colonic mucosa.

    b. *Phenolphthalein* (Ex-Lax) produces laxation principally by interfering with H_2O absorption. It imparts a red color to an alkaline urine and may cause pigmentation of the dermis.

    c. *Bisacodyl* (Dulcolax) interferes with H_2O absorption in the small and large intestine and may stimulate peristalsis.

4. **Stool softeners**

These are substances that specifically act within the stool to soften it, thus reducing any need for straining at defecation.

    a. *Mineral oil* is a non-digestible lipid that penetrates feces to soften them. Adverse effects include (a) impaired absorption of fat-soluble vitamins (D,E,A,K), (b) foreign body reaction, such as lipid pneumonitis, (c) impaired healing of anorectal lesions, like hemorrhoids, and (d) anal leakage.

    b. *Docusates* (Colace) are surfactants with a detergent action that emulsifies fat into the fecal mass, thereby increasing bulk.

Uses for laxatives

1. Functional constipation (natural fibers are best)

2. Counteraction of the constipating effect of a drug (e.g., combined MgO and CaO)

3. In geriatric and ill patients (hemorrhoids, post-surgery, post-infarct, congestive heart failure). Normal straining during defecation causes an increase of esophageal pressure and a collapse of venae cava. This reduces the return of venous blood to the heart, thereby producing a decrease in cardiac output and compensatory increase in peripheral vascular resistance (PVR).

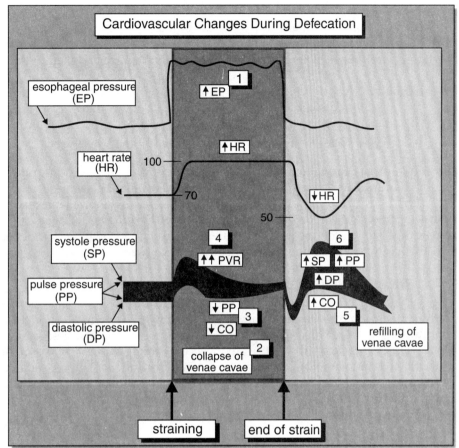

Figure 7.7

At the end of straining, cardiac output (CO) returns to control levels while there is still an increased PVR and elevated blood pressure. Weakened vessel walls could be ruptured **(Figure 7.7)**. To prevent this cascade of events in the at-risk group, laxatives are indicated.

Contrainications for laxatives include any undiagnosed abdominal pain, appendicitis and intestinal obstruction. Dangers of laxative abuse include spastic colon, rupture of an inflamed appendix; excess loss of fluid, electrolytes and nutrients, including vitamins.

CATHARTICS

Cathartics are used specifically to evacuate all contents from the lumen of the GI tract, under the following conditions:
1. in chemical poisoning, to evacuate the toxin from the bowel;
2. with an anthelmintic, to expel worms and the toxic drug from the bowel;
3. before radiologic exams;
4. before surgery.

ENEMAS

Enemas may consist of virtually any non-toxic liquid, although there must be caution to prevent excess loss of body electrolytes. Substances include salts (as per saline cathartics), isosmotic electrolyte solution (Golytely, Colyte), sorbitol, soaps, detergents (Docusate), mineral oil, glycerin, etc.

Uses for enemas
1. Before radiologic exams
2. Before surgery
3. Pre/post-partum

Prevention of gas formation. High fiber diets (especially whole grains, nuts, peas, cabbage family) contain oligosaccharides with α-galactosidase linkages which cannot be cleaved by human intestinal enzymes. These oligosaccharides are thus carried to the large intestine where colonic bacteria metabolize them to gases (hydrogen, methane, carbon dioxide) which produce bloating and flatulence. The fungal enzyme, α-galactosidase, can be taken with meals to facilitate the digestion of oligosaccharides in the upper intestine, thereby reducing gas formation in the colon. Trisilicates have been used to reduce gas formation.

C. TREATMENT OF DIARRHEA

Diarrhea is a symptom of some disorder, not an entity of its own. The underlying basis of diarrhea should be determined. The objective for antidiarrheal drugs is prevention of excess loss of body water (dehydration) and electrolytes. In undeveloped countries, diarrhea is a leading cause of childhood death. The most common causes of diarrhea include amoebial, bacterial, viral and parasitic infections; inflammatory bowel syndrome (e.g., Crohn's disease and ulcerative colitis). Diet adjustment, without drugs, is usually all that's needed for diarrhea in self-limiting infections in adults. Because total body water and electrolyte stores are much less in children, there is more justification for these drugs in children.

1. **Opiates and analogs.** Actions of the opiates are shown in **Figure 7.8**.
   a. *Loperamide* (Imodium) is poorly absorbed, acting locally in the lumen of the GI tract. The abuse potential is nil, since high amounts produce dysphoria.

   b. *Diphenoxylate* and *Atropine* (Lomotil). Diphenoxylate is a prodrug, converted to the active diphenoxylic acid, which acts primarily in the lumen of the GI tract. In high amounts, diphenoxylate could produce euphoria. However, being water-insoluble, it cannot be injected. Atropine, an antimuscarinic anti-diarrheal drug which produces unpleasant systemic effects when taken in large amounts, is added to supplement the effects of diphenoxylate and reduce the abuse potential.

   c. *Native opiate preps* include opium tincture (10%) (Laudanum), paregoric (camphorated opium tincture) and codeine. The abuse potential is higher than for the above opiate anti-diarrheal preps.

2. **Adsorbent Powders.** These drugs adsorb water in the lumen of the GI tract, increasing bulk and slowing passage of lumen contents. Causative bacteria and toxins may also be adsorbed.

   a. *Kaopectate* (kaolin and pectin)
   Kaolin powder is hydrated aluminum silica clay. Its actions are analogous to that of aluminum antacids. Pectin is a complex carbohydrate. Its actions are analogous to that of bulk laxatives, but the effect is anti-diarrheal.

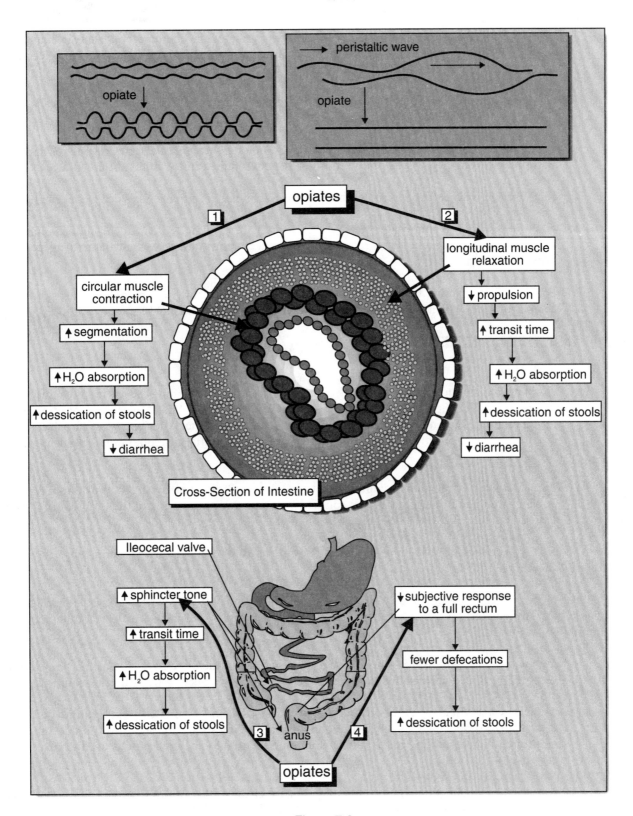

Figure 7.8

b. *Bismuth subsalicylate* (Pepto-Bismol) adsorbs water, while liberated salicylate is claimed to inhibit PG synthesis in the GI tract to further inhibit diarrhea. This drug is also an anti-ulcer drug. In high amounts it could cause tinnitus.

c. *Dietary fibers*, such as methylcellulose, psyllium, etc. are used for diarrhea as well as a laxative effect—the result depending on initial GI status.

3. *Broad spectrum antibiotics*, such as tetracyclines, quinolones, etc. are *cidal* for many microorganisms that produce diarrhea. Taken orally, they are largely retained in the lumen of the GI tract.

4. Rehydration with electrolyte replacement symptomatically improves recovery from diarrhea.

5. Glucocorticoids and aminosalicylates (Sulfasalazine) are sometimes used for chronic inflammatory bowel syndrome. Salicylic acid is released in the lumen of the intestine with the latter drug, possibly affecting PG synthesis.

6. Selective Ca^{++}-channel blockers are being studied as new drugs for irritable bowel syndrome. Peppermint oil, which is partially effective, contains the Ca^{++}-channel blocker, menthol.

D. EMETICS AND ANTIEMETICS

Emetics

1. *Apomorphine* produces emesis by virtue of its agonist action at dopamine D_2 sites in the chemoemetic trigger zone (CTZ; floor of 4th ventricle). This facilitates the flow of impulses to the vomiting center in the medulla. Patients are often titrated with i.v. apomorphine. After emesis occurs, infusion is stopped. Recovery from nausea and vomiting is rapid, as residual systemic apomorphine is rapidly metabolized.

2. *Ipecac* is a plant extract that contains several alkaloids, including emetine. After oral administration, a local irritation of the stomach wall coupled with alkaloid stimulation of the CTZ, leads to rapid emesis. Emetine has a potential of cardiac toxicity.

Emetics should be taken with large volumes of water. They are used for oral poisoning by non-caustic and non-volatile hydrocarbons.

Antiemetics. These drugs are sometimes used for people with motion sickness, vertigo and other entities that induce emesis. Antiemetics are often combined with anticancer drugs that induce emesis. Antiemetics act predominately at the CTZ.

1. *Dopamine D_2-blockers* like promethazine (Phenergan) are antiemetic by virtue of their block of D_2 receptors at the CTZ. Side effects are common **(see Chapter 9, Section I, Antiparkinsonian Drugs)**.

   Metoclopramide (Reglan) blocks D_2 receptors at the CTZ and facilitates ACh-induced peristaltic activity in the upper GI tract, while simultaneously reducing sphincter tone. This promotes stomach emptying, thereby adding to the central antiemetic action. Metoclopramide is used to treat reflux esophagitis.

2. *Antimuscarinics* like scopolamine are antiemetic by virtue of their block of muscarinic receptors at the CTZ. Scopolamine is frequently administered via a dermal patch behind the ear, to limit side effects.

3. *Histamine H_1-blockers*, like diphenhydramine (Benadryl) are weakly antiemetic, probably by virtue of their additional block of muscarinic receptors. This class of drugs is useful for prevention of motion sickness.

4. *Ondansetron (Zofran) is a 5-HT_3 receptor antagonist*, approved for preventing chemotherapy-induced emesis.

5. *Marijuana* and its active principle tetrahydrocanabinol (THC) are also useful antiemetics. The analog, dronabinol (marinol), is approved for preventing chemotherapy-induced emesis.

CHAPTER 8
PAIN AND INFLAMMATION

BACKGROUND ON PAIN

Basis of pain (Figure 8.1). When intensely stimulated, sensory receptors transmit nociceptive stimuli to the spinal cord via large myelinated delta A fibers and small unmyelinated C fibers. The endings of these sensory nerves contain opioid mu receptors, but release substance P and/or other neurotransmitters in laminae I, II (substantia gelatinosa) and V of the spinal cord. Secondary neurons convey the nociceptive impulses primarily through the lateral spinothalamic tract to the thalamus and higher brain centers. The sensation that is perceived is known as pain. Opioid mu receptors are located in many of the central pathways for pain.

Integumental pain versus visceral pain. Integumental pain represents the pain in integumentary structures such as the dermis, mucosa, skeletal muscle and joints. However, headache and uterine pain (i.e., dysmenorrhea) comprise other kinds of integumental pain. Often, integumental pain is sharp and easily localized (e.g., tooth ache).

Cutaneous and mucocutaneous pain is effectively treated with local anesthetics. Muscle and joint aches, as well as dysmenorrhea and headache, are effectively treated with nonsteroidal antiinflammatory drugs (NSAIDs) which primarily act locally at sites of inflammation to reduce synthesis of prostaglandins (PGs) and inhibit release of the algesic autacoid, bradykinin **(Figure 8.1)**. These and other autacoids have major roles in the pain and edema that occur with inflammation. Corticosteroids (CS), specifically glucocorticoids, act differently than NSAIDs but produce a similar and more potent attenuation of inflammation and associated pain. Integumental pain is generally better controlled by NSAIDs and CS than by opioids. NSAIDs and CS are non-addictive, although treatment with CS may be life-long.

Visceral pain generally represents pain within body cavities such as the thorax and abdomen. Angina, pleural and intestinal pain are representative types of visceral pain. Characteristically, visceral pain is diffuse, hard to localize and often referred to distal sites. For example, pain from a cardiac infarct is often referred to the left arm. Visceral pain is best treated with opioids which exert a major action on the pain pathway in the spinal cord, reticular formation, periaqueductal gray, thalamus and many other sites within the central nervous system. Visceral pain is often not effectively treated with NSAIDs. Opioids, unless administered intrathecally, are addictive.

1. **Background on inflammation.** An inflammatory reaction is usually caused by infection or irritation. Mobile and non-mobile cells of the host are influenced by the irritation and release a variety of autacoids which act to "wall off" the source of inflammation. The resulting edema and pain is the effect brought about by the different autacoids which include histamine, kinins (bradykinin, kallikrein is lys-bradykinin), a variety of eicosanoids and other substances such as proteolytic enzymes. Eicosanoids are thought to have the most prominent role in an inflammatory reaction and these autacoids are discussed in this section.

A. EICOSANOIDS AND THE INFLAMMATORY PROCESS

1. Eicosanoids are acid lipids derived from polyunsaturated (greater than 2 double bonds) fatty acids with a length of 18-22 carbons ("eicosa" is 20) having biological activity (auta and "coid"). There are several major groups of eicosanoids:
    Prostaglandins (PGs)—prostanoic acid nucleus (originally thought to be derived from the prostate gland, which accounts for the name).

Leukotrienes (LTs)—3 double bonds (tri and ene), first obtained from *leuk*ocytes.
Thromboxanes (TXs)—made primarily in platelets.
Prostacyclins (PGIs)—made primarily by vascular endothelium.
HPETES (hydroperoxyeicosatetraenoic acids)—chemotactic substances.
HETES (hydroxyeicosatetraenoic acids)—chemotactic substances.
EETES (epoxyeicosatetraenoic acids)—actions still being defined.
Lipoxins (LXs)—actions still being defined.
Platelet activating factor (PAF)—a platelet-derived factor that has been implicated in acute asthma and other conditions.

2. **Synthesis of eicosanoids: chemistry and nomenclature.** Eicosanoids are natural products that are normally produced by cells in small amounts. Macrophages are the only cells able to synthesize all of the eicosanoids. Most cells are able to make some but not all of the eicosanoid products, according to the enzymes present in the cells. T and B lymphocytes lack the enzymes needed to make eicosanoids. Platelets normally produce TXs because TX synthase is present in large amounts in

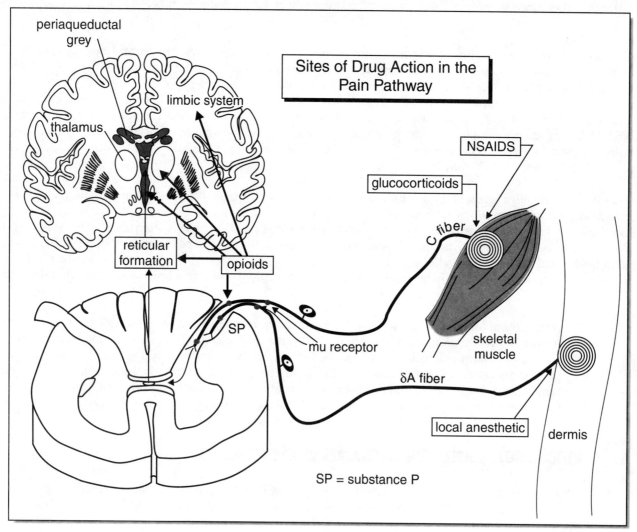

Figure 8.1

these cells. Conversely, endothelial cells normally produce PGI_2 because PGI_2 synthase is present in large amounts in these cells. Generally, those cells producing PGs have large amounts of a single PG synthase, so that only a single PG is made in large amounts. LTs

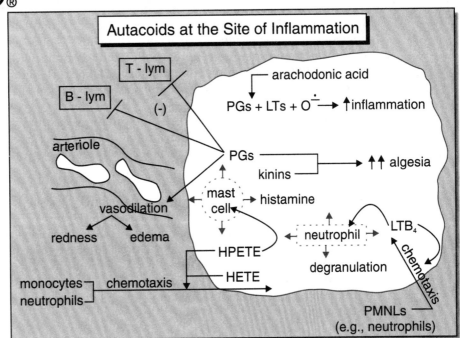

Figure 8.2

are produced primarily in mobile cells: PMNLs(PolyMorphoNuclear Leukocytes), monocytes, etc. PAF is produced by platelets and other mobile cells (macrophages, PMNLs, etc.).

The eicosanoids are not stored, but are released after de novo biosynthesis. They are stable in blood but their $t_{1/2}$ is short (0.5 to 5 minute). The lung is the major organ involved in this metabolism, but the spleen, kidney, gastrointestinal tract and adipose tissue also play substantial roles. PGs are degraded mainly by a dehydrogenase (15-hydroxy-PG-dehydrogenase) to 15-keto derivatives.

The nomenclature follows certain principles. For example, for $PGF_{2\alpha}$ the main letter "F" refers to a particular ring or structural configuration; the digit "2" refers to the number of double bonds; and the greek letter "α" refers to the planar configuration. In humans PGE_2 and PGF_2 are the most abundant PGs.

Precursors for PGs
PGs with 2 double bonds (e.g., $PGE_{2\alpha}$, $PGF_{2\alpha}$) arise from arachidonic acid (5,8,11,14-eicosatetraenoic acid.
PGs with 1 double bond (e.g., $PGE_{1\alpha}$) arise from 8,11,14-eicosatrienoic acid.
PGs with 3 double bonds (e.g., $PGE_{3\alpha}$) arise from 5,8,11,14,17-eicosapentaenoic acid.

3. **Actions of eicosanoids**. There are many actions attributable to the eicosanoids as described below. However, these autacoids are described in the section on "Pain and Inflammation" because of their prominent actions in these processes.

   a. **Inflammation (Figure 8.2)**. Inflammatory exudates have high concentrations of PGs, LTB_4 and ETEs. Free radical production during PG and LT synthesis may contribute to the inflammatory process. PGEs, particularly PGE_1, have prolonged vasodilatory effects on microcirculatory vessels, thereby contributing to redness and swelling. Algesic actions of kinins (bradykinin and lys-bradykinin) are potentiated by PGs. LTB_4 is chemotactic for PMNLs, particularly neutrophils.

In neutrophils LTB_4 also promotes adherence and degranulation. Both HETE and HPETE are chemotactic for monocytes and neutrophils; 5-HPETE enhances release of histamine from mast cells. At the same time PGs suppress both T and B lymphocytes, thereby depressing an immune response and enhancing the inflammatory reaction. (The immunosuppression that occurs in some carcinomas is associated with increased production of PGs.)

Glucocorticoids promote formation of the protein lipocortin which inhibits phospholipase A_2 activity and thereby suppresses formation of all of the eicosanoids. Nonsteroidal antiinflammatory drugs (NSAIDs) inhibit cyclooxygenase activity and subsequent formation of PGs but not LTs or EETES.

b. **Allergy**. The difficulty in breathing in acute asthma is largely attributable to bronchoconstrictor and pro-bronchosecretory effects of LTs. LTC_4, LTD_4 and LTE_4, which are considered to be the "slow reacting substance of anaphylaxis" (SRS-A), also increase microvascular permeability and consequent edema. $PGF_{2\alpha}$ and TXA_2 are also potent bronchostrictors. Histamine is a minor component of acute asthma. Histamine H_1-blockers are not effective for acute asthma, since these drugs do not counteract the actions of LTs. In contrast, glucocorticoids, by (a) preventing formation of LTs, PGs and PAF, and (b) inhibiting histamine release, are effective in acute asthma.

In non-asthmatic allergy PGs and PGI_2 are produced by mast cells, along with histamine. In this instance PGs and/or PGI_2 replicate many of the actions of histamine, including vasodilatation of arterioles and venules with an increase of capillary permeability, stimulation of nerve endings which produces itching, and algesia. At the same time PGE suppresses histamine release. LTB_4 and LTC_4 are found in high amounts in psoriatic lesions.

c. **Gastrointestinal (GI) patho/physiology**. PGE_2 and PGI_2 suppress gastric acid formation and pepsinogen secretion, while promoting mucus secretion particularly in the small intestine. It is notable that all PG synthesis-inhibitors like anti-inflammatory drugs and glu-

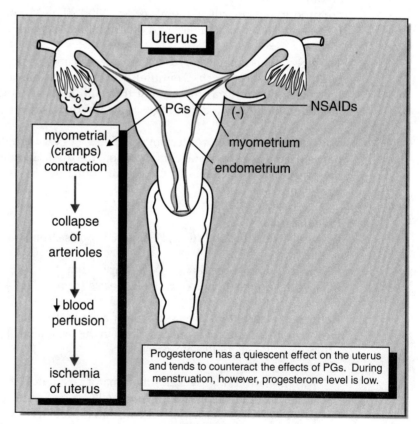

Figure 8.3

cocorticoids are ulcerogenic. Misoprostol (Cytotec), a PGE_1 analog that promotes mucus formation, is specifically approved for preventing induction of stomach ulcers by NSAIDs.

PGE_2 and PGI_2 promote pancreatic secretion. PGE_2, $PGF_{2\alpha}$ and LTs stimulate contraction of longitudinal muscles in the GI tract, decreasing transit time and producing consequent diarrhea. In ulcerative colitis high amounts of LTB_4 are found in GI mucosal samples. Menstrual diarrhea is probably related to increased uterine production of PGs which circulate in blood to act on the GI longitudinal muscle. When PGE_2 and $PGF_{2\alpha}$ are administered to humans, colicky cramps and bronchospasm occur.

d. **Reproductive function**. PGs are produced by the endometrium in large amounts during menstruation. $PGF_{2\alpha}$ acts locally to stimulate uterine contractions, particularly when progesterone levels are low, producing menstrual cramps **(Figure 8.3)**. *Dysmenorrhea* is accompanied by elevated levels of PGF_2 and is effectively treated with PG synthesis-inhibitors.

Because PGs produce contractions of even a gravid uterus, while dilating and softening the cervix, they have been used to induce labor and to produce abortions (Dinoprost is Prostin $F_{2\alpha}$; Dinoprostone is Prostin E_2; Carboprost is 15-methyl-$PGF_{2\alpha}$) when administered into the amnionic sac, vaginally or i.m. Generally diarrhea also occurs because of effects of absorbed PGs on the GI tract. At term, the level of PG in amnionic fluid is elevated.

A low level of PG in semen is associated with reduced fertility. Since PGs induce luteal degeneration, they are used to synchronize estrus in farm animals.

e. **Regulation of thrombus formation (Figure 8.4)**. *Endothelial cells* lining blood vessel walls normally produce small amounts of PGI_2 which produce vasodilation and inhibit platelet aggregation. Platelets produce TXA_2 which promotes vasoconstriction and platelet aggregation. When a blood vessel is broken, the effects of TXA_2 override the normally predominant effects of PGI_2, thereby promoting platelet aggregation which would stop the loss of blood. Aspirin is effective in preventing a second myocardial infarct (MI) and in preventing transient ischemic attacks (TIAs), because of preferential inhibition of TXA_2 formation. Blood clotting is thus impaired so thrombus formation is less likely to occur. PGI_2 (Cyclo-Prostin) is useful in hemodialysis partly because of its vasodilatory action and inhibitory effect on platelet aggregation.

f. **Cardiovascular effects**. PGs are responsible for maintaining a patent ductus arteriosus in the fetus (PGE_1 is Alprostadil; PGI_2 is Prostin VR). The PG synthesis-inhibitor indomethacin (Indocin) is used to promote closure of ductus arteriosis in newborns.

TXA_2 levels are elevated during pulmonary hypertension. Reduced circulating levels of PGI_2 and PGE_2 have been reported in primary hypertension. TXA_2 is more effective in producing venoconstriction than arterial constriction. LTs have a negative inotropic effect.

g. **Body temperature**. PGE_1 and PGE_2 produce fever. Antipyretics inhibit pyrogen-induced production of PGE_2.

h. Renally-formed PGE_1, PGE_2 and PGI_2 increase glomerular filtration and promote diuresis. Furosemide, a potent diuretic, promotes renal PG formation by stimulating cyclooxygenase activity.

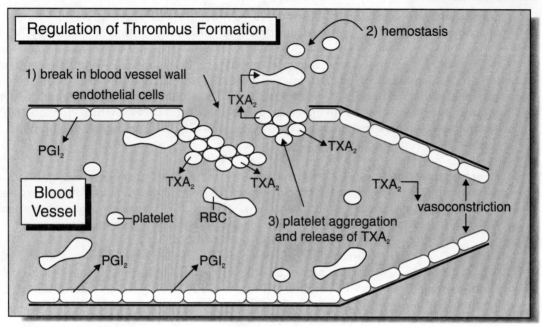

Figure 8.4

Summary. The prominent involvement of the eicosanoids in many patho/physiological processes is recognized but still not well understood. Agonist and antagonist analogs of the eicosanoids are still actively being investigated for their potential therapeutic utility.

B. DRUGS FOR INTEGUMENTAL PAIN OR INFLAMMATION

1. ANALGESIC/ANTIPYRETIC DRUG LACKING ANTIINFLAMMATORY ACTIVITY

A. **Acetaminophen (Tylenol)**. This drug is often used in place of aspirin for mild to moderate pain and fever, particularly in children. This drug is not anti-inflammatory.

Pharmacokinetics of acetaminophen. Absorption is rapid (less than 1 hour) and extensive (60-90%). Peak level occurs in less than 1 hour. Distribution is to all tissues and fluids, including brain. There is some protein binding (25-50%), but protein-bound drugs are not displaced by acetaminophen.

Metabolism occurs by cytochrome P_{450} enzymes **(Figure 8.5)**. In *toxic* amounts, excess N-acetyl-p-benzoquinone is formed and this generates free radicals which deplete liver-SH groups and thereby produce centrolobular necrosis of the liver (approximately 10-15 g, or 120 µg/ml) and death (about 20-25 g, or about 200 µg/ml). Liver enzymes, SGOT and SGPT, become elevated, with or without jaundice. Symptoms of liver damage may not occur for 2-3 days, by which time it is too late to administer the antidote, *acetylcysteine* (preferably within 4 hours, but up to approximately 24 hours). Effective treatment of toxicity, in addition to the antidote, consists primarily of removal of acetaminophen (gastric lavage, emesis).

Excretion occurs via the kidneys, with less than 5% being excreted unchanged.

Mechanism and pharmacological actions. Acetaminophen acts in the brain to produce an antipyretic action. Analgesia is produced by an unknown mechanism. Since acetaminophen is only a weak inhibitor of PG synthesis, this drug is not ulcerogenic, does not cause stomach bleeding, does not inhibit platelet aggregation and does not increase bleeding time. There are no anti-inflammatory actions.

Use of acetaminophen. Acetaminophen is used for fever, headache, myalgia and post-partum pain. It is preferred over aspirin for those with ulcers, bleeding tendencies, aspirin-allergy and in gout patients treated with uricosurics (2 g aspirin enhances excretion of uricosurics, but acetaminophen does not interfere with uricosurics).

Children have a limited capacity for excreting acid aspirin metabolites, but acetaminophen metabolites are readily excreted. Acetaminophen is not associated with Reye's Syndrome in children with viral infections (e.g., chicken pox).

2. ANALGESICS WITH ANTIINFLAMMATORY ACTIVITY

A. ASPIRIN

Chemistry and activity. Aspirin (acetylsalicylic acid) (pKa is 3.5) and its analogs, like the major metabolite salicylic acid (pKa is 3.0), are acidic and irritating substances. Acid environs (e.g., stomach acid) inhibit dissociation and thereby promote formation of the non-ionized form which is more readily absorbed **(Figure 8.6)**. Most absorption, nonetheless, occurs in the alkaline upper small intestine because of its greater surface area.

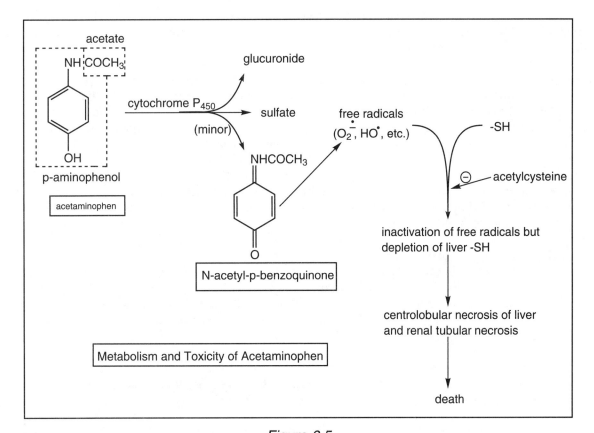

Figure 8.5

PHARMACOLOGY

The irritant property of aspirin is associated with destruction of gastric mucosal cells, erosion of the stomach mucosal barrier and formation of stomach *ulcers*. Aspirin is sufficiently irritating to be useful as a keratolytic for warts and corns.

The acetyl moiety (-COCH$_3$) combines with some proteins, including enzymes like thromboxane synthetase and cyclooxygenase, producing permanent inactivation. Consequently, some effects, like inhibition of platelet aggregation, persist for 3 days (i.e., the time required for new enzyme synthesis).

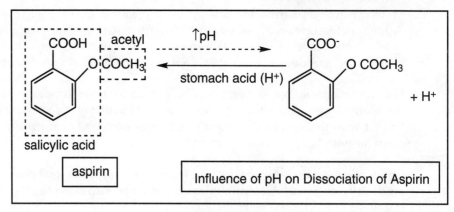

Figure 8.6

Pharmacokinetics of aspirin. Absorption is rapid (1-2 hours) and nearly complete (greater than 90%), occurring initially in the stomach and later in the duodenum. Increasing the tablet dissolution rate, as in buffered and effervescent preps, increases the rate of absorption. Peak levels occur in approximately 2 hours after an analgesic dose of aspirin (650 mg).

Distribution is to all tissues and fluids including brain, milk and fetus. Aspirin accumulates in synovial fluid of inflamed joints. Most (greater than 80%) salicylate binds to plasma albumin. Interactions occur with other protein-bound drugs (e.g., phenytoin, warfarin, hypoglycemics).

Metabolism occurs primarily via hepatic cyt P$_{450}$ enzymes **(Figure 8.7)**. First order kinetics apply with analgesic doses (i.e., rate of metabolism is proportional to blood level). Percentages of the different metabolites are shown for the 650 mg dose **(Figure 8.7)**. With anti-inflammatory doses (2000-6000 mg per day), the enzyme complexes are saturated, so zero order kinetics apply (i.e., enzymes are saturated, so rate is independent of blood level). The t$_{1/2}$ is about 4 hours (analgesic dose) to over 15 hours (anti-inflammatory dose).

Excretion occurs primarily via the kidneys. Because aspirin is secreted by acid-carrier sites in the nephron, low dose aspirin interferes with uric acid excretion and promotes gouty attacks. Renal excretion of acidic aspirin and analogs is enhanced by alkalinizing the urine (increased pH with bicarbonate (HCO$_3^-$) etc.). This is often used for aspirin intoxication.

Pharmacological actions of aspirin (Figure 8.8)
1. *GI ulcers* occur because aspirin erodes the stomach mucosal barrier by (a) damaging mucosal cells and (b) inhibiting formation of cytoprotective PGs. GI bleeding is a consequence, particularly since aspirin reduces platelet aggregation by inhibiting TXA$_2$ synthesis. Coated aspirin preps are less ulcerogenic. Misoprostol (Cytotec) is a PGE$_1$ analog that can be taken concurrent with aspirin to inhibit ulcerogenesis **(see Chapter 7)**.

2. *Bleeding time* increases twofold after aspirin because of the effect on TXA$_2$. Anti-inflammatory doses of aspirin also reduce RBC t$_{1/2}$. Nonacetylated salicylates (e.g., sodium salicylate, salsalate) do not inhibit platelet aggregation but retain antiinflammatory activity.

3. *Tinnitus* occurs with excess anti-inflammatory doses, due to adverse effects on hair cells in the cochlea. This can be a gauge of toxic dosing.

4. *Liver damage* occurs with toxic doses of aspirin: fatty infiltration, increased SGOT, increased SGPT and increased alkaline phosphatase.

5. *Respiratory actions* and alterations in acid-base balance occur. Aspirin uncouples oxidative phosphorylation in skeletal muscle causing excess CO_2 and stimulation of chemoreceptors which leads to increased depth of respiration. This means there is no change in P_{CO_2}. (A respiratory depressant, like morphine, inhibits this compensation and causes a large increase in P_{CO_2}—respiratory acidosis.)

Mechanism of salicylates. Aspirin inhibits
    a. Cyclooxygenase, causing decreased PG synthesis, inflammation and fever.
    b. Thromboxane synthetase, causing decreased TXA_2 synthesis with anti-coagulation.
    c. Conversion of prekallikrein to kallikrin, causing decreased bradykinin synthesis, resulting in diminished local pain, decreased capillary permeability, decreased migration of PMNLs and macrophages into inflammatory sites, as well as increased lysosome stability.

Toxicity
1. Acute salicylate intoxication. Very high amounts of aspirin directly stimulate the respiratory center causing:
    a. greatly increased respiratory minute volume with

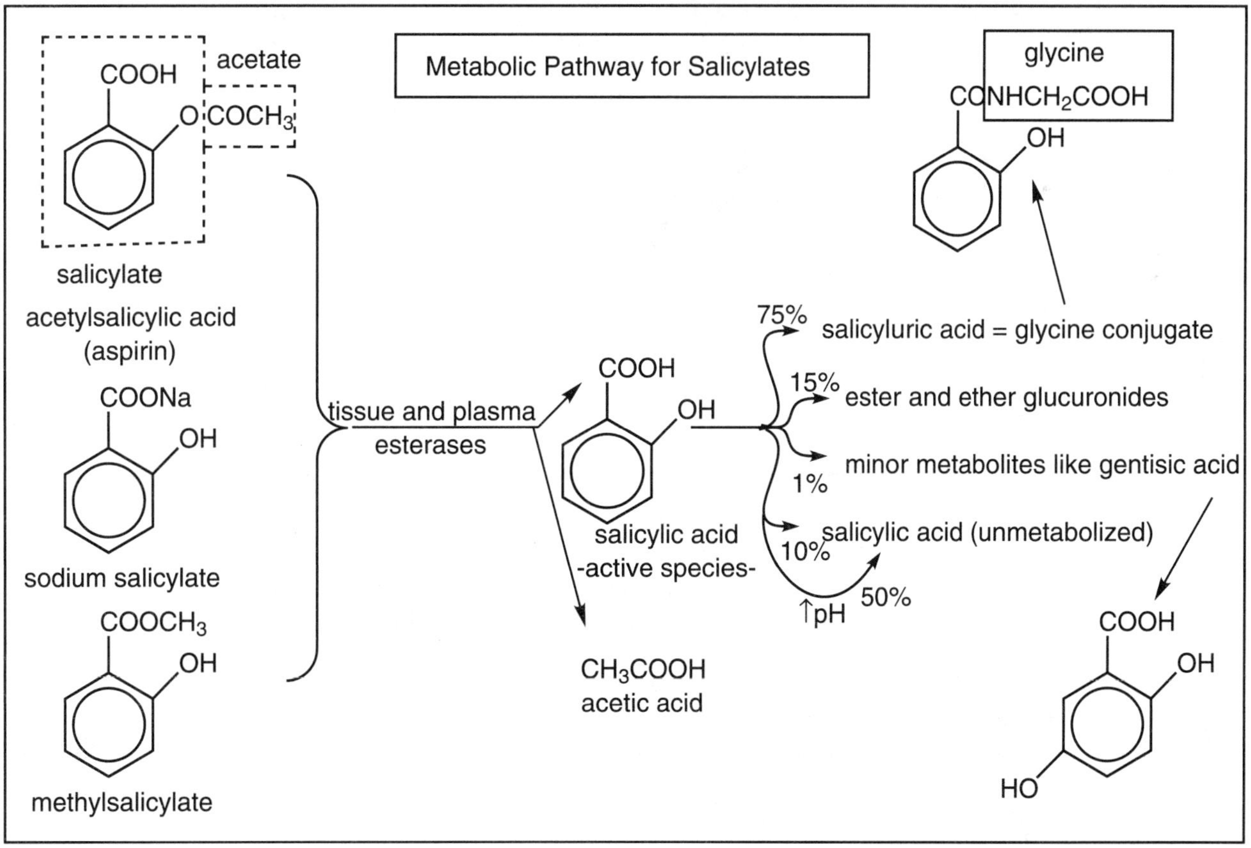

Figure 8.7

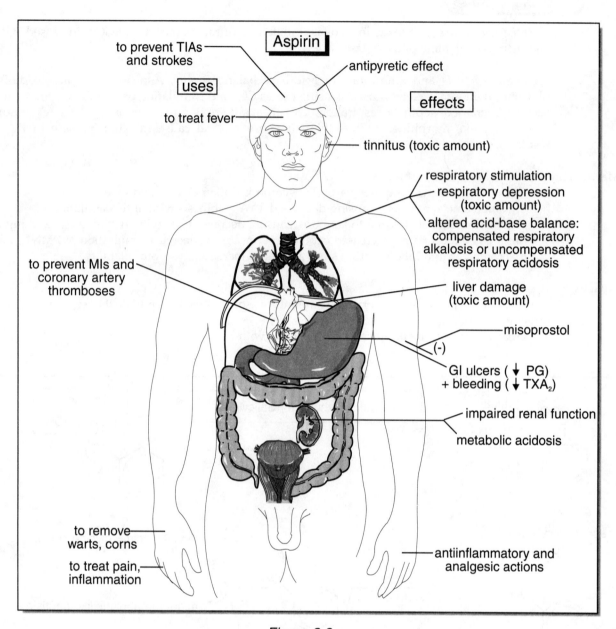

Figure 8.8

b. decreased P_{CO_2} (respiratory alkalosis) plus
c. enhanced renal excretion of HCO_3^- (compensated respiratory alkalosis).

Aspirin also produces metabolic acidosis, which reflects
    a. the presence of salicylic acid in plasma;
    b. decreased BP with decreased renal function and an accumulation of metabolically-derived sulfuric and phosphoric acid;
    c. increased carbohydrate metabolism, causing increased formation of pyruvic, lactic and acetoacetic acids and
    d. hyponatremia and hypokalemia (i.e., Na^+ and K^+ accompany HCO_3^- excretion).

Finally, dehydration may be present from
   a. salicylate-induced sweating;
   b. increased respiration and increased insensible respiratory H_2O elimination and
   c. increased Na^+, K^+, HCO_3^- excretion and increased renal H_2O excretion.
   Renal papillary necrosis can occur (decreased PG-mediated vasodilation).

2. Higher amounts of aspirin (around 500 µg/ml) cause
   a. respiratory center depression, which leads to an increased P_{CO_2} with
   b. low plasma HCO_3^- levels (uncompensated respiratory acidosis).

Mild, chronic salicylate intoxication (salicylism)
   CNS: headache, confusion, drowsiness.
   Auditory: tinnitus, hearing loss, dizziness.
   Respiratory/acid base: hyperventilation, sweating, dehydration.
   Intestinal: vomiting, diarrhea.

Severe chronic salicylate intoxication
   CNS: salicylate jag, delirium, hallucinations, incoherent speech, EEG abnormalities (tremors, convulsions), cardiovascular collapse.
   Respiratory/acid-base: hyperpnea, acidosis, dehydration (fever), hypoglycemia, respiratory depression, death.
   Intestinal: vomiting, diarrhea, abdominal pain.
   Dermal: petechial hemorrhages.

Methylsalicylate poisoning
   CNS: excitation.
   Respiratory/acid-base: hyperpnea, dehydration (fever).
   Intestinal: vomiting, diarrhea, abdominal pain.

Treatment of salicylate poisoning
1. Remove/inactivate salicylate in GI tract (especially methylsalicylate)
   a. gastric lavage or emesis and
   b. activated charcoal and a cathartic.

2. Enhance renal salicylate elimination by (a) alkalinizing urine with HCO_3^- or (b) dializing.

3. Correct H_2O, electrolyte and acid/base imbalances.

4. Reduce fever with external sponging.

Use of Salicylates
1. Low dose aspirin (325-1000 mg) is analgesic and antipyretic (plasma concentration 20-50 µg/ml) and is used for headache, toothache, joint pain and fever. Aspirin is not so useful for dysmenorrhea or migraine.

2. High dose aspirin (3.6-6 g) (approximately 4,000 mg) is anti-inflammatory (plasma concentration is about 200-300µg/ml) and is useful for chronic inflammation (e.g., rheumatoid arthritis, osteoarthritis, rheumatic fever). Aspirin is as effective as other NSAIDs and is much cheaper, but many people cannot tolerate the GI effects of aspirin.

3. Very low dose aspirin (160 mg) inhibits platelet aggregation, is used to prevent transient ischemic attacks (TIAs), stroke, coronary artery thrombosis and myocardial infarcts (MIs). One aspirin every other day is adequate. Beware of aspirin in patients with bleeding tendencies (hemophilia, hypoprothrombinemia, vitamin K deficiency).

4. Topically, salicylates in cremes are keratolytics for warts and corns.

5. Methylsalicylate is used as an ointment for muscle-joint inflammation. Toxicity is too great for systemic use.

Precautions for aspirin
1. Aspirin intolerance/hypersensitivity (0.3%) is associated with urticaria, rhinorrhea, asthmatic reaction and anaphylaxis. This is more frequent in people with allergies. There may be cross-sensitivity with other NSAIDs and the yellow dye tartrazine in many drug preps.

2. Aspirin should be avoided in peptic ulcer disease.

3. Aspirin (no more than 2 g per day) reduces uric acid secretion and should be avoided in gout.

4. Aspirin should not be used for certain viral infections like influenza and chicken pox because of the risk of inducing Reye's Syndrome (causes encephalopathy, hepatic damage, death).

5. Aspirin should be avoided in children with hypertension, heart disease, diabetes and thyroid disorders.

6. Aspirin, in moderation, is not contraindicated in pregnancy. However, prolonged high-dose aspirin is associated with prolonged gestation and labor, low birth weight, increased post-partum hemorrhage, and increased perinatal mortality.

B. **NONSTEROIDAL ANTIINFLAMMATORY DRUGS (NSAIDS)**

These NSAIDs have antipyretic and analgesic activity (especially integumental versus visceral pain). NSAIDs are commonly used for rheumatoid-, osteo- and psoriatic arthritis, as well as ankylosing spondylitis and Reitter's syndrome. All resemble aspirin to the extent that they are well-absorbed after oral administration; act by inhibiting cyclooxygenase; produce GI irritation, ulcers and bleeding (inhibit platelet aggregation); bind extensively to plasma albumin and displace other drugs; are metabolized in liver and excreted via the kidneys.

Acute side effects. These are less frequent and less severe than for aspirin. Long-term effects may be more severe than aspirin (e.g., blood dyscrasias like agranulocytosis and aplastic anemia; renal toxicity like interstitial nephritis and nephrotic syndrome). These NSAIDs are 5-20 times more expensive than aspirin.

A. Propionic acids are also used to treat dysmenorrhea.

1. *Ibuprofen* (Motrin) **(Figure 8.9)** does not displace anticoagulants from plasma binding sites. Now available OTC (over-the-counter or non-prescription), ibuprofen is used for headache, myalgia and joint pain. In low dose, there are few adverse effects.

2. *Naproxen* (Naprosyn) has a long $t_{1/2}$ (12 hours) and can be taken twice a day. It also is available OTC.

Adverse effects. These include jaundice, rash and pruritus, peripheral edema, tinnitus, CNS excitation and dizziness.

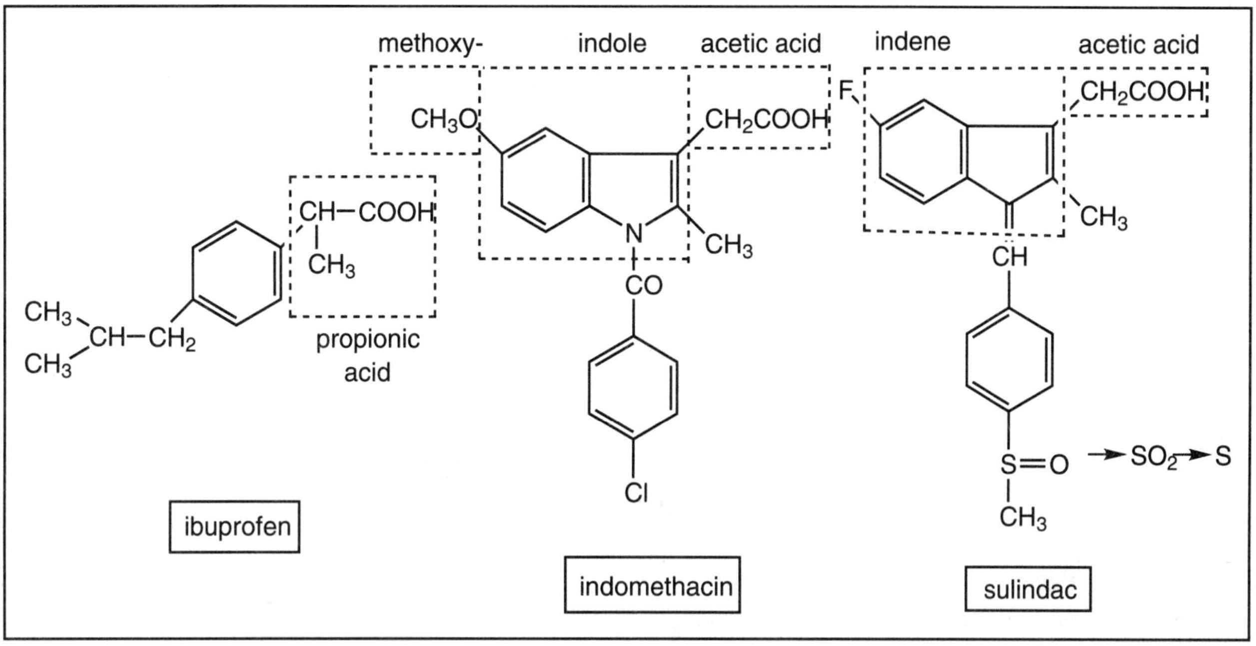

Figure 8.9

B. Indoleacetic acids and analogs.

1. *Indomethacin* (Indocin) **(Figure 8.9)** produces overdose tinnitus and headache like high dose aspirin. Side effects are generally greater with indomethacin overdose than other NSAIDs, especially CNS effects (headache, depression, disorientation). Indomethacin is preferred over colchicine for acute gout and for inducing closure of a patent ductus arteriosus in premature infants. (The ductus arteriosus is normally kept patent by local PG synthesis. Indomethacin inhibits PG synthesis.)

2. *Sulindac* (Clinoril) **(Figure 8.9)**, a sulfoxide (SO) prodrug, is sequentially metabolized by the liver to a sulfone ($-SO_2$) then the active sulfide (S). Since the inactive form is taken orally, gastric PG synthesis is not inhibited, so gastric irritation is not produced. Accordingly, the incidence of ulcers is less. The $t_{1/2}$ is long (about 18 hours) because of enterohepatic circulation. About 25% of this drug is excreted in feces.

3. Tolmetin (Tolectin) does not displace anticoagulants from plasma binding sites.

C. Oxicam Analog

Piroxicam (Feldene) **(Figure 8.10)** undergoes enterohepatic circulation which imparts a long $t_{1/2}$ (approximately 2 days). Dosing is once a day as per sulindac. GI symptoms occur in 20% of patients.

D. Newer NSAIDs

1. *Nabumetone* (Relafen) is a prodrug that is absorbed mainly in the duodenum, then metabolized primarily in the liver to the active 6-methoxy-2-naphthylacetic acid (6-MNA), which has a $t_{1/2}$ of about 24 hours. This drug is approved for rheumatoid arthritis and osteoarthritis.

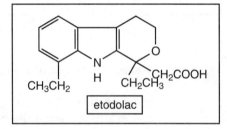

Figure 8.10

The incidence of GI ulcers should be low since the prodrug does not inhibit gastric PG synthesis.

2. *Etodolac* (Lodine) is a pyranocarboxylic acid that is approved for osteoarthritis, not rheumatoid arthritis. It does not produce stomach ulcers (less than 1%), probably because it does not inhibit gastric PGE_2 production.

Less common NSAIDs. These drugs are very potent, but adverse and potentially life-threatening side effects often limit their use to short-term relief of acute inflammation.

A. Fenamates

Meclofenamate (Meclomen) **(Figure 8.10)** is effective for dysmenorrhea and is still occasionally used for arthritis. Diarrhea is common and often severe.

B. Pyrazolones

Phenylbutazone (Butazolidin) **(Figure 8.10)** is very potent and an excellent drug for short-term relief of inflammation, including acute gout. Serious adverse effects include agranulocytosis and aplastic anemia, hepatic and renal necrosis, nephrotic syndrome, severe *hypertension* and exfoliative dermatitis. An active metabolite, oxyphenbutazone has the same spectrum of activity as the parent drug.

Drug interactions with NSAIDs. The NSAIDS attenuate the diuretic and natriuretic effects of diuretic drugs, by interfering with renal PG synthesis. The NSAIDs also tend to displace drugs which are highly protein bound (e.g., anticoagulants, oral hypoglycemics), thereby enhancing their effect.

C. **REMITTIVE AGENTS FOR RHEUMATIC DISEASES**

Aspirin and NSAIDs are used for long-term management of chronic inflammatory disorders. Those drugs abate the inflammatory response, but do not alter progression of the disease. The drugs to be described are used in an attempt to retard the progression of the inflammatory disorder (remission). The previously described NSAIDs are taken concurrent with remittive drugs that might take months to produce their effect (exception, methotrexate). Often, hydrochloroquine is tried initially, then gold or methotrexate. There must be regular monitoring for adverse effects.

A. *Hydrochloroquine* (Plaquenil) and Chloroquine (Aralen) **(see Chapter 11, Section 5, Antimalarials)**.

Use. These antimalarials are taken daily for several months, for treating rheumatoid and juvenile arthritis (not psoriatic arthritis) and systemic lupus erythematosus. Their *immunosuppressant* activity is associated with a decline in rheumatoid factor titer. Improvement is seen in 75% of patients. However, the progression of arthritis may not be abated as with gold compounds. Hydrochloroquine is preferred. The antimalarials should not be given with gold.

Adverse effects. These drugs accumulate in the lungs, liver, spleen and kidneys; adverse reactions are seen at these sites. Corneal deposits, retinopathy, leukopenia, dermal lesions and EKG changes can also occur. Vision should be monitored at 6 month intervals.

B. Gold (Au) Therapy (Chrysotherapy)

Chemistry is shown in **Figure 8.11**.

Pharmacokinetics. Auranofin is taken orally. About 25% is absorbed. Aurothioglucose and gold sodium thiomalate are administered i.m. Peak plasma levels are attained at 4 hours. Au becomes concentrated five to tenfold in inflamed joints. Other major sites of accumulation are liver, spleen, kidneys, bone marrow, lymph nodes, reticuloendothelial cells and macrophages. Gold compounds are highly bound to plasma protein (greater than 90% initially). The $t_{1/2}$ is long, greater than 1 week after a single dose. Gold is still present for over 1 year after ceasing a course of therapy. About 60% is excreted by urine, 40% in feces.

Adverse effects. These occur in 25-50% after 200-400 mg of gold over a period of 3-6 months, and 10% may discontinue gold therapy. Mortality was as high as 0.4% at one time. Pruritus is sometimes used as a gauge for regulating the dosage.

Since 75% of *oral gold* (Auranofin) is unabsorbed, GI symptoms (e.g., diarrhea) are greater versus i.m. gold. Adverse systemic effects are greater after i.m. gold versus oral gold, possibly because the total gold dosage is much higher by the i.m. route (25 mg of Au weekly *per se* versus less than 1 mg weekly of absorbed Au with auranofin).

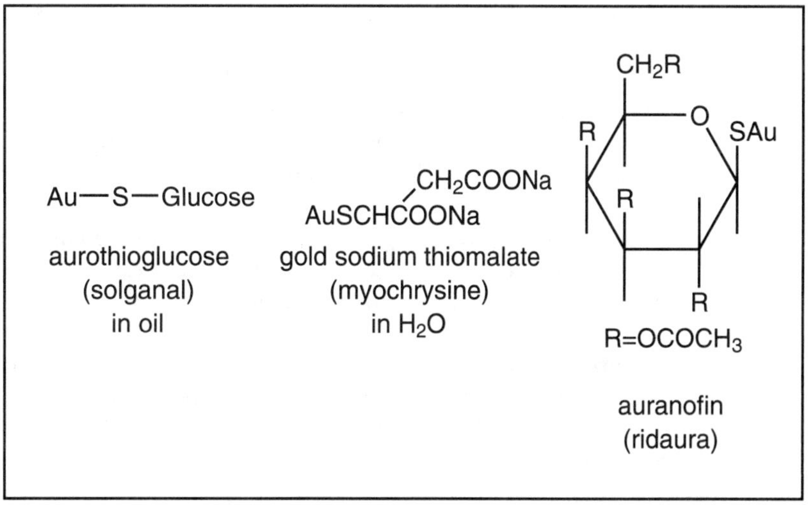

Use and effectiveness. Gold compounds are used for progressive rheumatoid arthritis. These drugs are taken for several month to greater than 1 year. There

Figure 8.11

is controversy about effectiveness but symptoms tend to be reduced; complete remission may occur in greater than 10%. The severity of relapses may be less than before gold therapy.

After gradually increasing weekly doses, a weekly maintenance dose of 6 mg of auranofin or 50 mg of the other gold compounds is used.

Treatment of gold toxicity. Glucocorticoids as well as the metal chelators, penicillamine (Cuprimine) and dimercaprol (BAL) are used for acute gold toxicity.

Contraindications to use of gold include
    a. previous toxicity with gold compounds;
    b. renal/liver disease and blood dyscrasias and
    c. cardiovascular disease.

C. Penicillamine (Cuprimine)

Chemistry. This penicillin analog is described in **Chapter 13, the Section on Toxic Agents**.

Pharmacokinetics. About half of oral penicillamine is absorbed, peak levels occur at 1-2 hours, most is protein bound (approximately 80%) and most (60%) is excreted in less than 24 hours.

Use/Mechanism and Actions. Penicillamine is used for rheumatoid arthritis, usually if gold therapy was unsuccessful. After gradually increasing the dose, a daily amount of 250-750 mg is taken for several months. The immunosuppressive action of penicillamine is associated with a decline in rheumatoid factor titer.

Adverse actions. These include many like that with gold compounds: dermatitis and stomatitis, nephritis and blood dyscrasias including aplastic anemia. Also, autoimmune disorders occur, including lupus erythematosus. A metallic taste or loss of taste (*ageusia*) is characteristic. Penicillin allergy is not a contraindication for penicillamine.

D. Immunosuppressive Drugs

1. *Methotrexate* (Rheumatrex) (MTX) is a chemotherapeutic agent that inhibits folic acid synthesis. MTX is also used for rheumatoid arthritis. Weekly doses may produce desired effects in 4-6 weeks. MTX is generally well tolerated. Blood count and hepatic function should be monitored every 4 weeks; renal function, less often. Folic acid supplements (1 mg daily) may reduce the severity of mucosal ulceration and cytopenia without altering the anti-rheumatoid effect. MTX is described more completely in **Chapter 12, the Section on Cancer Chemotherapy**.

2. *Azathioprine* (Imuran) **(Figure 8.12)**, an analog of mercaptopurine, inhibits purine incorporation into RNA and DNA. This is S-phase specific. Azathioprine is most effective after immunological challenge (i.e., after grafting or transplantation).

   Actions. T-lymphocytes are suppressed more than B-lymphocytes. As expected, synthesis of red and white blood cells is impaired but total bone marrow suppression is not common. Severe toxicity is uncommon. Blood counts and liver function tests should be done regularly. GI toxicity often produces diarrhea.

   Uses. Azathioprine is a mainstay in graft and transplantation procedures, but is also used in rheumatoid arthritis and systemic lupus erythematosus.

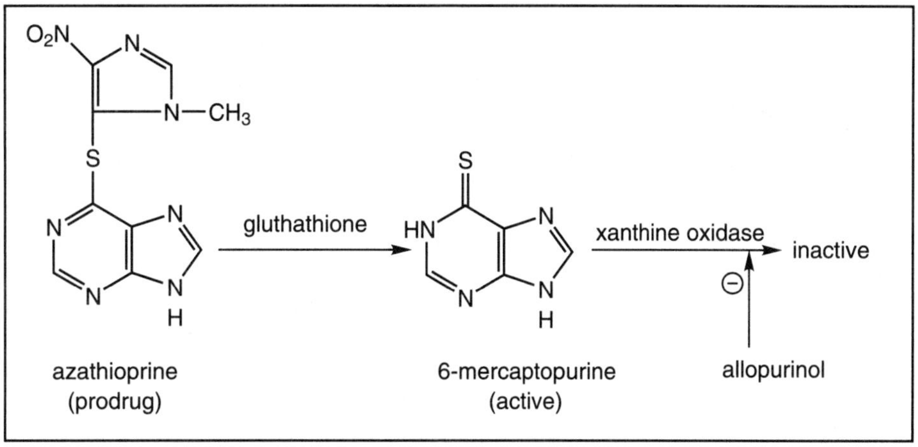

In patients treated with allopurinol (Zyloprim), the dose of azathioprine should be reduced 75% (see above).

Figure 8.12

3. *Cyclophosphamide* (Cytoxan) is an anticancer prodrug that is activated by cytochrome P_{450} enzymes mainly in the liver to alkylating agents **(see Chapter 12)**. B-lymphocytes are suppressed more than T lymphocytes. This drug is used for rheumatoid arthritis and other autoimmune disorders. Blood dyscrasias, diarrhea and alopecia occur.

4. *Chlorambucil* (Leukeran) is another alkylating agent.

E. Immunopotentiating Drugs

1. *Levamisole*, an anthelmintic, has been used for rheumatoid arthritis. Levamisole enhances T lymphocyte activity and macrophage function. This drug is taken daily for several months. Adverse effects include rash (most common), stomatitis, nephritis and blood dyscrasias.

D. STEROIDAL ANTIINFLAMMATORY DRUGS (SAIDS)

Glucocorticoids are often used to treat chronic inflammatory disorders, particularly in patients unresponsive to common NSAIDs or unable to tolerate NSAIDs. There are many disadvantages to glucocorticoids, including adrenal suppression and a variety of disturbing, serious side effects. The actions of glucocorticoids are described in **Chapter 10** on Adrenocortical Hormones.

Orally-active SAIDs

A. Short-acting SAIDs ($t_{1/2}$ approximately 8-12 hours).

1. Hydrocortisone (Cortisol) has potent, undesired, Na^+-retaining action.
2. Cortisone, has the same activity spectrum as Cortisol, but is only about 80% as potent.

B. Intermediate-acting SAIDs ($t_{1/2}$ approximately 1 day).

1. *Prednisone* and *Prednisolone* have about 5 times the antiinflammatory potency of cortisol and less than half the salt-retaining activity.
2. *Meprednisone* and *Methylprednisolone* have about 5 times the antiinflammatory potency of cortisol and no salt-retaining activity.
3. *Triamcinolone* has antiinflammatory potency similar to that of prednisolone.

C. Long-acting SAIDs ($t_{1/2}$ about 2 days) lacking salt-retaining activity.

1. *Betamethasone* and dexamethasone have about 5 times the potency of prednisolone.

E. ANTI-GOUT DRUGS

Gout is a special form of arthritis, produced by crystals of uric acid (UA) and sodium urate (NaU) in extracellular spaces like synovial fluid of joints and subcutaneous tissue. Primary gout can be due to either (a) excess UA production or (b) inadequate UA excretion. Gout can be secondary to a specific disorder in which high amounts of nucleic acids are broken down (e.g., leukemia, lymphoma, polycythemia) or in which UA excretion is impaired (e.g., renal damage); or secondary to therapy with drugs that are acidic or have acidic metabolites that compete with UA for renal excretion (e.g., salicylates, alcohol).

Hyperuricemia represents a serum NaU concentration that exceeds the serum NaU solubility (7 mg%). Serum NaU levels are low in children, increase during puberty and remain stable in men (5-6 mg%) and women (4 mg%) during the lifespan, except for a 1 mg% increase in women after menopause. There are 6 men versus 1 woman with serum NaU greater than 7 mg%; there are 6 times as many men with gout. Most gout is primary. Mean age of onset of gout in men is 47 years; gout in women is usually postmenopausal. Enzyme defects account for a small percentage of hyperuricemia, but the disease is familial, transmitted as an X-linked recessive character. Renal NaU/UA stones occur in nearly half of men when their serum NaU level is greater than 9 mg%, but NaU tends to deposit in joints and subcutaneous sites rather than kidneys.

Formation of UA. UA is the final product of purine metabolism in humans. Whereas most catabolic pathways generate metabolites that are increasingly more soluble, UA is peculiarly less soluble than its precursors, hypoxanthine and xanthine. Allopurinol, an inhibitor of the xanthine oxidase (XO) enzyme involved in UA formation, is accordingly useful for treating gout. Note that while most sodium salts of organic acids are quite soluble, NaU is not very soluble.

Purines are normally formed by (a) *de novo* synthesis and (b) nucleic acid catabolism **(Figure 8.13)**. Excess purine formation can occur by virtue of excessive activity of the enzymes 5-phosphoribosyl-1-pyrophosphate (PRPP) synthetase or PRPP amidotransferase.

Excess uric acid formation can also be due to defects in the purine reutilization pathway, which would normally recycle purines into new nucleic acid synthesis. The defect here is usually abnormally low activ-

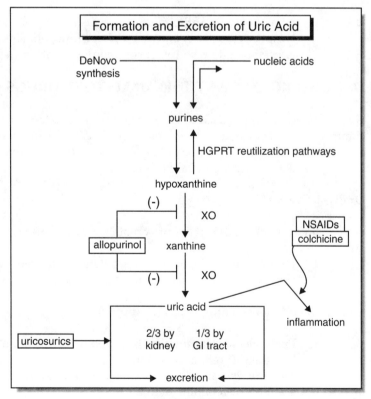

Figure 8.13

ity of the enzyme hypoxanthine guanine phosphoribosyltransferase (HGPRT) **(Figure 8.13)**. Dietary purines are not an important source of UA.

Excretion of UA. About two-thirds of UA is excreted by the kidneys and one-third is excreted via secretion into the gastrointestinal (GI) tract. In the kidneys most UA enters the nephron via glomerular filtration of serum; some UA is secreted by an organic acid transport system of low capacity. From the lumen of the nephron, most UA is reabsorbed by an organic acid transport system of high capacity. Uricosuric drugs block the secretory and reabsorptive sites, with the net effect being greater UA excretion **(Figure 8.14)**. Because the secretory system may be blocked to a greater extent than the reabsorptive system by low amounts of uricosurics, initiation of uricosuric therapy often precipitates gout. It is standard procedure to give anti-inflammatory drugs when starting uricosuric therapy.

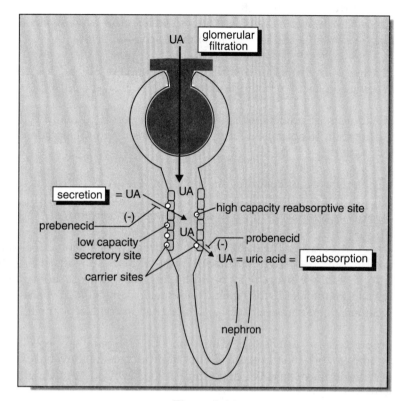

Figure 8.14

High serum levels of UA lead to a high level of UA in tissues and joints, predisposing to precipitation of UA crystals at these sites (tophi). The ensuing inflammation leads to granulocyte infiltration and phagocytosis of the UA crystals. As part of this process, lactic acid is produced, while lysosomal enzymes, kinins and chemotactic factors are released. The resultant acidic pH tends to precipitate more UA crystals. As shown in **Figure 8.15,** this is a self-propagating process.

Sites of tophi formation. Because joints in the limbs have a low pH, UA tends to precipitate there.
It precipitates in the large toe, instep, ankle and heel more than the knee, elbow, wrist and fingers. Less commonly, UA precipitates in subcutaneous tissue and kidneys.

Treatment of acute gout. To relieve the pain and to treat the inflammation of acute gout, colchicine or an NSAID is commonly given for a week or more.

General steps in long-term treatment of gout
1. Restrict dietary purines.
2. Increase fluid intake (increase urine volume).
3. Elevate urine pH greater than 6.5 ($NaHCO_3$; K•citrate).
4. Inhibit UA synthesis with allopurinol.
5. Increase renal excretion of UA with uricosurics.
6. Give anti-inflammatory drugs like colchicine and ibuprofen.

An anti-inflammatory drug is commonly given for a week or more, when instituting therapy with uricosurics or allopurinol, since these latter agents initially tend to precipitate gout or make it worse.

Uricosuric drugs (Figure 8.16). *Probenecid* (Benemid) and *sulfinpyrazone* (Anturane) are older agents. Benzbromarone is a newer more-potent uricosuric, effective even in patients with renal dysfunction. Benzbromarone is not yet available and is not part of the discussion below.

Pharmacokinetics of uricosurics. Absorption is rapid and peak plasma levels occur in about 2 hours. Binding to plasma proteins is extensive (90%, probenecid; 99%, sulfinpyrazone).

Metabolism of probenecid is by glucuronidation ($t_{1/2}$ is about 10 hours); sulfinpyrazone, by hydroxylation ($t_{1/2}$ greater than 10 hours). Sulfinpyrazone itself is a natural metabolite of phenylbutazone and has anti-inflammatory activity.

Mechanism of uricosurics. As shown in **Figure 8.14**, uricosurics act mainly by competing with UA for reabsorptive sites in the lumen of the nephron, thereby enhancing excretion of UA. Greater UA excretion leads to lower serum levels of UA and ultimately to resorption of UA tophi.

Use of uricosurics. The main use of these drugs is to enhance renal elimination of UA. Uricosurics are sometimes used in combination with other acidic drugs, to block secretory sites in the nephron and thereby increase the plasma half life of these drugs (e.g., penicillin, cepahlosporins). Alternatively, acidic drugs like aspirin compete with uricosurics for sites in the kidneys and can block their effectiveness. Probenecid and sulfinpyrazone can be given together for an additive effect, or can be used in combination with allopurinol.

Side effects of uricosurics. The main side effects of probenecid are allergic reactions (rash) and GI upset. Sulfinpyrazone mainly causes GI upset. By increasing the renal excretion of UA, both drugs elevate renal levels of UA and may cause UA deposition in the kidneys. Accordingly, uricosurics should not be given to patients with renal stones.

A. Allopurinol (Zyloprim)

Pharmacokinetics of allopurinol. Absorption is rapid and peak plasma levels occur in about 1 hour. Metabolism occurs by xanthine oxidase (XO), which converts allopurinol ($t_{1/2}$ approximately 2 hours) to an active and long-lived metabolite, oxypurinol ($t_{1/2}$ about 20 hours).

Mechanism of allopurinol (Figure 8.17). Allopurinol is an inhibitor of XO (competitive

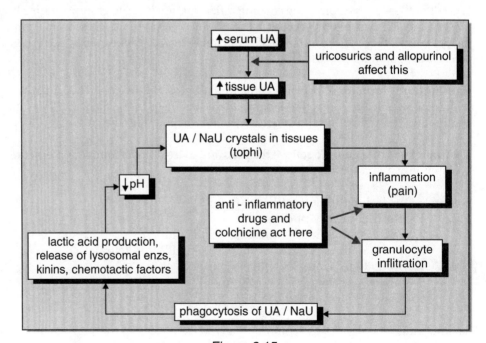

Figure 8.15

at low concentrations; non-competitive at high concentrations). Oxypurinol is a non-competitive inhibitor of XO. The anti-gout effect is largely due to oxypurinol. By inhibiting XO, there is less conversion of hypoxanthine (HX) and xanthine (X) to UA. This reduces plasma UA levels which enhances the resorption of UA crystals. Also, HX, X and UA have independent solubilities, so purine metabolites can be excreted in larger amounts and with less risk of crystal formation in the kidneys.

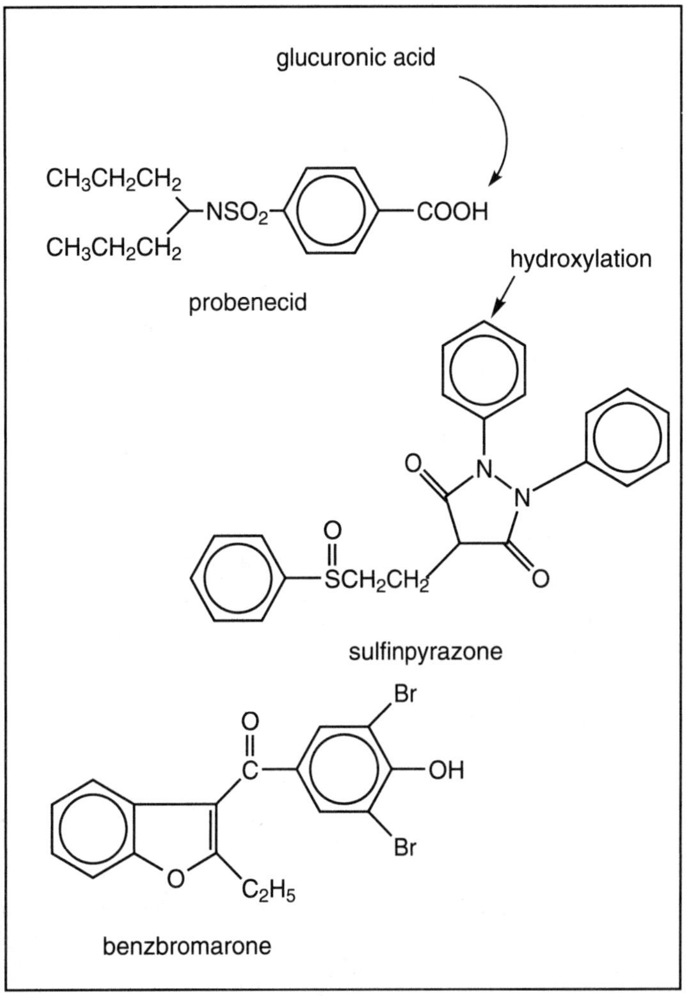

Uses of allopurinol. This drug is used for gout patients that are producing large amounts of UA; or have renal tophi; or are unresponsive to uricosurics. Allopurinol is also used to inhibit UA synthesis, when beginning chemo- or radiation therapy in patients with leukemias and malignancies. Allopurinol is commonly used for hyperuricemia of Lesch-Nyhan disease (genetic deficiency of HGPRT).

Side effects of allopurinol. Allopurinol causes allergic responses (rash), GI upset (vomiting, diarrhea), hepatotoxicity, depression of bone marrow (leukopenia and leukocytosis), neuritis and cataracts.

Figure 8.16

Drug interactions of allopurinol. Allopurinol, an inhibitor of XO, increases the effectiveness and toxicity of drugs that are metabolized by XO (e.g., the antineoplastic drugs, mercaptopurine and azathioprine).

B. Colchicine

Colchicine is an anti-inflammatory drug that is effective only for gout.

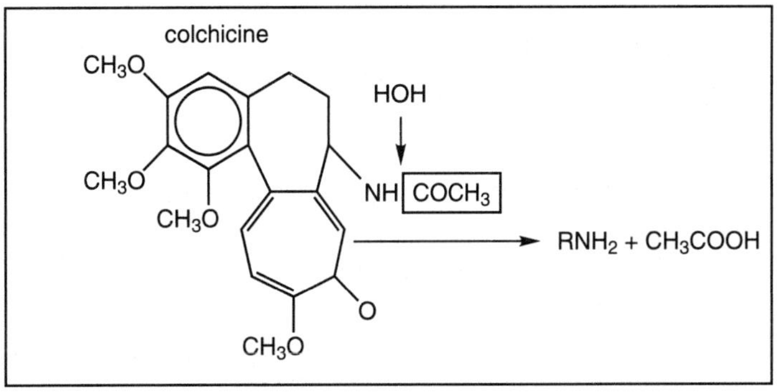

Pharmacokinetics of colchicine. Absorption is rapid and peak levels occur in less than 2 hours. Distribution is mainly to viscera (intestine, spleen, liver, kidneys) and leukocytes. Metabolism is mainly by deacylation in the liver. Excretion is mainly into bile, then feces (80%); 20% in urine.

Mechanism of colchicine. Colchicine binds to tubulin in leukocytes and thereby interferes with leukocyte migration and phagocytosis of UA **(see Figure 8.15).** This interrupts the self-propagating cycle of inflammation in gout.

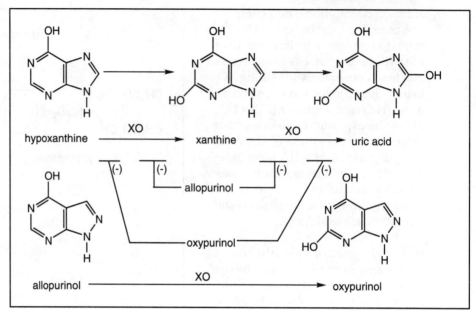

Figure 8.17

Use of colchicine. Colchicine is used to diagnose gout. If the pain is relieved by colchicine, then gout is considered to be present (95% of gouty patients respond). Colchicine is not effective in other anti-inflammatory disorders.

Colchicine is used prophylactically to prevent recurrence of painful gout (0.5-2 mg/day or every other day). Colchicine is commonly given at the start of uricosuric or allopurinol therapy. Colchicine is used to treat acute gout. Orally, 1 mg of colchicine is often given, followed by 0.5 mg every 2-3 hours (6-7 mg maximum), until pain is relieved. By an i.v. route, 2 mg (less than 4 mg max) is often given.

Adverse effects of colchicine. Colchicine stops mitosis of rapidly dividing cells. This leads to side effects in tissues in which colchicine is largely distributed:
    a. Intestinal pain and diarrhea (80% of patients)
    b. Liver damage
    c. Bone marrow depression with hematopoietic changes
    d. Hematuria
    e. Alopecia

Other effects include vascular damage with shock; hypertension from stimulation of the vasomotor center; and fever. The intestinal pain is used as a gauge of colchicine's adverse effects, so this symptom should not be treated.

C.   Other Anti-inflammatory Drugs for Gout

All anti-inflammatory drugs and glucocorticoids can be used for gout. Their action is illustrated in **Figure 8.3**.

3. **TREATMENT OF HEADACHE**

   A. Occasional Headache

   Occasional headaches are routinely treated with aspirin, acetaminophen (Tylenol), ibuprofen (Motrin) or naproxen (Naprosyn). These drugs have already been discussed. All of these are also used for migraine and cluster headaches (vascular headaches), but other drugs are more effective.

   B. Migraine Headache

   1. **Abortive treatment**. The headache phase of migraine is associated with dilation of extracranial blood vessels.
      a. Aspirin, acetaminophen, ibuprofen or naproxen (all in analgesic doses).

      b. Ergotamine (Ergomar) and ergotamine with caffeine (Cafergot) are the drugs of choice. Usually 1 or 2 mg is administered, followed by smaller supplements at intervals, up to a maximum of 6 mg or less.

      Ergotamine acts directly on vascular smooth muscle to produce a *vasocontrictor* effect which supersedes the potential vasodilator effect of α-receptor block.

      c. Dihydroergotamine (DHE 45), ergonovine (Ergotrate) and methysergide (Sansert) act similarly to ergotamine.

      d. *Sumatriptan* (Imitrex) is a serotonin 5-HT_{1D} agonist that relieves migraine by constricting large intracranial blood vessels and inhibiting neural release of algesic peptide neurotransmitters. Sumatriptan does not cross the blood-brain barrier and is not an analgesic.

      Adverse effects. These include (a) transient hypertensive response, (b) angina and (c) EKG abnormality in patients with coronary artery disease or Prinzmetal's angina.

      e. Dexamethasone, a corticosteroid, is sometimes used.

   2. **Prophylactic treatment**. The aura of migraine is associated with constriction of extracranial blood vessels.
      a. *Propranolol*, a β-adrenoceptor antagonist, is the drug of choice.
      b. Clonidine, an $α_2$-agonist.
      c. Phenelzine, a monoamine oxidase inhibitor.
      d. Amitriptyline, a blocker of NE transporters.
      e. Caffeine and papaverine, vasodilators.
      f. *Cyproheptadine* (Periactin), an H_1/5-HT receptor blocker. This is the drug of choice in children.
      g. Methysergide and ergotamine.

      Ergotamine, dihydroergotamine, ergonovine and methysergide are classified as *ergot alkaloids*. Their pharmacology is discussed in the chapters on (a) the sympathetic nervous system and (b) uterine drugs. These are unpleasant drugs that frequently produce (a) nausea and vomiting with diarrhea, (b) numbness in fingers and toes (vasoconstriction), (c) uterine contractions and accompanying pain, (d) hallucinations and retroperitoneal fibrosis. These drugs cannot be taken during pregnancy (abortions occur) or with hypertension (acute hypertension occurs).

C. Cluster Headache

1. **Abortive treatment**
   a. 100% oxygen, inhaled for 5-10 minutes, is very effective for cluster headache.
   b. *Sumatriptan* has the same effectiveness as oxygen.
   c. Ergotamine is often used.
   d. Dihydroergotamine.

2. **Prophylactic treatment**
   a. *Methysergide* (Sansert) is the drug of choice.
   b. Ergotamine
   c. Indomethacin
   d. Lithium

4. LOCAL ANESTHETICS

Local anesthetics (LAs) are drugs that produce reversible topical (local) anesthesia or nerve block by virtue of *sodium channel block,* thereby reducing the ability of excitable membranes (e.g. nerve axons) to conduct impulses (i.e., action potentials).

A. Esters **(Figure 8.18)**

1. Procaine (Novocain)
2. Tetracaine (Pontocaine)
3. Cocaine
4. Benzocaine does not have a tertiary amine group. It is water-insoluble and cannot be given parenterally. Being soluble in oil, benzocaine can be dissolved in cremes and lotions for topical anesthesia. Benzocaine penetrates the skin.

These first three LAs are effective when applied topically to mucous membranes. Cocaine is rarely used as a LA today because of its abuse potential.

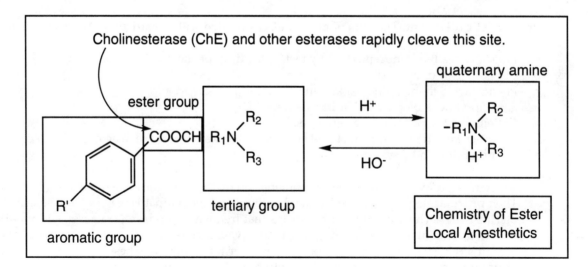

Figure 8.18

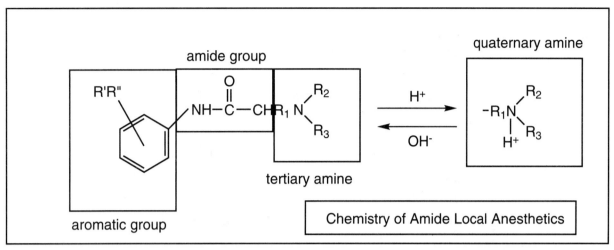

Figure 8.19

B. Amides **(Figure 8.19)**

Lidocaine (Xylocaine) is one of the most commonly used LAs.

LAs have the suffix *-aine* (e.g. lido*caine*). Ester LAs have a much shorter $t_{1/2}$ than amide LAs.

Pharmacokinetics of LAs. LAs are usually injected into the site to be anesthetized (e.g. procaine into the gum, for dentistry) or into the region of the nerves to be anesthetized **(see Figure 8.20)**.

Anesthetic action is terminated primarily by removal of the LA from its site of action via absorption into the vasculature. The richer the blood supply, the quicker the anesthesia "wears off." Except for cocaine, a vasoconstrictor, LAs generally produce vasodilation or no effect on vascular tone. Epinephrine (Epi) (1:200,000), a vasoconstrictor, is often added to the LA to prolong its duration of anesthesia. Epi should not be administered when LAs are used in digital extremities. Epi should be used cautiously if the patient has cardiovascular disease. Esterases rapidly metabolize ester LAs **(Figure 8.18)**. Liver microsomal enzymes metabolize both ester and amide LAs. However, LA action is terminated by absorption from its site of action into the bloodstream—not by metabolism.

Mechanism of action of LAs. The pKa of LAs is about 8.5. As shown in **Figures 8.18 and 8.19**, the N group of LAs will be largely nonionized at a pH greater than 8.5; largely ionized at a pH less than 8.5. The nonionized form diffuses most easily through tissues and across membranes **(Figure 8.21)**. [In an acid environ, as in inflammation, the LA would not be effective since the ionized form of the LA would persist and not diffuse to the nerve or through the nerve membrane.] However, it is the ionized form that is active. This species of LA is formed in the acidic environs inside the nerve from any of the LA that diffuses across the nerve membrane. The ionized LA blocks sodium (Na) channels from inside the nerve. Action potentials, due to Na^+ flux across the nerve membrane, cannot occur. Accordingly, the nerve is "blocked." This is the basis of LA action.

As shown in **Figure 8.22**, Na channel block adversely affects the generation of an action potential (AP) in nerves. There is (1) a reduced rate of rise in the excitatory postsynaptic potential (EPSP), (2) an elevation in the threshold potential (TP), (3) a reduced rate of rise in the AP and (4) a reduced amplitude or overshoot of the AP. The net effect of these actions is (5) a reduced rate of propagation of nerve impulses, (6) a reduction in the number of nerves sending impulses and eventually (7) complete block

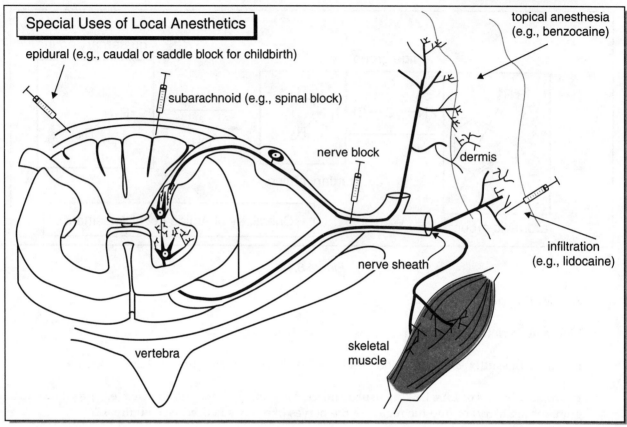

Figure 8.20

of conduction of impulses. The resting membrane potential (E_m) is not appreciably altered, since K^+ currents predominately determine the E_m and these are generally not altered by LAs.

Susceptibility of nerves to LAs. Rapidly firing nerves are most easily blocked by LAs, because LAs have greatest affinity for receptors of activated Na channels (open Na channels). Therefore, Na channel block is voltage- and time-dependent.

Small nonmyelinated fibers (e.g., dorsal root type C pain fibers) are most easily blocked by LAs, then lightly myelinated delta A pain fibers. Less heavily myelinated α and β A sensory fibers are next most susceptible to: LAs, followed by gamma-A myelinated motor fibers.

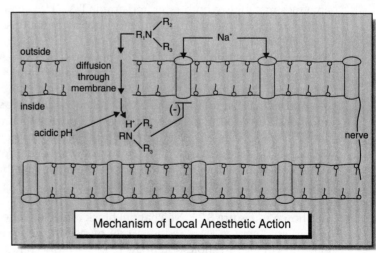

Figure 8.21

As many have experienced in dentist office procedures, sensations of touch or pressure are still noticeable when temperature and pain sensations have been obtunded. The sensation of pain is last to recover after an LA. It should be appreciated that in large nerve trunks, the LA would first affect the nerves in the outer portions of the trunk, so motoneurons could be blocked before smaller deeper sensory fibers in the nerve trunk.

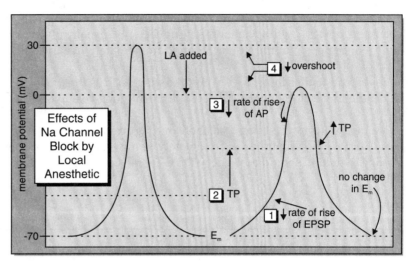

Figure 8.22

Potency of LAs. Tetracaine is more potent than Prilocaine (Citanest), and these are more potent than Lidocaine and Procaine.

Actions and special uses of LAs. These are outlined in **Figure 8.20**.

C. DRUGS FOR VISCERAL PAIN: NARCOTIC ANALGESICS

Chemistry. *Opiates* are "phenanthrenes" isolated from the opium poppy plant **(Figure 8.23)**. *Opioids* are compounds that have opiate-like activity, but not necessarily the phenanthrene nucleus. Many opioids are synthetic and relatively small molecules (e.g., methadone and meperidine). Morphine, hydromorphone and oxymorphone are potent opiates that differ from one another only by the number of H atoms or an added -OH group **(see Figure 8.23)**. Their respective CH_3- derivatives (codeine, hydrocodone and oxycodone) are much less potent opiates. Heroin, a diacetylated derivative of morphine, is not approved in the United States and is only a drug of abuse. Heroin is metabolized to morphine.

When alkyl groups are substituted for the methyl (CH_3-) group on the N atom of opiates, opiate antagonists are produced (e.g., nalorphine, naltrexone, naloxone).

Collectively, the above drugs are known as narcotics and narcotic antagonists. *Narcotic* (narcolepsy means sleep) is an old generic term that is now loosely used to refer to drugs of abuse, even non-opioid chemicals.

Endogenous opioids are made primarily in the brain, pituitary gland and gut from one of three proteins (257 to 267 amino acids)

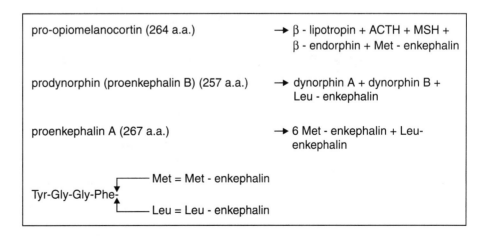

182 PHARMACOLOGY

Pharmacokinetics of morphine. Absorption is rapid but erratic by oral and rectal routes of administration. Therefore, opiates (with a phenanthrene nucleus) are usually given parenterally. Peptide opioids cannot be taken orally, since pepsin in the stomach would immediately inactivate the peptide.

Opioids tend to distribute largely in soft tissues such as the lungs, liver, spleen and kidneys. However, opioids distribute in varying degrees to all tissues including skeletal muscle and fat.

There is significant first-pass metabolism **(Figure 8.24)**. In a naive patient a therapeutic dose of morphine (approximately 10 mg i.v. or approximately 30 mg oral) has a $t_{1/2}$ of about 2 hours and duration of about 5 hours. Dosage reduction should be made when there is liver damage. Opioids are excreted primarily in urine, but also in bile.

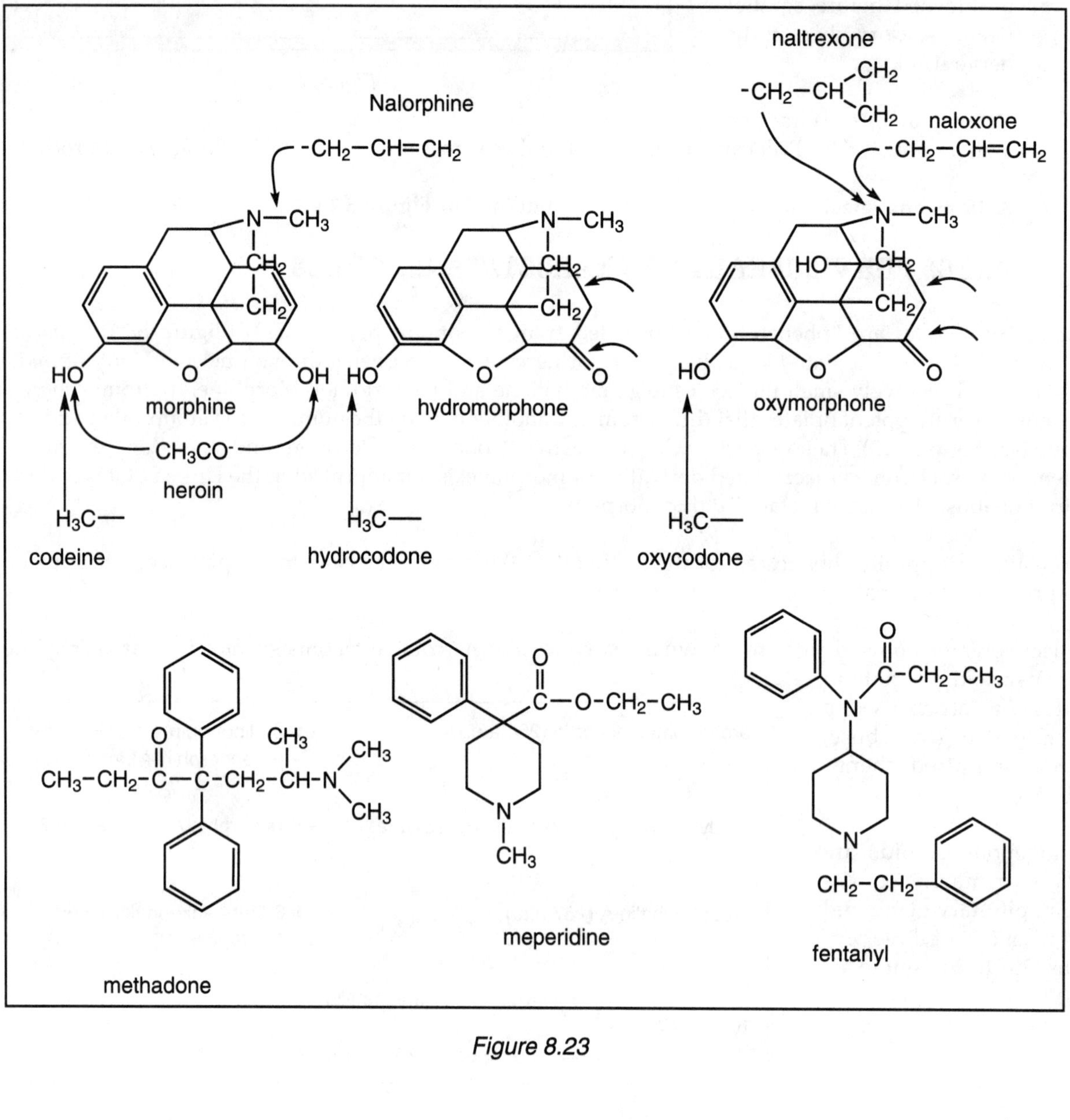

Figure 8.23

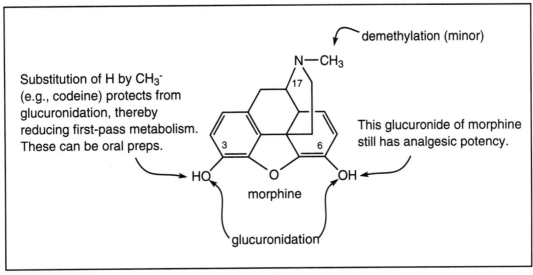

Figure 8.24

Opioid actions and receptors. Opioids produce effects on many organ systems in the body, while acting on several receptor types (mu, kappa (κ), sigma (σ), and delta (δ)) and subtypes (mu_1, mu_2, etc.). Their involvement in opioid actions is outlined in **Figure 8.25**.

Common opioid analogs
1. Meperidine (Demerol) has high potency but shorter duration than morphine. Antimuscarinic effects produce tachycardia, etc. Meperidine does not suppress the cough reflex. Because it has little effect on uterine tone versus the relaxing effects of other opioids, meperidine is used for labor pain.

2. Methadone (Dolophine) is an orally-active drug with a spectrum of action similar to morphine. Tolerance and physical dependence develop slowly, while withdrawal signs are less severe. For these reasons methadone is used to wean addicts off of heroin.

3. Fentanyl (Sublimaze) and analogs are potent very short-acting opioids that are used in "balanced anesthesia" **(see Chapter 9, General Anesthetics)**.

4. Lower potency phenanthrenes (codeine, hydrocodone, oxycodone) are commonly used for analgesia with aspirin and acetaminophen, not alone. Codeine was used in the past as sole therapy for diarrhea; today, for cough.

5. Propoxyphene (Darvon) is a low potency analgesic (less potent than codeine).

6. Loperamide (Imodium) and Diphenoxylate with atropine (Lomotil) are used for diarrhea **(see Chapter 7, Treatment of Diarrhea)**.

Mixed Opioid Agonist-Antagonists. These drugs act partly as opioid agonists and partly as opioid antagonists. Consequently, when administered to an addict in place of a full agonist, dysphoria and opiate withdrawal may be produced.

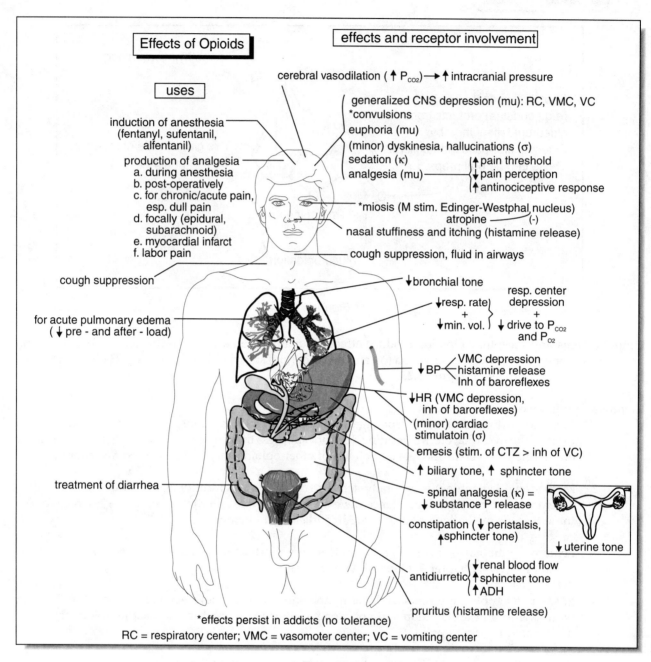

Figure 8.25

A. *Phenanthrenes* used for pain.

1. Nalbuphine (Nubain) is a potent κ-agonist and mu-antagonist. There is less respiratory depression versus morphine—a "ceiling" effect. Naloxone cannot reverse nalbuphine actions, since both drugs are mu-antagonists.

2. Buprenorphine (Buprenex) is a potent partial mu-agonist that dissociates slowly from the receptor, thereby imparting a long duration and resistance to naloxone-reversal. This drug also exerts a "ceiling" effect on respiratory depression.

B. Non-Phenanthrenes. Butorphanol (Stadol) and pentazocine (Talwin) are κ-agonists like nalbuphine.

Antagonists at mu-receptors. *Naloxone* (Narcan) has a $t_{1/2}$ approximately 1-2 hours; *naltrexone* (Trexan) has a $t_{1/2}$ about 10 hours. Both are used to produce immediate (less than 3 minutes) reversal of opioid toxicity, which simultaneously produces withdrawal. Naloxone (0.2-0.4 mg) must be injected at intervals, since most opiate drugs of abuse have a longer $t_{1/2}$.

Uses of Opiods. These are outlined in **Figure 8.25**.
1. Analgesia: Morphine, meperidine and hydromorphone.

2. Cough suppression. Opioids depress the medullary cough center at a dose less than that required for analgesia.
    a. Codeine, 15 mg
    b. Dextromethorphan
    c. Levopropoxyphene
    d. Benzonatate—a local anesthetic

    Dextromethorphan and levopropoxyphene stereoisomers are devoid of analgesic and addictive properties.

3. Pulmonary edema: Morpine is often given to exert a generalize calming effect and allay the anxiety the accompanies respiratory difficulty.

4. Diarrhea: Imodium and diphenoxylate are most commonly used.

Contraindications
1. **Cholelithiasis**. Opiates should not be used for cholelithiasis, since the usual spasmogenic action on the bile duct would increase pain.
2. **Respiratory depression**. Opiates would further depress respiration.
3. **Hypotension**. Opiates release histamine which further reduces blood pressure.

Addiction potential of opioids. Tolerance begins to develop after the first dose, occurring most rapidly with (a) high doses and (b) close dosing intervals. Addicts can tolerate greater than 25-times the therapeutic dose. Notably, tolerance does not develop to miosis, constipation and convulsant actions.

Cross-tolerance is prominent for all opioids except the antagonists (e.g., naloxone) and antagonist actions of partial agonist-antagonists. Also, for the agonist-antagonists, full tolerance to agonist opioids does not develop. Physical dependence accompanies development of tolerance.

The *withdrawal syndrome* is characterized by (a) cold clammy skin ("cold turkey") with chills, despite hyperthermia and (b) muscle spasm ("kicking the habit") cramps and tremors. Other signs are lacrimation, mydriasis, rhinorrhea, yawning, diarrhea and a general feeling of anxiety and irritability. This syndrome occurs at about 8 hours after the last dose of morphine (or heroin), peaks at 1-2 days and largely subsides in less than 1 week. With methadone, withdrawal is less severe but it persists longer. This syndrome is instantaneous after naloxone, persisting for about 1 hour. Withdrawal from opioids is extremely unpleasant, but generally not life-threatening. Symptoms can be lessened by administering opiates periodically, to wean the addict. A recent approach involves administering naltrexone to an addict that is unconscious and artificially respired, so that withdrawal symptoms are not experienced.

CHAPTER 9
PHARMACOLOGY OF THE CNS

A. GENERAL ANESTHETICS

General anesthesia is a state of unconsciousness, accompanied by:
  a. analgesia
  b. anterograde amnesia
  c. skeletal muscle relaxation and
  d. inhibition of sensory reflexes.

It is intentionally induced for invasive procedures such as surgery and catheterization. The depth of anesthesia is ideally regulated to coincide with the degree of discomfort caused by the procedure being performed.

Conventional anesthetics are inhaled or administered by i.v. infusion. All inhalant anesthetics in the United States today are gases. In other countries ether vapor may still be used, and it is discussed briefly to impart some features related to anesthetic action. Drugs that are commonly used as adjuncts for anesthesia (e.g., analgesics, muscle relaxants) are likewise discussed.

Stages of anesthesia (Guedel's signs). These can all be seen when ether is the anesthetic. However, none of today's anesthetics produce all of the stages and/or planes of anesthesia. In fact, most of the anesthetics are used with adjuncts, which would otherwise eliminate some of the signs and stages. For example, when scopolamine is used as a preanesthetic to inhibit bronchosecretion, pupil diameter cannot be an index (scopolamine causes mydriasis). Opioids, as analgesic adjuncts, depress the respiratory center. The neuromuscular blockers, d-tubocurarine and succinylcholine, paralyze respiratory muscles. Accordingly, changes in respiration cannot be gauged in the presence of these drugs since a ventilator would be required and is now routine.

In fact, the concurrent use of antimuscarinics, analgesics and muscle relaxants with anesthetics is intended to optimize the actions of each while minimizing the risk of excess CNS depression that might be engendered with use of anesthetic alone. This is called "balanced anesthesia."

With these limitations, and for the purpose of indicating the expected physiological changes that accompany gradually increasing anesthesia, the stages of anesthesia are shown in **Table 9.1**. It is desirable to pass through stage 2, an excitement phase, as rapidly as possible. Stage 3 corresponds with surgical anesthesia, but not until plane 3 of stage 3 is optimal skeletal muscle relaxation achieved. The depth of anesthesia should not proceed beyond this level.

Mechanism of action. Anesthetics are thought to "dissolve" in membranes and disorder the structure of nerve membranes so that ion fluxes are disrupted. This is illustrated in **Figure 9.1**.

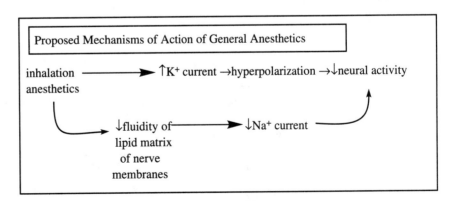

Figure 9.1

Stages of Anesthesia (Table 9.1)

Stage	Begins with...	Features	Lost reflex	Pupil size
1	analgesia*	irregular respiration		
2	loss of consciousness*	amnesia and excitement (movement, coughing, vomiting)	eyelid	
3	rhythmic respiration*	surgical anesthesia		
plane 1		moist and mobile eyes	conjunctival	
plane 2	initial signs of paralysis of intercostals	dry and immobile eyes more diaphragmatic breathing	skin incision	
plane 3	full paralysis of intercostals	skeletal muscle relaxation* full diaphragmatic breathing	corneal larynx peritoneal	*
plane 4	paralysis of diaphragm*	apnea	light, anal	
4	depression of respiratory center and vasomotor center	apnea circulatory failure death		

Major anesthetic effects
1. **Analgesia**

$$\boxed{\text{ether, nitrous oxide}} > \text{methoxyflurane} > \text{all others}$$

Except for N$_2$O and ether, an analgesic must be added to all anesthetics.

2. **Skeletal muscle relaxation**

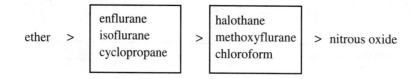

Except for ether, cyclopropane, enf- and iso-flurane, a muscle relaxant must be added to the common anesthetics for a balanced anesthesia.

3. **Blood pressure and respiration**
   a. Ether and cyclopropane increase blood pressure and respiration.
   b. N_2O does not change the blood pressure.
   c. All other anesthetics reduce blood pressure.

4. Myocardial sensitization to circulating epi is produced by halothane.

Other anesthetic effects
1. Increased cerebral blood flow causes increased cerebral volume, which elevates intracranial pressure. The effect can be fatal in a trauma patient with head injury.
2. Increased renal vascular resistance causes reduced renal blood flow and reduced GFR.
3. Increased hepatic vascular resistance causes reduced hepatic blood flow.
4. Reduced uterine tone is produced by halothane, enf-, iso- and sevo-flurane.

A. Inhalant Anesthetics

1. **Dosing**
The rate of delivery is proportional to the concentration of anesthetic in inspired air (gas). Often, a concentration higher than the maintainance concentration is used to induce anesthesia, to speed induction. The percent of each anesthetic in inspired gas will reflect relative differences in potency/efficacy. The MAC (minimum alveolar concentration) is a gauge of this. MAC is defined as the minimum concentration (partial pressure in percentages) of a gas at equilibrium that is needed to inhibit reflex movement to a skin incision in 50% of patients. Nearly 100% efficacy is achieved at about 1.3 times the MAC value.

Adjuncts like an analgesic, muscle relaxant, sedative or another anesthetic will reduce the MAC. The MAC is inherently higher in infants and is elevated by sympathomimetics like amphetamine.

In **Table 9.2** it can be seen that methoxyflurane is extremely potent, since such a small amount of this gas produces anesthesia. In contrast, nitrous oxide is a low-potency anesthetic, since 100% would be needed to produce anesthesia. Since 20% oxygen must be maintained, 100% nitrous oxide would produce fatal anoxia.

2. **Speed of induction** is accelerated when
   a. The concentration of anesthetic is high—faster rate of delivery of anesthetic.
   b. The ventilation rate is high—faster rate of delivery of anesthetic.

MAC of General Anesthetics (Table 9.2)

Anesthetic	MAC (%)
Nitrous oxide	greater than 100 (least potent)
Cyclopropane	about 10
Ether	about 2
Enf-, Iso-, Sevo-flurane	1-2
Halothane	less than 1
Methoxyflurane	0.2 (most potent)

This is especially important for anesthetics that are very soluble in blood, since rapid dissolution in blood would effectively reduce the alveolar anesthetic concentration and accordingly reduce anesthetic concentration in inspired air, slowing delivery to the body.

c. The cardiac output is low—delay in reducing alveolar concentration of anesthesia. Obviously, a low cardiac output is undesired. However, it is important to appreciate that a high cardiac output (e.g., with hyperthyroidism) would lengthen the induction time, especially for gases highly soluble in blood.

Table 9.3

Anesthetic	Blood/Alveolus Partition Coefficient
Cyclopropane	0.4-0.5 = Low solubility in blood Rapid induction Rapid recovery
Nitrous Oxide	
Isoflurane	1.0-2.0
Enflurane	
Halothane	2.0-2.5
Ether	greater than 10 = High solubility in blood Slow induction Slow recovery
Methoxyflurane	

d. The solubility of anesthetic in blood is low. If an anesthetic is insoluble in blood, only a few molecules are needed to substantially increase its partial pressure in blood. The anesthetic would similarly pass more readily from blood to brain—the target tissue for anesthetics. This is highlighted in **Table 9.3**.

e. The A-V gradient is high—more anesthetic being delivered to brain than removed from brain. The rank order of tissue saturation by an anesthetic is:

| brain, heart, liver, kidneys, spleen
high blood flow | > | muscle, skin
intermediate blood flow | > | fat, connective tissue, bone
low blood flow |

The above features related to induction are illustrated below:

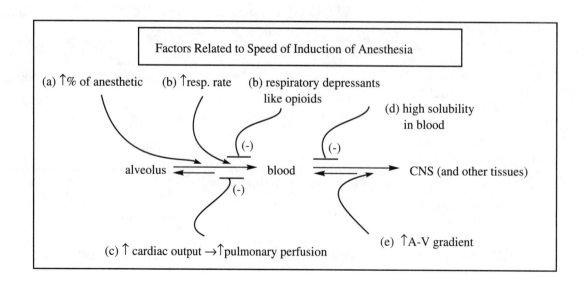

3. **Termination of anesthesia**
   Anesthetics "wear off" because of redistribution to tissues high in lipids, like skin and fat. These sites serve as a sink for anesthetics. The relatively poor blood perfusion of skin and fat restricts initial delivery of anesthetic to these tissues. Even for the few anesthetics that are partly metabolized, redistribution is still the major means of terminating anesthesia.

4. **Elimination of anesthetics**
   Except for halothane and methoxyflurane, only small amounts of inhaled anesthetics are metabolized:

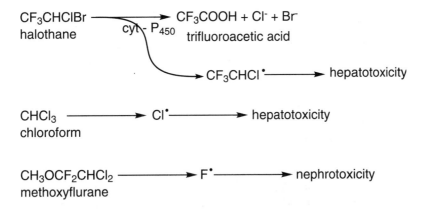

   Inhaled anesthetics are excreted mainly by the lungs in expired gases.

PROPERTIES OF INHALED ANESTHETIC DRUGS

1. *Nitrous oxide* (N_2O), in low concentration (less than 25% N_2O), is an excellent analgesic. It is used alone for this purpose in labor or for brief procedures. At 60-80% N_2O, unconsciousness is produced without muscle relaxation. N_2O is unsuitable as sole agent for surgical anesthesia since stage 2 is the maximal depth of anesthesia. However, N_2O is a good analgesic supplement to a potent anesthetic. Despite direct myocardial depression, blood pressure is maintained by reflex sympathetic stimulation. N_2O is not flammable but it supports combustion.

2. *Ether* is largely obsolete because it is flammable and explosive. Induction with ether is extremely unpleasant and is associated with bronchial irritation, bronchosecretion, vomiting and sympathetic stimulation. However, it will produce good analgesia, skeletal muscle relaxation and surgical anesthesia. If used as the sole anesthetic, ether would be unsafe compared to standard anesthesia today.

3. *Cyclopropane* is rarely used because it is also flammable and explosive. Unlike ether, induction with cyclopropane is rapid, pleasant and unaccompanied by bronchial irritation. However, sympathetic stimulation occurs as with ether.

4. *Halothane* (Fluothane) is nonflammable and produces rapid, pleasant induction. Skeletal muscle relaxation is insufficient when halothane is used alone, but halothane potentiates skeletal muscle relaxants. The myocardium is sensitized to epinephrine-induced arrythmias. A rare hepatotoxicity occurs (about 1:10,000). Halothane has a narrow margin of safety.

5. *Enflurane* (Ethrane) is a commonly used anesthetic. It is nonflammable but pungent. Induction is rapid and the level of analgesia is good. Some skeletal muscle relaxation occurs. Adverse effects include myocardial depression, hepatitis (rare) and overdose-tonic-clonic movements.

$CH_3CH_2\text{-}O\text{-}CH_2\text{-}CH_3$	Diethyl ether
$CHF_2\text{-}O\text{-}CF_2\text{-}CHFCl$	Enflurane
$CHF_2\text{-}O\text{-}CHCl\text{-}CF_3$	Isoflurane
$CH_3\text{-}O\text{-}CF_2\text{-}CHCl_2$	Methoxyflurane
$CHF_2\text{-}O\text{-}CHF\text{-}CF_3$	Desflurane

6. *Isoflurane* (Forane), *desflurane* (Anaquest) and *sevoflurane* have efficacies similar to enflurane. None of these sensitizes the myocardium to epinephrine-induced arrhythmias.

7. *Methoxyflurane* is the most potent anesthetic (lowest MAC value) so it induces anesthesia rapidly. (The high solubility of methoxyflurane in blood, from **Table 9.3**, tends to slow induction, but this is only one of the many factors related to induction.) Metabolism of methoxyflurane generates F• which produces nephrotoxicity. For this reason, this anesthetic is unpopular.

B. Intravenous Anesthetics

Properties of I.V. anesthetics
1. **Ultrashort-acting barbiturates**
   Thiopental (Pentothal), *Methohexital* (Brevital) and *Thiamylal* (Surital)

   Induction is rapid (less than 1 minute) since these highly lipid soluble drugs readily enter the brain. Cerebral blood flow may be reduced, an advantage in head injury patients with elevated intracranial pressure. Although virtually all of these drugs are ultimately metabolized, the rate of metabolism is relatively slow. The anesthetic action is terminated by redistribution from brain to peripheral tissues like skin and fat.

   Adverse effects. These include respiratory depression (reduced CO_2 response), dose-dependent depression of the myocardium with arterial and venous dilation. Barbiturates are hyperalgesic, not analgesic, and skeletal muscle relaxation is not produced. Accordingly, reflex responses to painful stimuli would be enhanced. For these reasons the thiobarbiturates are used only for induction of anesthesia, not as single anesthetic agents.

2. **Ketamine** (Ketalar)
   Ketamine produces *dissociative anesthesia*, a catatonic-like state accompanied by anterograde amnesia and cutaneous- but not visceral-analgesia. Ketamine acts as an antagonist at the N-methyl-D-aspartate (NMDA) receptor, blocking actions of the endogenous neurotransmitters glutamate and aspartate. Hallucinations and terrifying dreams occur in adults, but not so frequently in children. Benzodiazepine pretreatment reduces the incidence of this effect.

   Ketamine produces central vagal inhibition and central sympathetic stimulation with epi release from the adrenal medulla, so cardiac stimulation with bronchodilitation occur. Skeletal muscle relaxation is not produced.

3. *Etomidate* (Amidate) acts similarly to thiobarbiturates. Induction is rapid and redistribution accounts for termination of anesthesia. Cerebral blood flow is reduced, but analgesia is not produced.

   Etomidate is eventually metabolized by cyt P_{450} enzymes and blood esterases. Blood pressure is well-maintained. Peculiarly, etomidate can produce myoclonic activity and inhibition of steroid synthesis in the adrenal (Addison-like). Vomiting is a common post-anesthetic event.

4. *Propofol* (Diprivan) acts similarly to etomidate. Induction is rapid and cerebral blood flow is reduced. Respiratory and cardiovascular activities are well-maintained. However, post-anesthetic nausea is uncommon.

5. *Opioids: Fentanyl* (Sublimaze), *sufentanil* (Sufenta), *alfentanil* (Alfenta)

These short-acting analgesic synthetic opioid agonists are used for short anesthetic procedures. They do not release histamine. However, they are not amnestic and do not produce skeletal muscle relaxation. Respiratory depression can be rapidly reversed with naloxone, an opiate receptor antagonist.

Opioids are preferred in cardiac patients, particularly as a component of balanced anesthesia with a skeletal muscle relaxant, benzodiazepines (BZDs), etc. Combined opioid and BZD makes "conscious sedation" for endoscopy and brief surgery. *Neurolepanalgesia* (opioid plus droperidol, a butyrophenone like haloperidol) is used to produce a compliant patient that can respond to verbal commands and be physically maneuvered (e.g., for endoscopy).

Longer-acting morphine is often administered to relieve pain before or after surgery.

6. *Benzodiazepines*: *Diazepam* (Valium), *lorazepam* (Ativan) and *midazolam* (Versed)

Because BZDs produce an anxiolytic effect, anterograde amnesia, skeletal muscle relaxation and have a high therapeutic index, they are widely used as preanesthetic medication or as a component of balanced anesthesia. *Midazolam* is the most water-soluble and is able to be administered i.v. It produces the most rapid induction. Overdose toxicity can be rapidly reversed with flumazenil, a BZD receptor antagonist.

B. ANXIOLYTICS

BENZODIAZEPINES
Benzodiazepines are the principle anti-anxiety (*anxiolytic*) drugs in use throughout the world. This class is among the most commonly prescribed drugs in medicine.

Chemistry and Pharmacokinetics of benzodiazepines. Benzodiazepines are heterocyclic molecules having two ring nitrogen amino groups ("diazo" and "amine") with an electronegative moiety at position seven (e.g., Cl⁻) and a benzene ring ("benzo") at position five **(see Figure 9.2, top right)**.

These bulky lipophilic molecules are rapidly and nearly completely absorbed. Clorazepate (Tranxene) is an inactive prodrug that is converted by stomach acid to an active metabolite. The onset of action of benzodiazepines is seen in less than 90 minutes. In emergencies i.m. or i.v. preps can be used (e.g., diazepam, lorazepam and midazolam). Although more than 80% of benzodiazepines are plasma protein bound, they generally do not interact with other protein-bound drugs **(Figure 9.3)**.

Chlordiazepoxide, diazepam, clorazepate and some other benzodiazepines are metabolized to the long-lived ($t_{½}$ equals 1 day to 1 week) active metabolite, nordazepam which is also known as desmethyldiazepam **(Figure 9.2)**. Alprazolam, midazolam and triazolam are metabolized to α-hydroxy metabolites which have short or intermediate durations of action **(Figures 9.2 and 9.3)**. Flurazepam is metabolized to active N-analogs. Hepatic cytochrome P_{450} enzymes play the major role in N-dealkylation, aliphatic hydroxylation and eventual glucuronidation, prior to excretion of benzodiazepines in urine **(Figures 9.2 and 9.3)**. The rate of metabolism is reduced in geriatric and liver-diseased patients, necessitating dosage adjustment of the benzodiazepines.

Triazolam has the shortest duration ($t_{1/2}$ less than 5 hours); alprazolam, lorazepam, oxazepam and temazepam have intermediate durations ($t_{1/2}$ equals 3 to 24 hours). Most other benzodiazepines have prolonged activity or are converted to active metabolites with long durations of action ($t_{1/2}$ equals 1 day to 1 week). If taken daily, these long-lived benzodiazepines would accumulate and produce cumulative (residual) effects.

Subjective effects to benzodiazepines diminish within a few hours of dosing, even though high blood levels are maintained long after this time. Therefore, it seems that redistribution of benzodiazepines from the CNS to adipose and other tissues accounts for the loss of effect **(Figure 9.3)**.

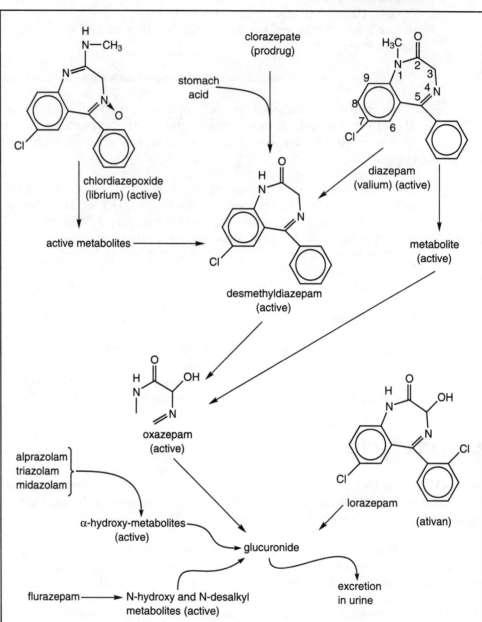

Figure 9.2

Mechanism of benzodiazepine action. Benzodiazepines act as agonists at the α subunit of GABA-gated Cl channels but at sites distinct from those that bind GABA or barbiturates **(Figure 9.4)**. Benzodiazepines potentiate GABA binding to the GABA receptor, increasing the frequency of Cl channel opening and thereby producing greater hyperpolarization and inhibition of nerve firing. Benzodiazepines have no action in the absence of GABA.

Pharmacological activity. The desired anxiolytic effect of benzodiazepines is the consequence of an action predominately on the limbic system (amygdala, hippocampus, etc.). This effect occurs in the relative absence of generalized CNS depression. Therefore, the degree of drowsiness is less than that observed with barbiturates, histamine H_1-blockers and tricyclic antidepressants. Nevertheless, sedation is the most frequently reported adverse effect of benzodiazepines. Benzodiazepines do not seem to produce overdose lethality (high therapeutic index), unless alcohol or other CNS depressants are taken concurrently.

Adverse effects of benzodiazepines
1. *Sedation* occurs in about half of patients treated with benzodiazepines. Patients may fall or have temporary memory impairments.
2. Paradoxical excitement, observed as anxiety and overactivity, is occasionally seen.
3. CNS depression is more severe when benzodiazepines are used with other CNS depressants such as alcohol or histamine H_1-blockers.
4. *Cimetidine* (Tagamet) increases the $t_{1/2}$ of diazepam. Monoamine oxidase inhibitors (e.g., tranylcypromine) enhance the CNS depressant effect of benzodiazepines.
5. Tolerance, psychological dependence and withdrawal symptoms can occur with benzodiazepines, particularly shorter-acting preparations. (With long-acting preps, high and stable levels of drug are maintained in the blood, so that the gradual metabolic decline in blood level is not readily perceived.) Cross-tolerance develops with alcohol and barbiturates.

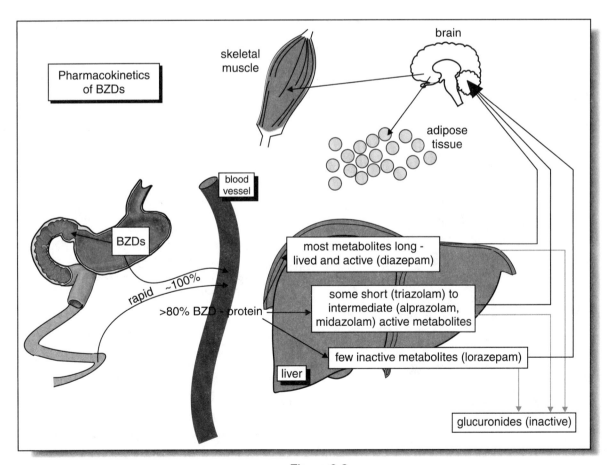

Figure 9.3

Uses of benzodiazepines

1. **Anxiolytic effect.** This is the major use for benzodiazepines. Diazepam (Valium) and chlordiazepoxide (Librium) are most often used for this purpose. Tolerance does not develop to this effect.

2. **Induction of sleep (hypnosis).** Alprazolam (Xanax), flurazepam (Dalmane), and temazepam (Restoril) and other short- to intermediate-acting preps are used to induce sleep. Since the duration of action of these particular drugs is less than 24 hours, cumulative effects should not occur.

3. **Preanesthetic medication.** Benzodiazepines are used for this purpose partly because they produce the desired anxiolytic effect, anterograde amnesia and muscle relaxation. However, their long duration of action is associated with prolonged postanesthesia depression.

4. **Endoscopic and short surgical procedures.** Midazolam (Versed), a water-soluble benzodiazepine, is sometimes administered i.v., when it is beneficial to have a conscious compliant patient that can follow instructions during these invasive procedures, while being amnestic to the experience.

5. **Muscle relaxation.** Diazepam is used to treat muscle spasms.

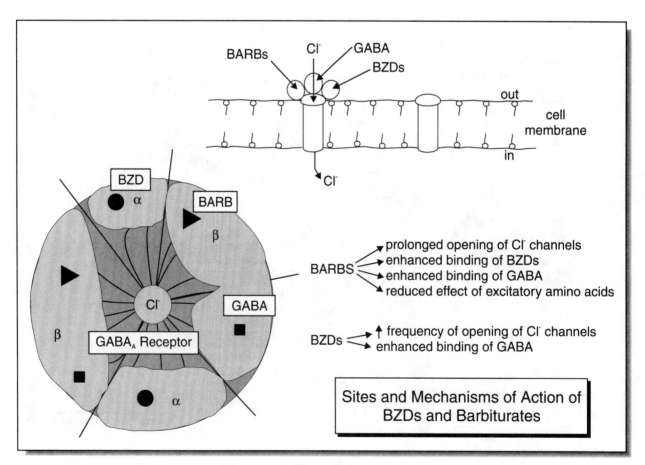

Figure 9.4

6. **Anticonvulsant effect**. Diazepam is used for status epilepticus and seizures induced by local anesthetics. Clonazepam is used as an antiseizure drugs. Benzodiazepines owe this action to a suppression of post-tetanic potentiation **(see later section on Anticonvulsants.)**

7. **Drug withdrawal.** Long-acting benzodiazepines are use to attenuate the symptoms of drug-dependent withdrawal from barbiturates, alcohol and short-acting benzodiazepines.

8. **Panic attacks.** These are effectively treated with alprazolam (Xanax).

Benzodiazepine antagonist. *Flumazenil* (Mazicon) binds to the benzodiazepine receptor on the a subunit of the $GABA_A$ receptor. Flumazenil displaces benzodiazepine agonists, but produces no effect of its own. This drug is used to treat benzodiazepine overdose and to reverse the sedation of benzodiazepine preanesthetic medication. Because of its short duration of action (less than 4 hours) repeated i.v. injections of flumazenil are likely to be needed. Patients may experience anxiety, panic attacks or seizures.

Benzodiazepine inverse agonist. Some β-carbolines bind to the benzodiazepine receptor and produce effects opposite to those observed with benzodiazepines. (Antagonists block an action and do not produce opposite effects.) These compounds are classified as inverse agonists (inverse is opposite to the action of agonists). Humans experience anxiety, panic and/or convulsions. Since these are not desired actions, there is no clinical use for these substances. They are of value for better understanding drug and receptor mechanisms.

TREATMENT OF ANXIETY

1. *Benzodiazepines* have been the drugs of choice for treating anxiety. Diazepam (Valium) and chlordiazepoxide (Librium) are the most frequently used agents. All CNS depressants also have an anxiolytic action. However, benzodiazepines produce an anxiolytic action with minimal sedation. Nevertheless, sedation is the most frequently reported adverse effect of benzodiazepines.

2. *Buspirone* (Buspar) is a non-benzodiazepine that is preferred by many for treating anxiety. Buspirone acts as a partial agonist at serotonin $5-HT_{1A}$ receptors, not benzodiazepine receptors. Buspirone is well-absorbed, undergoes extensive hepatic first-pass metabolism (95%) and is excreted in urine (2/3) and feces (1/3). A major advantage of buspirone is that it does not produce sedation. Like benzodiazepines, buspirone has a high therapeutic index and produces few adverse effects. Unlike benzodiazepines, buspirone does not enhance the CNS depressant actions of sedative-hypnotics like alcohol and barbiturates and does not induce tolerance or dependence. There is a latency of 1 to 3 weeks for the anxiolytic action of buspirone, in contrast to the more immediate action of benzodiazepines.

TREATMENT OF INSOMNIA

1. *Benzodiazepines* have been the drugs of choice for sleep-induction. *Alprazolam* (Xanax), *lorazepam* (Ativan) and *temazepam* (Restoril) are frequently used, because of their relatively short duration of action. This averts cumulative effects with daily dosing. Benzodiazepines increase the time spent in the stage of sleep characterized by non-rapid eye movement (non-REM sleep), so that sleep may not be as satisfying.

2. *Zolpidem* (Ambien) is a new non-benzodiazepine that acts at benzodiazepine receptors (mainly the BZ_1 subtype) to exert a sedative-hypnotic effect. This drug is well absorbed, rapidly metabolized in the liver, and excreted in urine. A short $t_{1/2}$ (2.5 hours) ensures that the drug does not accumulate with daily dosing. Since high dose zolpidem induces vomiting, there is little abuse potential. Like benzodiazepines, zolpidem produces few adverse effects and has a high therapeutic index and high safety margin. In high dose, zolpidem produces anterograde amnesia, as per benzodiazepines.

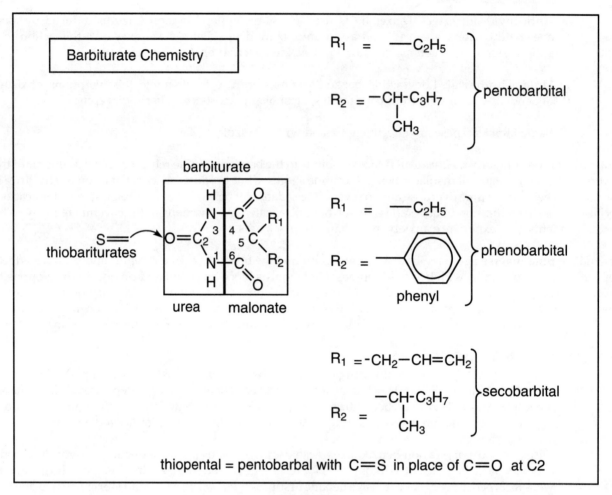

Figure 9.5

Flumazenil is an effective antidote for zolpidem. Zolpidem does not interact with cimetidine, imipramine, haloperidol or warfarin. Unlike benzodiazepines, zolpidem does not have anxiolytic, anticonvulsant or muscle relaxing properties. Zolpidem does alter the time spent in different stages of sleep. Not altering the sleep cycle, zolpidem may be a better agent than benzodiazepines for inducing sleep. Psychotic reactions have been reported with higher dose zolpidem.

3. Barbiturates and barbiturate-like drugs (e.g., meprobamate) were once widely used to promote sleep. These drugs have been supplanted by benzodiazepines. Tolerance and dependence develops to barbiturates, so there is a tendency to gradually increase the dosage. Also, the margin of safety for barbiturates is much less than for benzodiazepines.

C. BARBITURATES

This once, widely-used hypnotic class of drugs has been largely supplanted today by safe and effective benzodiazepines. However, barbiturate drugs have a niche in medicine, so it is important to understand their pharmacology.

Chemistry and its influence on activity. Barbituric acid can be visualized as a condensation product of urea and malonic acid **(Figure 9.5)**. Substitutions at R_1 and R_2 result in the common barbiturate drugs. These lipid-soluble drugs readily enter the CNS and produce generalized CNS depression. Substitution of S for O at C2 produces more highly lipid-soluble compounds that have short latency times (seconds to minutes). These latter drugs are often used to induce anesthesia (e.g., thiopental). They are rapidly converted to C=O, so that the parent C=S drug has a short duration, while the metabolite is active.

Pharmacokinetics of barbiturates. All are well-absorbed, protein-bound and readily distributed throughout the body. Thiobarbiturates, being highly lipid-soluble, rapidly enter the brain and produce anesthesia—an effect that is short-lived, owing to rapid redistribution from brain to adipose and other tissues. Oxybarbiturates have long half lives and tend to accumulate when taken daily for a sedative-hypnotic effect.

Barbiturates compete with some drugs for metabolism (e.g., tricyclic antidepressants). The more common effect is an enhancement of metabolism of many drugs, owing to barbiturate-induction of hepatic microsomal cytochrome P_{450} enzymes. Rates of drug metabolism may be increased twofold, and blood levels of drugs may be reduced to ineffective concentrations (e.g., phenytoin, dicoumarol).

Most barbiturates are completely metabolized by processes shown below **(Figure 9.6)**. Phenobarbital is an exception, with about 25% being excreted unmetabolized.

Mechanism of action of barbiturates. Barbiturates act on the β subunit of GABA-gated Cl channels, but at sites distinct from those that bind GABA on this subunit; and at sites distinct from those for benzodiazepines on the α subunit **(Figure 9.3 of Benzodiazepine Section)**. The site of barbiturate action is identical to the site at which picrotoxin acts. Barbiturates increase the duration of opening of the Cl channel and thereby potentiate the hyperpolarizing effect of GABA. The frequency of nerve firing is thus reduced.

Pharmacological actions of barbiturates

A. CNS actions

1. Barbiturates reduce the amount of REM sleep, so that sleep is less satisfying. An undesired consequence is that patients may continually increase their dosage in order to achieve "good" sleep.

2. Barbiturates produce generalized CNS depression, so that their anti-anxiety effect would be

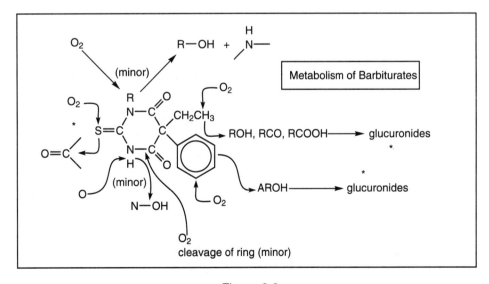

Figure 9.6

accompanied by a general feeling of tiredness. The CNS depression is enhanced by other drugs that depress the CNS, such as ethanol and H_1-blockers. (In this situation the lethal dose is greatly reduced.) Paradoxical excitation sometimes occurs, particularly in debilitated or geriatric patients, or in the presence of pain.
3. Pain perception is not suppressed by barbiturates. Small doses may produce hyperalgesia.

Tolerance. Barbiturates induce pharmacokinetic tolerance and pharmacodynamic tolerance. The former, of lesser magnitude, reflects the fact that barbiturates are potent inducers of cyt P_{450} enzymes that metabolize barbiturates. Thus, subsequent doses of barbiturates would be metabolized more rapidly. Pharmacodynamic tolerance reflects neuronal adaptations that render subsequent doses of barbiturates to be less potent. Because there is little tolerance to the lethal dose of barbiturates, the therapeutic index (TI = TD_{50}/ED_{50}) diminishes with continued use. Cross-tolerance to other CNS depressants occurs.

Adverse effects of barbiturates
1. Generalized CNS depression is associated with a feeling of drowsiness in most people treated with barbiturates.
2. Respiratory depression is due to
   a. suppression of the neurogenic drive in the reticular activating system,
   b. later suppression of the CO_2 drive (carotid body) and
   c. eventual suppression of the hypoxic (O_2) drive (medulla).

   Death occurs because of respiratory depression. In overdose toxicity respiration must be maintained artificially. Since barbiturates are weak acids, bicarbonate can be administered to alkalinize the urine and facilitate excretion of barbiturates. Hemodialysis would also enhance excretion.
3. Anesthetic doses of barbiturates enhance the actions of neuromuscular blocking drugs.
4. There is little effect of barbiturates on cardiovascular function.
5. Barbiturates induce γ-aminolevulinic acid (ALA) synthase, thereby increasing porphyrin synthesis. Consequently, barbiturates are contraindicated in acute intermittent porphyria (inability to inactivate porphyrins).
6. Allergic hypersensitivity, mainly dermal reactions can occur.
7. Tolerance, physical dependence and psychological dependence occurs and there is a tendency to gradually increase the dose. Barbiturate addiction is difficult to terminate, since withdrawal is severe. Particularly when taken with other sedative-hypnotics like ethanol, patients may inadvertently overdose and kill themselves.

Specific drugs and uses
1. **Anticonvulsant action.** *Phenobarbital* (Luminal) has a long duration of action and is used in low dose for prophylactic treatment of epilepsy. Although diazepam is the drug of choice for status epilepticus, phenobarbital could be used.
2. **Induction of anesthesia.** *Thiopental* (Pentothal) is ultrashort acting and is commonly used to induce anesthesia **(see Section A of this chapter)**.
3. *Secobarbital* (Seconal) has a long duration of action and was once a popular drug of abuse **(see section later in this chapter on Drugs of Abuse)**.
4. **Treatment of kernicterus.** Because phenobarbital potently induces hepatic glucuronyl transferase enzymes and bilirubin-binding protein, it is used to treat neonatal kernicterus and hyperbilirubinemia. Phethbarbital, a non-sedating barbiturate, is also used.

NON-BARBITURATE SEDATIVE-HYPNOTICS
1. Histamine H_1-blockers are occasionally used for sleep induction. Diphenhydramine (Benadryl) and hydroxyzine (Atarax, Vistaril) are among the more frequently used agents.

2. Meprobamate (Miltown) exerts depressant actions on the CNS, similar to that of the barbiturates. Tolerance, physical dependence and psychological dependence occur with this drug. This drug has largely been supplanted by benzodiazepines.

3. Chloral hydrate is a prodrug that is rapidly metabolized to the active species, trichloroethanol. Chloral hydrate has also been supplanted by benzodiazepines. Chloral hydrate was never popular, but was used to induce sleep. Like barbiturates, tolerance and dependence developed to chloral hydrate. A "Mickey Finn" is an alcoholic drink to which chloral hydrate has been added.

D. ETHANOL

Chemistry and Pharmacokinetics. Ethanol (EtOH) is a 2-carbon lipid-soluble alcohol that readily passes through lipid membranes. EtOH is well-absorbed from the GI tract. Absorption begins in the stomach, but most absorption occurs in the intestine because of its greater surface area. Peak levels occur in approximately 40 minutes, but food delays absorption and increases time to peak levels.

Distribution is to all body compartments, including brain and fetus. Because EtOH is a small molecule it is also water-soluble and its distribution follows body water.

Metabolism occurs mainly (about 95%) by liver dehydrogenases (deH$_2$ases) as shown:

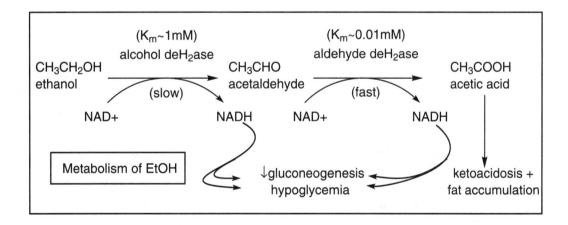

EtOH either saturates the deH$_2$ases or leads to depletion of NAD$^+$, since metabolism follows *zero-order kinetics* (i.e., constant rate, independent of serum concentration). The amount of deH$_2$ase is roughly proportional to liver size and lean body mass. Significantly, the deH$_2$ases are not induced.

There are important sex differences in alcohol metabolism. First, females have a greater proportion of body fat. Since EtOH distributes according to body water, the blood alcohol level would be higher in females versus males of the same weight. Second, females have approximately 50% less gastric alcohol deH_2ase, so there is less "first-pass metabolism" in females. This means that the same relative amount of EtOH would produce a disproportionately higher blood level in females versus males.

Only about 2% of alcohol is metabolized by cytochrome P_{450} enzymes. Although EtOH induces these enzymes, even a doubling of this rate would not substantially accelerate overall EtOH metabolism in regular drinkers. In regular drinkers EtOH metabolism by deH_2ase (e.g., about 95% metabolism of EtOH) would proceed at a rate identical to that in non-drinkers. Consequently, the overall rate of EtOH metabolism is similar in drinkers versus non-drinkers. EtOH-induction of cyt P_{450} enzymes translates into EtOH interactions with other drugs.

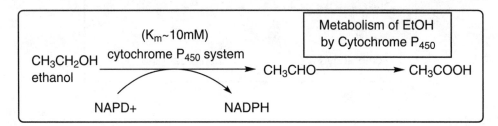

The urine concentration of EtOH is 1.3 times the blood concentration of EtOH. This is the basis of a urine sample sobriety test. The concentration of EtOH in expired air 0.05% times the concentration of EtOH in blood. This is the basis of the breathalyzer sobriety test. Both of these routes account for less than 3% of EtOH excretion.

Pharmacological actions of EtOH. Acutely, EtOH has only a few adverse effects. Chronically, EtOH has adverse effects on virtually all body systems.

Acute effects of EtOH
1. CNS effects are dose related, as outlined below:

mg% EtOH in blood	Effect
15	↓visual acuity
	↓motor coordination for fine skills
	euphoria, sedation, relief of anxiety, exaggerated emotions
	action on thermoregulatory center; flushing, warm skin, sweating, heat loss
100	legal definiton of inebriation in most places
	↑reaction time to light and sound (At 30 mph or 44 ft/sec this is a significant effect. After EtOH there is a six to sevenfold increased likelihood of an auto accident. Half of auto deaths are from drunk drivers.)
250	coma (Reduced analeptic effect of CO_2 on the respiratory center.)
400-900	Death (The lethal concentration of EtOH is much lower when benzodiazepines or other depressants have been taken.)

EtOH potentiates GABA inhibitory actions. Part of the CNS effect of EtOH may also be related to changes in nerve membrane fluidity, which then alters conduction and/or transmitter release.

2. **Liver**: Fat droplets appear in liver cells after one drink, due to an increased $NADPH/NADP^+$ ratio in liver cells and mobilization of fat from adipose tissue.

3. **GI tract**: EtOH increases histamine and gastrin secretion from gastric mucosa, thereby increasing gastric acid secretion and stomach ulcer formation.

4. **Kidneys**: Diuresis is related to reduced ADH release. (EtOH also suppresses oxytocin release from the posterior pituitary and will relax the uterus.) The acetic acid formed from EtOH metabolism competes with urate for secretory sites in renal tubules (urate excretion). EtOH may increase purine catabolism. These latter two actions promote acute gout.

5. **CDV system**: EtOH produces vasodilation and hypotension because of
    a. direct vasodilatory action of CH_3CHO on vascular smooth muscle and
    b. EtOH-induced depression of the vasomotor center. CH_3CHO produces arrhythmias and depression of myocardial contractibility.

    EtOH (1-2 drinks) causes an increase in HDL and a decrease in LDL (beneficial effects).

Chronic effects of EtOH

1. **CNS**: Demyelination, Wernicke's encephalopathy (related to thiamine deficiency), Korsakoff's psychosis, cerebellar degranulation

2. **Liver**: Hepatitis and cirrhosis occur and are likely to be related to
    a. increased O_2 consumption by liver cells and associated hypoxia at venus sinuses and
    b. CH_3CHO-enhanced lipid peroxidation and free radical production which depletes hepatic glutathione.

3. **GI tract**: Gastric ulceration may reflect the astringent action of EtOH on parietal cells. Pancreatitis also occurs. Constipation or diarrhea may reflect a poor diet.

4. **Kidneys**: Necrosis occurs.

5. **CDV/Skeletal muscle**: EtOH produces myopathies of skeletal and cardiac muscle. In moderation (1-2 drinks per day) EtOH is associated with a reduced incidence of coronary artery disease.

6. **Sexual function**: EtOH produces impotence and testicular atrophy by
    a. suppressing Leydig cell production of testosterone (i.e., reducing steroid hydroxylation), while enhancing hepatic catabolism of testosterone and
    b. damaging the liver, which reduces estrogen catabolism. Elevated plasma estrogen levels may induce gynecomastia.

7. **Immune system**: EtOH is associated with:
    a. decreased T cell number;
    b. decreased natural killer cell activity;
    c. decreased chemotaxis of granulocytes and decreased migration of WBC;
    d. decreased lymphocyte response to mitogens;
    e. vacuolization of precursors of RBC and WBC and
    f. thrombocytopenia.

8. *Longevity* is reduced in alcoholics, due in part to effects on the immune system. There is an increased mortality from cancers of mouth, larynx, pharynx, esophagus, lung, liver, breast, etc.

9. **Gestational effects of EtOH**: low birth weight, spontaneous abortions and stillbirths.

10. **Teratogenic effects** are displayed in the fetal alcohol syndrome (FAS):

a. microcephaly with retardation
b. flatten midfacial region: eyes wide apart, droopy eyelids, thin upper lip
A safe level of EtOH consumption by pregnant women has not been established.

Drug interactions with EtOH
1. Many drugs (e.g., sedatives/hypnotics, antidepressants, antipsychotics, opioids, etc.) potentiate the CNS depressant action of EtOH. A much lower amount of EtOH will produce coma and death.

2. Acutely, EtOH competes with many drugs for metabolism by the cytochrome P_{450} system (e.g., barbiturates, warfarin, phenytoin) and increases their $t_{1/2}$.

    Chronically, EtOH induces cytochrome P_{450} enzymes, thereby facilitating metabolism of many drugs and decreasing their $t_{1/2}$.

3. Some drugs have weak inhibitory actions on aldehyde deH_2ase, producing a disulfiram-like effect (oral hypoglycemics, cephalosporins, quinacrine).

4. EtOH potentiates effects of some drugs (e.g., vasodilators, antiplatelet action of aspirin).

Therapeutic uses of EtOH
1. **Topical uses**
    a. to cleanse the skin (e.g., ivy poisoning)
    b. to denature protein and toughen skin (50-70% EtOH) (to prevent bed sores)
    c. to reduce body temperature during high fever
    d. as an antiseptic (70%), (e.g., needle puncture)

2. **Systemic uses**
    a. as a mild sedative (e.g., an evening drink)
    b. as an aid to digestion (e.g., wine before dinner)
    c. to destroy nerves (e.g., trigeminal neuralgia)
    d. as an inhalant to collapse foam in airways during acute pulmonary edema

E. METHYL ALCOHOL (CH_3OH)

Chemistry and Pharmacokinetics. The pharmacokinetics of CH_3OH are like that of EtOH, but metabolism of CH_3OH occurs at about 1/7th the rate.

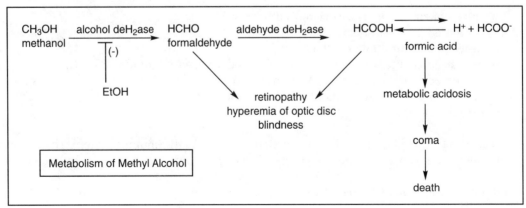

Metabolism of Methyl Alcohol

Acute toxic effects of CH_3OH
1. Metabolic acidosis causes CNS depression, dyspnea, coma and death. However, inebriation has a long latency (8 hours).
2. Retinal myopathy. (Diagnostic sign of CH_3OH poisoning is "snowstorm" vision.)
3. Pancreatitis with abdominal and back pain, diarrhea, vomiting.

Treatment of CH_3OH poisoning
1. Remove CH_3OH from the body by dialysis (especially if there is more than 50mg%) and gastric lavage.
2. Reduce the metabolism of CH_3OH by infusing EtOH which has higher affinity for alcohol deH_2ase and competes with CH_3OH as a substrate. This slows the rate of production of the toxic products of CH_3OH, namely HCHO and HCOOH.
3. Correct the acidosis. Infusion of $NaHCO_3$ (3g/hour) neutralizes H^+ and produces an alkaline urine which facilitates renal excretion of HCOOH.
4. Replace K^+.

F. OTHER ALCOHOLS

1. Long chain alcohols are more toxic than EtOH and produce greater CNS depression.
2. Ethylene glycol is the major component of antifreeze.

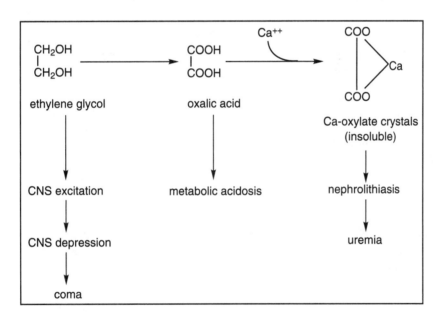

ESTIMATE OF BLOOD ALCOHOL LEVELS
1. A standard alcoholic drink has about 10g of EtOH.
   a. One oz of whiskey = (80 proof/2 = 40% EtOH)(30 ml) = 12 ml of 100% EtOH (12 ml) x (0.8 g/ml) = 10 g of EtOH.
   b. One mixed drink—same calculation.
   c. One 12-oz beer = (12 oz)(30 ml/oz)(3% EtOH)(0.8 g/ml) = 9 g of EtOH.
   d. Four oz wine = (4 oz)(30 ml/oz)(12% EtOH)(0.8 g/ml) = 11.5 g of EtOH.

2. In a 70kg (150 lb) lean healthy adult male, blood alcohol levels will be increased by about 20 mg % with each drink. (If he has 5 mixed drinks in one hour, his blood alcohol level will peak at about 100 mg %).

3. A 70kg (150 lb) lean healthy adult male or female metabolizes about 7g of EtOH per hour (i.e., 2/3 of a drink).
A 90kg (200 lb) lean healthy adult male or female metabolizes about 9g of EtOH per hour (i.e., 1 drink).
A 45kg (100 lb) lean healthy adult male or female metabolizes about 4.5g of EtOH per hour (i.e., 1/2 of a drink).

4. Total body water for lean males = 68%
Total body water for lean females = 55% (i.e., more weight is fat, proportionately.)

5. EtOH concentration in blood = $\dfrac{\text{Amount of EtOH}}{(\text{Body weight}) \times (\% \text{ of } H_2O)}$

6. Equation for calculating blood EtOH concentration in 90kg men:

$$\dfrac{(\text{Amount of EtOH consumed, g})}{(90 \text{ kg})(0.68)} - \dfrac{(9 \text{ g EtOH metabolized/hour}) \times (\text{hours})}{(90 \text{ kg})(0.68)} = ? \times 100 = \text{mg\%}$$

7. Equation for calculating blood EtOH conc in 45kg women:

$$\dfrac{(\text{Amount of EtOH consumed, g})}{(45 \text{ kg})(0.55)} - \dfrac{(4.5 \text{ g EtOH metabolized/hour}) \times (\text{hours})}{(45 \text{ kg})(0.55)} = ? \times 100 = \text{mg\%}$$

8. Sample problem. A lean healthy 200 pound man has 4 alcoholic drinks in 90 minutes time. What will be his blood alcohol level at the end of 2 hours?

$$\dfrac{(10 \text{ g of EtOH per drink})(4 \text{ drinks})}{(90 \text{ kg})(0.68)} - \dfrac{(9 \text{ g EtOH /hour})(2 \text{ hours})}{(90 \text{ kg})(0.68)} = 0.36 \times 100 = 36 \text{ mg\%}$$

9. His lean and healthy 100 pound girlfriend also has 4 alcoholic drinks at the same time. What is her blood alcohol level at the end of 2 hours? (Remember, the true blood alcohol level will be higher than your calculated answer, since women have less "first pass metabolism" than men.)

$$\dfrac{(10 \text{ g of EtOH per drink})(4 \text{ drinks})}{(45 \text{ kg})(0.55)} - \dfrac{(4.5 \text{ g EtOH /hour})(2 \text{ hours})}{(45 \text{ kg})(0.55)} = 1.25 \times 100 = 125 \text{ mg\%}$$

G. CLASSICAL CNS STIMULANTS

Mechanisms of action of these commonly abused drugs are outlined in **Figure 9.7**. Pharmacological actions are discussed more completely in the sections on Sympathomimetic Drugs **(Chapters 2 and 5)** and Drugs of Abuse **(later in this chapter)**.

A. Amphetamine, methamphetamine and methylphenidate

1. These displace norepinephrine (NE) from binding sites within sympathetic nerves, releasing excess NE which produces effects at most smooth muscles and glands in the body. The major adverse effects are (a) excess arterial constriction with markedly increased BP and (b) excess cardiac stimulation causing tachycardia or other arrhythmias.

2. These also displace dopamine (DA) from binding sites in DA nerves in the brain. Excess DA release produces euphoria.

Amphetamine was used in the past (a) in "diet pills" because of its appetite-suppressant property and (b) for narcolepsy, because of its anti-fatigue effect.

Amphetamine and methylphenidate are the principle drugs used to treat attention deficit hyperactivity disorder (ADHD), a childhood disorder characterized by overactivity and short attention span. The paradoxical effect of amphetamine to reduce general activity in children, contrasts with the stimulating effect of amphetamine in adults. This effect of amphetamine on activity is thought to involve DA release in the brain.

These three substances are drugs of abuse. In overdose, they produce a psychotic action with prominent paranoia.

B. *Cocaine* was once widely used as a local anesthetic agent. More recently it has become the most popular drug of abuse among the middle class. Cocaine is a potent and effective inhibitor of NE and DA transporters in nerves. The net effect is a block of uptake-inactivation of these neurotransmitters. The consequence is an effect similar to that of amphetamine:
1. Excess NE action on blood vessels and heart causes markedly increased BP and arrhythmias.
2. Excess DA action in brain causes euphoria, psychosis and paranoia.

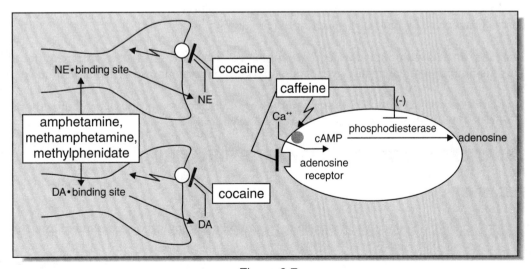

Figure 9.7

C. *Caffeine, theophylline* and *theobromine* are methylxanthines that are ingredients in coffee, tea and cola beverages. All stimulate or inhibit most smooth muscles in the body, by (a) blocking adenosine receptors, (b) inhibiting phosphodiesterase-inactivation of intracellular cAMP or (c) stimulating cellular Ca^{++} influx **(Figure 9.7)**. Tachycardia and other cardiac arrhythmias may also be produced.

These substances have weak CNS stimulatory effects which are sometimes manifested as mild tremor and anxiety.

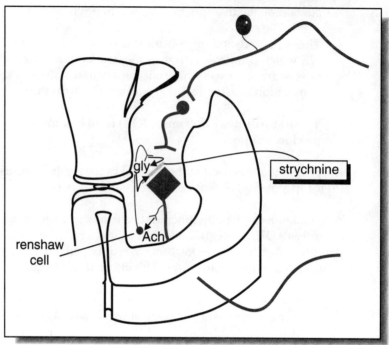

Figure 9.8

H. CNS TOXINS

The mechanism of action of this heterogeneous list of compounds is outlined below.

A. *Strychnine* is a glycine receptor antagonist that produces net excitation of neurons, consequent to disinhibition of glycine action **(Figure 9.8)**. This effect on spinal Renshaw cells is associated with violent *tonic convulsions* that may culminate in death (paralysis of diaphragm). Diazepam (i.v.) is a non-specific antagonist.

B. *Picrotoxin* is an antagonist at $GABA_A$ (gamma-aminobutyric acid) receptors **(Figure 9.9)**. This action produces net excitation of neurons, related to disinhibition of GABA action. *Tonic-clonic convulsions* ensue. Diazepam (i.v.) is a non-specific antagonist.

C. *Tetanus toxin* acts within glycine nerves to inhibit release of the glycine neurotransmitter **(Figure 9.10)**. This disinhibition at the spinal level results in tonic convulsions, as per the strychnine effect.

D. *Botulinum toxin* acts within cholinergic nerves to inhibit release of the neurotransmitter, *acetylcholine* **(Figure 9.11)**. Death is a consequence of respiratory paralysis.

I. TREATMENT OF NEUROLOGIC AND PSYCHIATRIC DISORDERS

1. EPILEPSY

Epilepsy (approximately 0.5% incidence) is the most common neurological disorder after stroke. About 75% of epilepsies can be effectively controlled by drugs.

Classification of epileptic seizures

A. *Partial (focal) seizures* are those in which a specific locus in the brain initiates seizure activity. This is correlated with the maximal electroshock (MES) test in animals.
   1. Simple
      a. no loss of consciousness (e.g., Jacksonian type)
      b. minimal spread of the discharge in brain
      c. a specific body region is affected

   2. Complex
      a. loss of consciousness
      b. spread of discharge throughout brain (bilateral)
      c. possible *aura*
      d. sometimes expressed as *automatisms* (tics, etc.) with retrograde amnesia

   3. Progressive, to generalized seizures

B. *Generalized seizures* are those produced by the discharge of multiple groups of neurons at no specific locus in brain. Loss of consciousness occurs. This is correlated with pentylenetetrazol (Metrazol) seizures in animals.

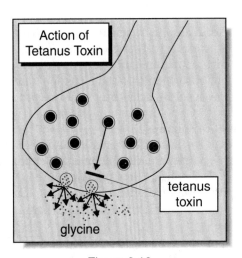

Figure 9.10

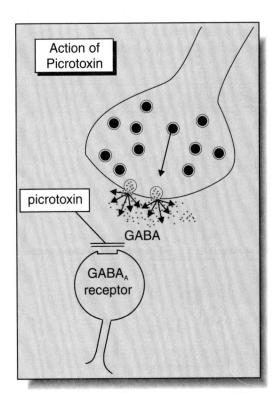

Figure 9.9

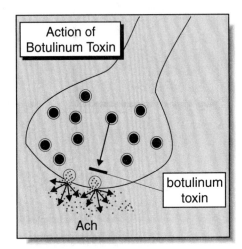

Figure 9.11

1. **Absence (Petit mal)**—brief (less than 45 seconds), accompanied by 3 Hz spike-wave EEG pattern. Sometimes clonic jerking or automatisms occur.

2. Tonic =

3. Clonic =

4. Tonic-Clonic(Grand Mal)

5. **Myoclonic**—brief burst of contractures of upper limbs and head

6. **Atonic**—brief loss of muscle tone (falling)

C. **Status epilepticus**—life threatening seizures, often tonic-clonic

General note. When antiseizure drugs are used in combination, they may compete for metabolism, so that excessively high levels of each drug may inadvertently occur. Conversely, phenobarbital or carbamazepine induce cytochrome P_{450} enzymes. When antiseizure drugs are combined with phenobarbital or carbamazepine, rates of drug metabolism would be increased, such that excessively low levels of each drug could be produced.

Treatment of seizures
1. **Partial**: Carbamazepine and phenytoin are often more effective than phenobarbital and primidone. Valproate, clonazepine and clorazepate are often less effective.
2. **Absence**: Ethosuximide is used over valproate and clonazepam, which are more widely used than trimethadione.

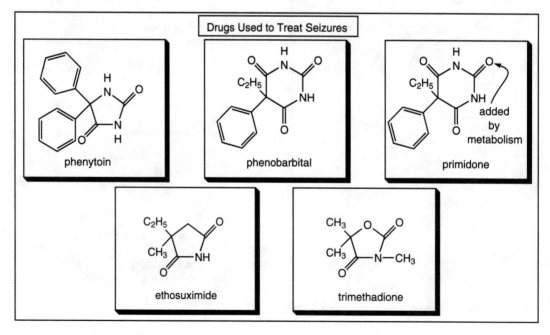

Figure 9.12

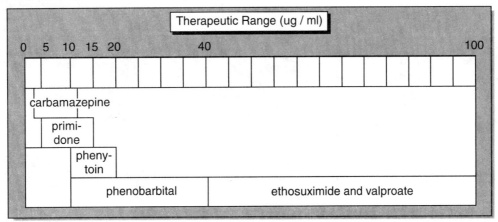

Figure 9.13

3. **Grand mal**: Carbamazepine, phenobarbital, phenytoin and valproate are often more effective than primidone.
4. **Myoclonic, atonic**: Clonazepam, clorazepate and valproate are used if there are concurrent absence seizures.
5. **Status Epilepticus**: Diazepam or lorazepam are more suitable than phenytoin or phenobarbital.

Summary of anti-epileptic drugs. Nearly 100% of these drugs are absorbed. Except for phenytoin, valproate and the benzodiazepines, anti-epileptic drugs are not appreciably bound to plasma proteins. All are metabolized by hepatic cyt P_{450} enzymes. Rates of biodegradation are slow for all of the drugs and zero order kinetics apply (except for phenytoin). Distribution follows total body water. Durations of action are less than 12 hours. There is a range of effective plasma concentration for each drug (lower limit is the threshold level; upper limit is an increase in adverse effects) **(Figure 9.13)**. The chemical structures are shown in **Figure 9.12**.

ANTICONVULSANTS

A. Phenytoin

Pharmacokinetics (Figure 9.15). Although about 90% of phenytoin is absorbed, the rate of absorption is highly variable and peak levels are attained between 3 and 12 hours. Phenytoin is bound to plasma proteins (90%) and displaces other bound drugs. Inactivation occurs via hydroxylation and glucuronidation **(see Figure 9.14)**. Virtually all of the drug is ultimately me-

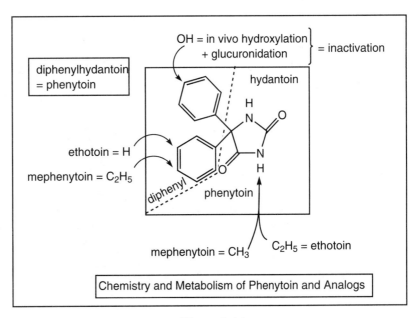

Figure 9.14

tabolized, initially by first order kinetics (rate is proportionate to amount); later, by zero order kinetics (enzymes are saturated, so rate is independent of amount) ($t_{1/2}$ about 1 day). Phenytoin induces cytochrome P_{450} enzymes, so metabolism of other drugs may be enhanced.

Mechanism (Figure 9.16). The effects of phenytoin on Ca, Na and K ionic fluxes, alter the nerve E_m and TP. These actions inhibit the spread of seizure activity. Since the seizure focus remains unaffected by phentoin, the aura may persist.

Adverse effects include
  a. gingival hyperplasia (about 20%) and hirsutism;
  b. diplopia and ataxia and
  c. cardiac arrhythmias (uncommon).

Mephenytoin (mesantoin) and ethotoin (peganone) have uses similar to phenytoin, but produce less gingival hyperplasia **(see Figure 9.14)**. However, they are less effective (ethotoin) or have more potential for adverse effects (mephenytoin).

B. Carbamazepine (Tegretol)

Chemistry. Carbamazepine is a *tricyclic* (3 rings) that resembles phenytoin and the antidepressant, imipramine. (Carbamazepine is likewise used for manic-depressive disorder.) This drug is oxidized *in vivo* as shown in **Figure 9.17**.

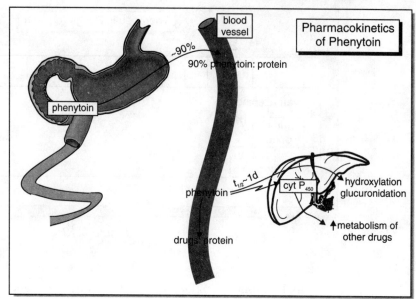

Figure 9.15

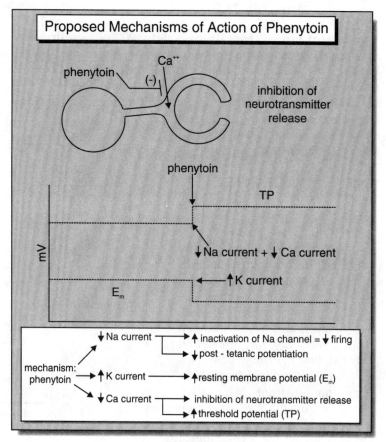

Figure 9.16

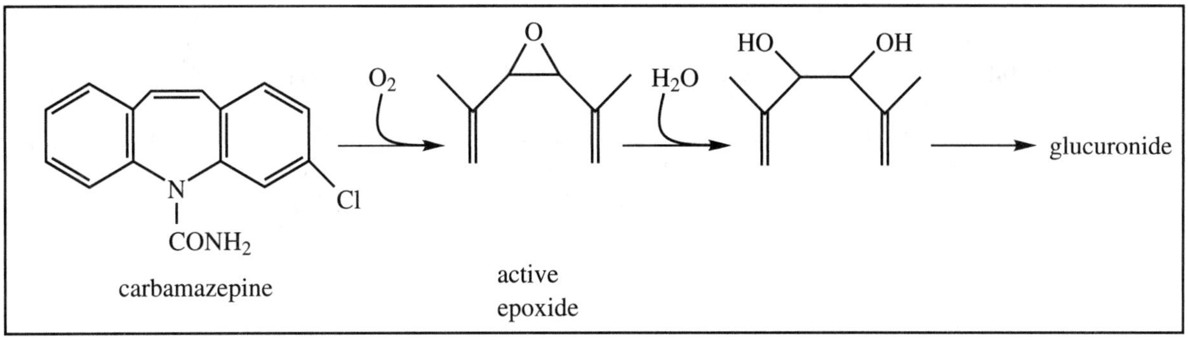

PHARMACOLOGY OF THE CNS

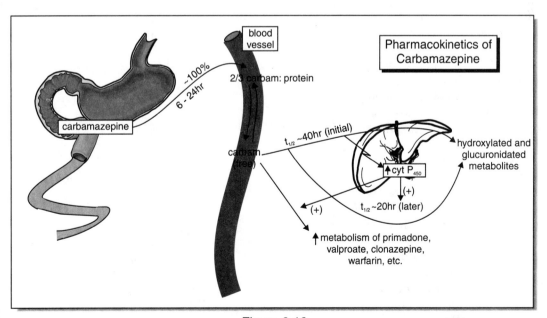

Figure 9.17

Pharmacokinetics. Carbamazepine is absorbed slowly at a variable rate. Peak plasma levels occur between 6 and 24 hours. About 2/3 is protein-bound, although interactions with other protein-bound drugs generally do not occur. Hydroxylated and glucuronidated metabolites are formed, initially with a $t_{½}$ approximately 40 hours and later with a $t_{½}$ approximately 20 hours, consequent to induction of cytochrome P_{450} enzymes **(Figure 9.18)**. This latter effect enhances clearance of other anti-seizure drugs (and many other drugs) that might be co-administered with carbamazepine. Drugs like cimetidine that inhibit cyt P_{450}, prolong the $t_{½}$ of carbamazepine.

Mechanism. Carbamazepine has an action on Na current and on post-tetanic potentiation like that of phenytoin.

Adverse effects include
a. diplopia and ataxia—like phenytoin;
b. idiosyncratic blood dyscrasias (agranulocytosis) and dermal reactions (exfoliative dermatitis) and

Figure 9.18

c. cardiac arrhythmias—uncommon, as per phenytoin.

C. **Phenobarbital.** Other actions are discussed earlier in this chapter **(see Hypnotics).**

Chemistry. The different anticonvulsant barbiturates are illustrated below. Phenobarbital is formed *in vivo* from both mephobarbital (by demethylation) and primidone (by oxidation). Note the similarity in chemical structure between phenobarbital and phenytoin **(see Figure 9.12).**

Pharmacokinetics (see Section C, Barbiturate Sedative-hypnotics).

Mechanism. Phenobarbital has effects on ion currents similar to that of phenytoin. The spread of seizure activity is inhibited. In addition, phenobarbital enhances GABA-mediated Cl channel opening while inhibiting effects of excitatory amino acids **(see Figure 9.4).**

Adverse effects. These are discussed in the section on Hypnotics, earlier in this chapter. Prominent adverse effects of antiseizure doses include
  a. *sedation* and
  b. exacerbation of petit mal and psychomotor seizures.

Mephobarbital (Mebaral) and metharbital (Gemonil) have actions and uses similar to phenobarbital.

D.   Primidone (Mysoline)

Chemistry. Primidone is converted *in vivo* to phenobarbital (by hydroxylation) and to phenylethylmalonamide (PEMA) (by dealkylation). Each of these compounds has anticonvulsant activity.

Pharmacokinetics. Primidone is well-absorbed, attains a peak plasma level at approximately 3 hours, and distributes according to body water. Primidone is largely unbound to plasma protein. Inactivation occurs via hydroxylation and eventual conjugation ($t_{1/2}$ approximately 8 hours).

Mechanism. Primidone has effects on ion currents similar to that of phenytoin. Adverse effects are similar to those of *phenobarbital*.

E.   Ethosuximide (Zarontin)

Chemistry. The structure of ethosuximide and analogs is shown below.

Pharmacokinetics. Ethosuximide is irritating to the GI mucosa and is usually administered in capsules to avoid this problem. Absorption is complete, peak levels occur in about 5 hours, and distribution follows body water. Protein binding does not occur. Metabolism is by hydroxylation and conjugation, with a $t_{1/2}$ of about 40 hours.

Mechanism. Ethosuximide reduces low-threshold (T-type) Ca current in pacemaker thalamic neurons, the origin of absence seizures. This elevates the threshold for seizure discharge.

Methsuximide (Celontin) is more toxic than ethosuximide; phensuximide (Milontin) is less effective than ethosuximide.

F.   Valproic acid and sodium valproate (Depakene)

Chemistry
$(CH_3CH_2CH_2)_2CHCOOH$ Valproic acid

$H_2NCH_2CH_2CH_2COOH$ GABA

Pharmacokinetics. Valproate is rapidly absorbed unless enteric-coated preps are used. Peak levels occur at about 3 hours. About 90% is plasma protein-bound, competing with phenytoin and other protein-bound drugs. Metabolism occurs slowly ($t_{1/2}$ approximately 12 hours) by oxidation and

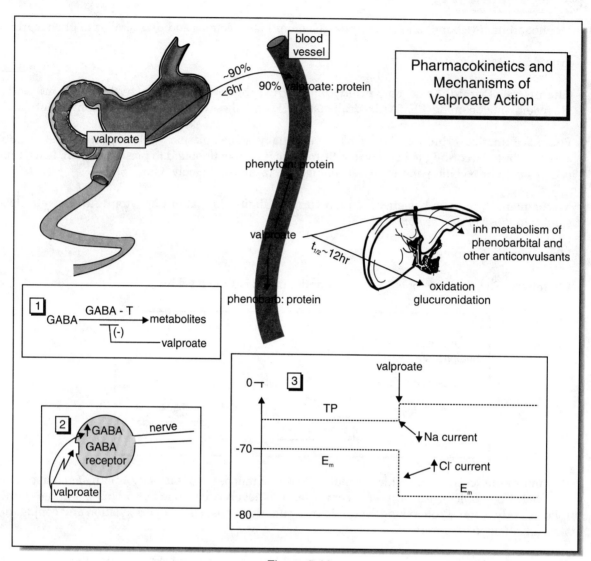

Figure 9.19

glucuronidation. Valproate inhibits the metabolism of phenobarbital, phenytoin, ethosuximide and carbamazepine **(Figure 9.19)**.

Mechanism. Valproate may mimic GABA, enhance its activity and/or elevate GABA levels by inhibiting the degradative enzyme, GABA-transaminase (GABA-T). Nerve membranes tend to become hyperpolarized and less active, and post-tetanic potentiation is inhibited (Na channels are inactivated) **(Figure 9.19)**.

Adverse effects
- a. GI pain, vomiting and diarrhea; tolerance occurs
- b. sedation, especially if combined with phenobarbital
- c. tremor and ataxia
- d. fatal idiosyncratic hepatotoxicity (1:25,000), especially if used with other anticonvulsants in young patients

G. Trimethadione (Tridione)

Chemistry

Pharmacokinetics. Trimethadione is rapidly absorbed and peak levels occur in less than 2 hours. Trimethadione is demethylated to the active metabolite, dimethadione, by hepatic cytochrome P_{450} enzymes. Trimethadione and dimethadione are not bound to plasma proteins. Dimethadione ($t_{½}$ approximately 2 weeks) is excreted unchanged.

Mechanism. Trimethadione, like ethosuximide, reduces T-type Ca current in pacemaker neurons, thereby elevating the threshold for seizure discharge.

Adverse effects
   a. sedation and ataxia
   b. ocular sensitivity to light
   c. dermal reactions (rash, dermatitis)
   d. blood dyscrasias (neutropenia, pancytopenia) and nephrotic syndrome

H. Benzodiazepines. These are discussed in detail earlier in the chapter **(see Anxiolytic Drugs)**. Only antiepileptic effects of benzodiazepines are described here.

Diazepam (i.v.) or lorazepam are drugs of choice for immediate control of status epilepticus. This is usually followed within 30 minutes by phenytoin or phenobarbital, to achieve stable control of seizures.

Diazepam, clonazepine and clorazepate are sometimes used to treat epilepsy, but often as an adjunct to other anti-epileptic drugs.

I. Felbamate is a newly approved drug for adults with partial seizures and children with Lennox-Gastaut syndrome. Its mechanism may involve reduction of Na current. This drug is rapidly absorbed, only 25% bound to plasma protein and has a long duration of action ($t_{½}$ about 1 day). Felbamate stimulates and inhibits cyt P_{450} enzymes, so there are significant interactions with other anti-epileptics. The drug produces few adverse effects, although blood dyscrasias have been reported.

Gabapentin and lamotrigine are expected to be approved as anti-epileptics. These act as GABA-mimetics. Gabapentin is not appreciably metabolized and does not interfere with metabolism of other drugs.

2. ANTIPARKINSONIAN DRUGS

The primary lesion in Parkinsonism is in the pars compacta of the substantia nigra (i.e., "black substance") (SN), where melanin-containing neurons degenerate. This deficit is correlated with a deficiency of dopamine (DA) in the basal ganglia. Lesions also occur elsewhere in the brain.

Etiology
1. Idiopathic for the majority of people.
2. Von Economo's encephalitis lethargica, pandemic from 1918-1926.
3. Atherosclerosis, stroke (often unilateral).
4. Drug-induced reversible forms (e.g., reserpine, phenothiazines, butyrophenones).
5. Poisoning by Co^{++}, Mn^{++}, Cu^+, CN^-, CO.
6. Possible environmental toxins that destroy the SN. MPTP (1-methyl-4-phenyl-1,2,3,6-tetrahydropyridine) was discovered as a contaminant in a drug of abuse, that produced irreversible Parkinsonism in humans. A single treatment of primates with MPTP can produce such overt destruction of cells in the SN that Parkinsonian symptoms can be manifest in hours. The mechanism is shown in **Figure 9.20**.

Objective of drug therapy today. Drugs are used to produce symptomatic relief by modifying the functioning of central dopaminergic or cholinergic neurons. No drug is curative.

The actions of the antiparkinsonian drugs are shown in **Figure 9.21**. As a review, DA is synthesized as a precursor of NE in sympathetic nerves. In DA nerves *per se*, only tyrosine hydroxylase (T-OH) and amino acid decarboxylase (AADC) enzymes are present, so DA is a final product that is stored in granules.

Treatment of Parkinsonism
1. *Levodopa* (Bendopa, Larodopa) is orally absorbed, reaching peak levels in less than 2 hours ($t_{1/2}$ approximately 1 hour). Because of extensive metabolism in peripheral tissues (liver, sympathetic nerves, etc.), less than 1% gets into the CNS. The objective is to load nigrostriatal DA nerves with L-DOPA, so that supernormal amounts of newly formed DA (from L-DOPA) will be released into

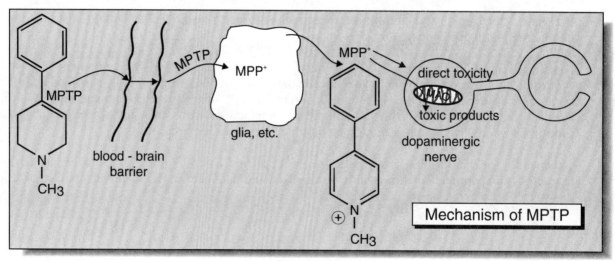

Figure 9.20

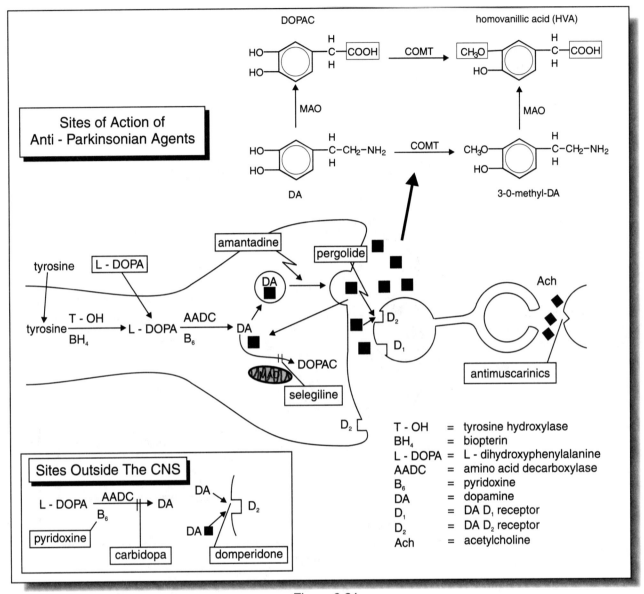

Figure 9.21

the synapse when these nerves are active. The site of L-DOPA action is shown in **Figure 9.21**, while effects are shown in **Figure 9.22**.

It is notable that excess *pyridoxine* (vitamin B_6) reduces the effect of L-DOPA by enhancing conversion of L-DOPA to DA at sites throughout the body, most of which are outside the brain (**Figure 9.21**, left corner inset). As a consequence, less L-DOPA is available for substantia nigra cells in brain.

2. *Carbidopa and L-DOPA* (Sinemet) is the most common form of L-DOPA today. Carbidopa blocks conversion of L-DOPA to DA in liver and other tissues outside the brain (**Figure 9.21**). Carbidopa cannot get into the brain, so L-DOPA is still converted to DA in DA nerves.

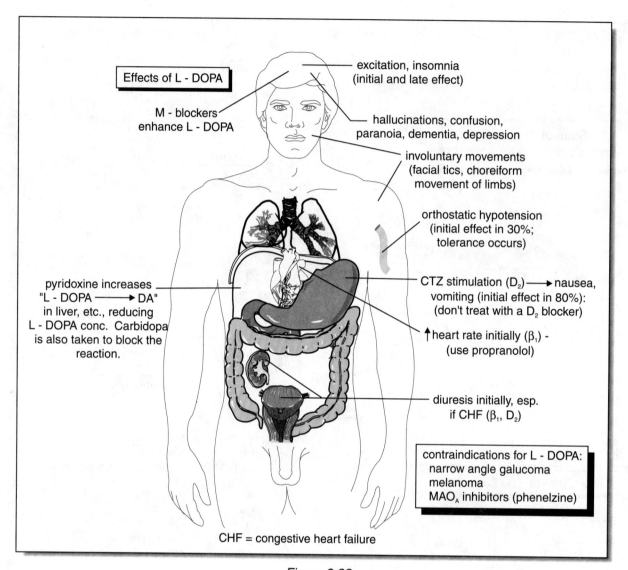

Figure 9.22

Advantages over L-DOPA alone
a. Since DA formation from L-DOPA is inhibited outside the brain there are less nausea, vomiting and CDV effects, as DA D_2 sites in the chemoemetic trigger zone (CTZ) and CDV system will not be stimulated.
b. Less L-DOPA is needed, so dosage can be reduced.
c. Pyridoxine excess does not interfere, since liver breakdown of L-DOPA is inhibited.

Limitations of L-DOPA
a. Initial benefits occur after several weeks since the desired dosage can only be attained gradually, as patients become tolerant of side effects. (Carbidopa shortens this period.)
b. Optimal benefits persist for only 1-5 years, for unknown reasons.
c. **Adverse effects** are more frequent with prolonged use. These include dyskinesias **(Figure 9.22)**, "end of dose akinesia" (a wearing off of L-DOPA effects) and "on-off phenomena" (unpredictable periods of akinesia between L-DOPA doses).

For these reasons L-DOPA is often reserved as the last treatment, even though it alleviates Parkinsonian symptoms better than any other drug.

3. *Amantadine* (Symmetrel) enhances DA release from DA nerve terminals. Amantidine is mildly effective and is good as initial treatment. Adverse effects are mild but include excitation, insomnia, confusion and hallucinations (like L-DOPA). Coma and seizures as well as "livido reticularis" can also occur. About 50% is excreted unmetabolized in urine, so excess blood levels occur if renal function is impaired. Amantidine is also used as an antiviral agent.

4. *Pergolide* (Permax) and *bromocriptine* (Parlodel) are agonists acting directly at the DA D_2 receptor (**Figure 9.21**). Bromocriptine seems to work only in patients that respond to L-DOPA. Bromocriptine can be used in patients unable to tolerate L-DOPA or can be used as an adjunct to L-DOPA, allowing L-DOPA dose to be reduced. Bromocriptine produces hallucinations, hypotension and other adverse effects. Generally, pergolide has fewer side effects.

    With the D_2 agonists and sometimes with other antiparkinsonian drugs, the DA D_2 receptor antagonist, domperidone, is sometimes co-administered to block D_2 sites in the CTZ, and thereby reduce the incidence of nausea and vomiting. Since block of D_2 receptors in the brain would induce parkinsonian symptoms, domperidone is useful only because it does not get into the brain.

5. *Selegiline* (Eldepryl) is a selective inhibitor of MAO_B, an enzyme that breaks down DA in the brain. By inhibiting DA metabolism, DA concentrations stay elevated. However, since this drug does not provide good symptomatic relief of Parkinsonism, this action may not be so important. Selegiline is the only drug reputed to slow the progression (worsening) of Parkisonism, presumably by preventing formation of free radicals (HO•, etc.) that arise from DA breakdown via MAO_B. It is also relevant that selegiline increases activities of superoxide dismutase and catalase, enzymes that degrade reactive oxygen species. This drug is widely used in combination with other Parkinsonian drugs.

6. *Antimuscarinics* have been used in Parkinsonism for over 100 years. They are useful, but produce predictable side effects in the elderly patients (e.g., confusion, anhydrosis, cycloplegia, urinary retention, constipation, etc.). Examples like trihexyphenidyl (Artane) and benztropine (Cogentin) are often used with other antiparkinsonian drugs.

7. H_1-blockers, like diphenhydramine (Benadryl) and orphenadrine (Disipal), have a spectrum of activity like the antimuscarinics. This latter action may account for their usefulness in parkinsonism.

Summary. All of the antiparkinsonian drugs are orally effective and are used in combination. Generally, L-DOPA should be reserved for more serious cases.

3. ANTIPSYCHOTICS (NEUROLEPTICS)

Nearly 1% of the global population has schizophrenia. It is estimated that a high percentage of homeless people have schizophrenia. This disorder is characterized by paranoia, psychosis and social withdrawal. In the past schizophrenics were often institutionalized, sometimes for life, because of their inability to cope in family and neighbor settings. Starting in the 1950s, antipsychotic drugs have provided the means for schizophrenics and psychotics to lead a more meaningful life.

Chemistry. Since antipsychotics (APs) act in the brain it is predictable that these drugs are highly lipid soluble molecules. Their general structure is outlined:

```
lipophilic end  ─────────────────────              R_2
(aromatic rings)              ↑            amine end (RNR_1 or RNR_1)
                       bulky lipophilic
                         groupings
```

Older classes of APs like phenothiazines have low potency and low receptor specificity. Newer classes like butyrophenones and benzamides have high potency, which is related to their high affinity and relative specificity for dopamine (DA) D_2 receptors. Relative potencies of the AP classes are illustrated in **Figure 9.23**.

[Chlorpromazine structure with metabolic pathways labeled: sulfoxidation, glucuronidation, hydroxylation, N-oxidation, deamination, demethylation]

Pharmacokinetics. Absorption is erratic but peak levels occur in 2-4 hours. Distribution is to all tissues and across the placenta. APs are highly protein-bound (90-99%), so the volume of distribution is high. Metabolism is extensive (90-99%) and occurs by a variety of enzymes, primarily in the liver. There is extensive first-pass metabolism of some of the parent drugs. However, the rate of metabolism is generally low, so that most antipsychotics have a long $t_{1/2}$. Some metabolites are also long-lived and active. Antipsychotics interfere with metabolism of many drugs. Excretion is in urine and bile.

Mechanism of action. All APs block the effects of DA at the DA D_2 receptor complex (D_2, D_3, D_4 subtypes). The dosage of any AP is inversely-correlated with its affinity for D_2 receptors.

Drugs like amphetamine release excess DA and produce a paranoid state. Drugs like reserpine attenuate DA nerve activity by depleting DA and produce an antipsychotic effect. Numbers of D_2 receptors are increased in the brain of some schizophrenics. For all of these reasons DA nerves are thought to have a major role in psychosis and schizophrenia (i.e., DA hypothesis of schizophrenia).

Although APs block most D_2 receptors after the first dose, the antipsychotic effect with altered thought process occurs only after 10-14 days. In this time interval neural adaptations to D_2 receptor block is thought to occur. Initial AP doses cause DA release; prolonged AP therapy inhibits DA release. The *limbic system* represents the site that is most important for AP action.

Significance of AP therapy. These drugs make it possible for psychotics to be managed as outpatients, thereby keeping the number of institutionalized patients to a minimum. However, when patients do not take their

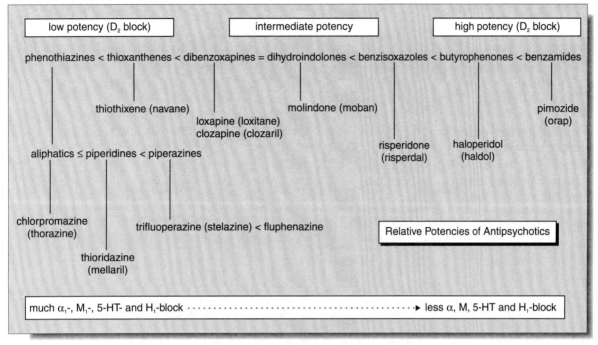

Figure 9.23

medication they revert to their psychotic state and become a threat to society. An unfortunate overuse of antipsychotics has occurred in some geriatric nursing homes.

Limitation of first generation APs (Classical or conventional APs). Positive symptoms like paranoia, aggression, delusions and hallucinations can be reasonably controlled by these APs. However, negative symptoms like emotional blunting and apathy are not relieved by the first generation APs.

Desired effect of APs. The benefit of APs is related to DA D_2 block in the mesolimbic system. Patients become less paranoid, less agressive and hallucinate less.

Adverse effects of first generation APs caused by DA D_2 block.
1. **Mesocortical system**. Thinking is not impaired, but sedation is produced.
2. **Tuberoinfundibular system**. Undesired endocrine effects including hyperprolactinemia occur (Figure 9.24).
3. **CTZ**. An antiemetic effect occurs.
4. **Nigrostriatal system**. Undesired Parkinsonian symptoms occur:
   tremor
   rigidity   These can be eliminated by reducing the
   akathisia   AP dose or adding an M-blocker.
   catatonia

Also, *dystonias* and *dyskinesias* occur: torticollis and tardive dyskinesia.

Tardive dyskinesia is perhaps the most dreaded adverse effect of APs. This is a choreoathetoid movement disorder, usually long-lived and most-commonly is characterized by oral dyskinesias like lip smacking and tongue thrusting. The risk is thought to be related to high dose and prolonged therapy (greater than 1 year). There is approximately a 5% incidence with each year of continuous treatment. For example, there is a 15% risk factor for patients treated for 3 years with first generation APs.

Adverse effects of first generation APs not related to DA D_2 block (Figure 9.25). First generation APs block receptors for several neurotransmitters, thereby producing an assortment of adverse effects. This spectrum of unpleasant effects is a factor in patients becoming non-compliant (not taking their medication). The neuroleptic malignant syndrome is thought to be related to elevated Ca^{++} concentration in skeletal muscle, which causes muscle rigidity, hyperpyrexia resulting from excess actin-myosin interactions, leukocytosis and autonomic instability. This may occur early in treatment or even

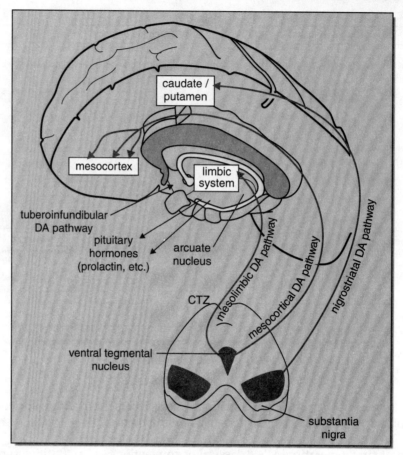

Figure 9.24

in patients that have been responding well to long-term AP therapy. The reason for the onset of neuroleptic malignant syndrome is not well-understood.

Drug interactions of first generation APs
1. Additive CNS depression with sedatives, etc.
2. Additive muscarinic block with tricyclic antidepressants, etc.
3. Counteraction of L-DOPA and other anti-Parkinson drugs.
4. Competition with many drugs for protein-binding sites, with activation of those drugs (e.g., NSAIDs, oral anticoagulants, oral hypoglycemics, etc.).
5. The quinidine-like effect of thioridazine is additive with other such drugs.

Second generation APs. In addition to their effects on positive symptoms of schizophrenia, second generation APs have a beneficial effect on negative symptoms. Patients feel much better after second generation versus first generation drugs. Another important feature is that second generation APs, in usual doses, do not produce a Parkinsonian syndrome.

The second generation drugs potently block DA D_2 and $5-HT_2$ receptors, and this combination is thought to be important for the beneficial profile. In addition, some second generation APs are not potent inhibitors of H_1, M and α receptors. Accordingly, there are fewer adverse effects of second generation APs. Hyperprolactinemia (D_2 block) is still a problem.

1. *Clozapine* (Clozaril) is the most potent antagonist of DA D_4 receptors, a subtype of the D_2 receptor. However, this may not be important in its antipsychotic effect.

   The major disadvantage of clozapine is the high incidence of *agranulocytosis* (1% of patients). For this reason, clozapine is approved only for patients refractory to other APs or patients unable to tolerate other APs. Clozapine is effective in a high percentage of patients that are refractory to first generation APs. Required routine blood monitoring, and high drug cost, make clozapine treatment expensive.

   Other adverse effects of clozapine include seizures (3%) and antimuscarinic actions: tachycardia (25%), sialorrhea (30%) and sedation (40%).

2. *Risperidone* (Risperdal) and its longer-lived ($t_{1/2}$ approximately 1 day) 9-hydroxy metabolite are active. In poor metabolizers, the $t_{1/2}$ of risperidone is approximately 1 day; in rapid metabolizers, about 3 hours.

   Adverse effects. These include initial orthostatic hypotension and reflex tachycardia. Risperidone, like all other antipsychotics, causes sexual dysfunction. Unlike clozapine, risperidone does not produce agranulocytosis.

Uses of APs
1. Psychoses: schizophrenia, schizoaffective disorders, acute mania
2. Tourette Syndrome
3. Senile dementia of the Alzheimer Type
4. Vomiting is effectively treated by prochlorperazine and promethazine (Phenergan), owing to their block of D_2 receptors in the CTZ. Neither of these drugs is used as an AP.

Summary. Unlike antiparkinsonian drugs, APs are not combined with one another. All of the first generation APs are about equally efficacious. The second generation drugs are more efficacious than first generation drugs, which accounts for the new designation as "second generation." A patient is generally given an AP for 6 months or longer. Some patients failing to respond to one drug may respond well to another.

4. TREATMENT OF DEPRESSION

There is a 10% lifetime risk for clinical depression. At any time approximately 1% of the global population suffers depression. About 15% ultimately commit suicide.

Depression is classified in several different ways:
1. *Endogenous depression* is insidious in onset and is seemingly unrelated to any identifiable event in one's life.
2. *Reactive depression* represents a mood change subsequent to a specific identifiable event in one's life (e.g., death in a family).

Depression may be subcategorized as follows:
1. *Agitated depression* is associated with overactivity, fidgety actions with purposeless unproductive tasks.
2. *Retarded depression* is associated with social withdrawal and inability to undertake even routine tasks.

In either case there is little reaction to other people or events. Both cases involve emotional blunting.

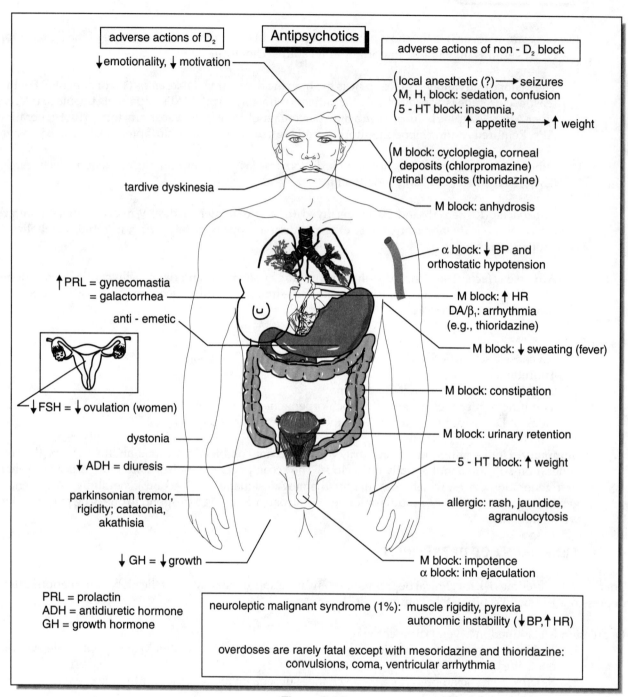

Figure 9.25

Endogenous depression is either *unipolar* or *bipolar*. In unipolar depression the person has long cycles (many months) between bouts of depression and a normal mood. In bipolar depression the person cycles between bouts of depression and bouts of mania (manic-depressive disorder).

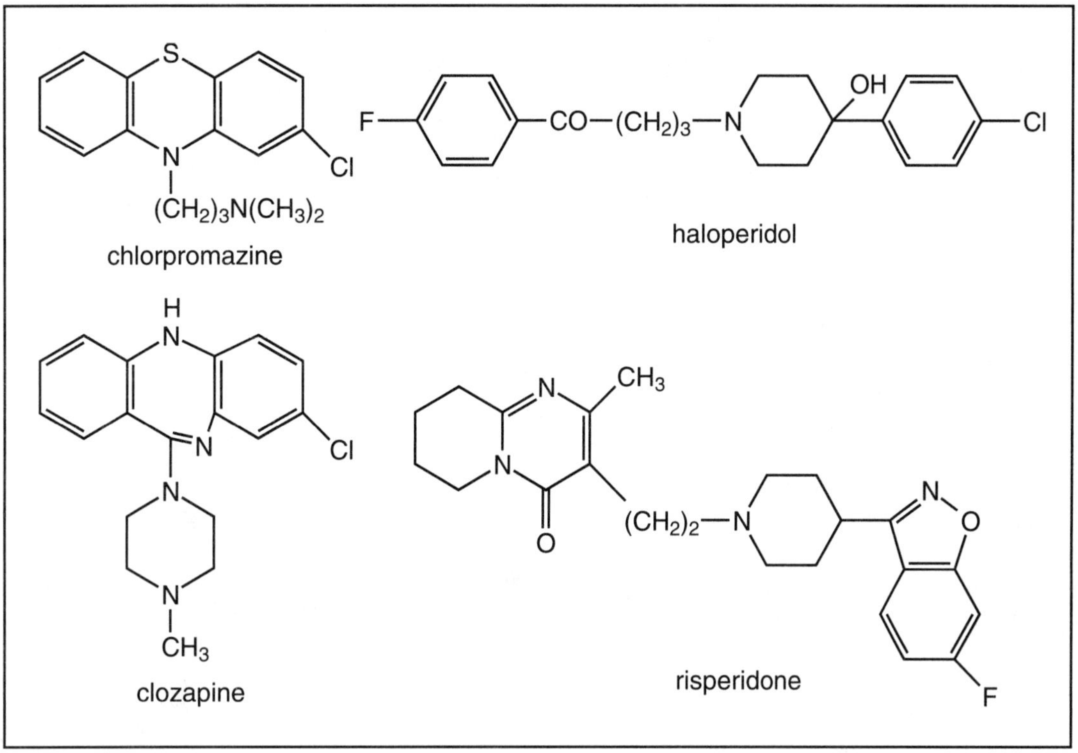

Figure 9.26

Mania represents a state of elevated mood. Manics tend to be unproductive in the sense that they undertake more than they can possibly accomplish. *Hypomania* is an elevated mood state, not quite as agitated as mania but moreso than normal.

Theory on origin of depression. The basis of depression in unknown but is thought to reflect altered interactions between different nerves and nerve regions in the brain. The limbic system is intricately associated with mood level.

Nerves containing the neurotransmitters norepinephrine (NE) and serotonin (5-HT) are thought to have prominent roles in regulating mood, as expounded in the "biogenic amine hypothesis of depression." This is based on the following:

1. Drugs that inactivate NE/5-HT nerves tend to produce depression, as seen after
   a. reserpine which depletes both NE and 5-HT and
   b. α-methyl-*p*-tyrosine which depletes NE.

2. Drugs that activate NE/5-HT nerves tend to produce mood elevation, as seen by
   a. amphetamine which releases NE and 5-HT and
   b. MAO-inhibitors which increase intraneuronal levels of NE and 5-HT.

3. There is a reduced concentration of 5-HT and its major metabolite 5-HIAA in brains of people with depression. The CSF concentration of 5-HIAA is reduced too.

4. First generation antidepressants increase synaptic content of NE and 5-HT by blocking NE and 5-HT transporter sites (TCADs **Figure 9.28**) or by increasing intraneuronal levels of NE and 5-HT (MAO-inhibitors).

5. Selective serotonin reuptake inhibitors=SSRIs, second generation antidepressants, increase synaptic content of 5-HT by blocking 5-HT transporter sites **(Figure 9.28)**.

All antidepressants produce their pharmacological effect after the first few doses (e.g., block of NE reuptake). However, the desired mood change does not occur until about 14 days or later. Therefore, it is emphasized that neural adaptations (e.g., changes in receptor numbers, etc.) must occur subsequent to the specific drug action, in order for an antidepressant effect to occur.

Antidepressants. First and second generations.
All antidepressants, including second generation drugs, are effective in about 2/3 of depressed patients. The "second generation" drugs are accorded that designation on the basis of their improved safety profile.

FIRST GENERATION ANTIDEPRESSANTS

A. Tricyclic Antidepressants (TCADs)

TCADs have been the mainstay for treating endogenous depression. TCADs are so-named because they consist chemically of 3 rings ("tri" "cyclic") **(Figure 9.27)**.

Chemistry and activity of TCADs
1. TCADs are remarkably similar in chemical structure **(Figure 9.27)**.
   a. In the upper part of the second ring there are 2 "C"s except for doxepin which has an "O" in place of one "C."
   b. With all TCADs, except protriptyline, the second ring is saturated.
   c. The lower part of the second ring has either a "C" or "N" that is attached by a carbon chain to a secondary (-NHCH$_3$) or tertiary [-N(CH$_3$)$_2$] amine.

2. Tertiary amines [-N(CH$_3$)$_2$] are active and are metabo-

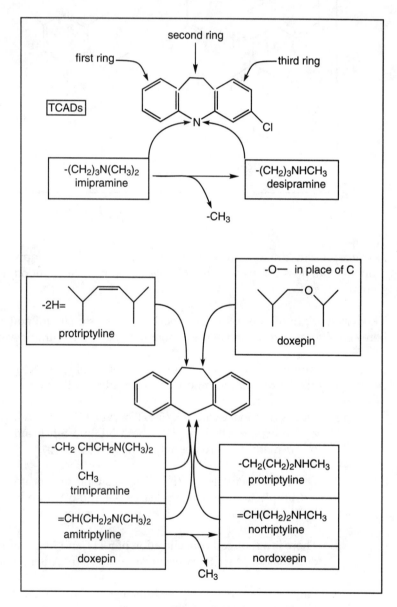

Figure 9.27

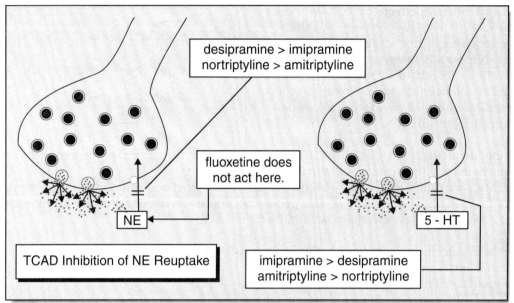

Figure 9.28

lized to secondary amines [-NHCH$_3$] which retain activity. Some of these metabolites are marketed separately as antidepressants (e.g., *desipramine, nortriptyline*). The demethylation may occur slowly, as with amitriptyline, so that most of the drug can persist for a long time as the parent tertiary amine.

3. All TCADs inhibit NE reuptake: Secondary TCADs more than Tertiary TCADs **(Figure 9.28)**.

4. All TCADs inhibit 5-HT reuptake: Tertiary TCADs more than Secondary TCADs **(Figure 9.28)**.

5. All TCADs block muscarinic receptors: Tertiary TCADs more than Secondary TCADs **(Figure 9.29)**.

6. All TCADs block histamine H$_1$ receptors: Tertiary TCADs more than Secondary TCADs **(Figure 9.29)**.

7. All TCADs block α-adrenoceptors: Tertiary TCADs more than Secondary TCADs **(Figure 9.29)**.

Pharmacokinetics of TCADs. TCADs, being drugs that act in the brain, are lipid-soluble and readily cross lipid barriers (if pH is greater than 7). Peak levels occur at 8 hours. All TCADs are highly protein-bound, so (a) toxic doses cannot be removed by dialysis and (b) TCADs displace other protein-bound drugs (e.g., NSAIDs, phenytoin) and enhance their effects (and vice-versa).

Metabolism occurs via cyt P$_{450}$ enzymes that
    a. hydroxylate the ring (add -OH)
    b. glucuronidate the -OH group and
    c. N-demethylate tertiary TCADs to active secondary TCADs.
There is extensive first pass metabolism. However, the t$_{½}$ for all TCADs is long (e.g., imipramine 12 hours; protriptyline 120 hours). TCADs compete with other drugs for metabolism and accordingly, may enhance their effects (e.g., antipsychotics).

Pharmacological effects of TCADs (Figure 9.30). TCADs do not exert an antidepressant effect by producing euphoria. In fact, dysphoria is observed. Many unpleasant effects are related to inhibition of NE/5-HT reuptake or blockade of receptors for neurotransmitters. Patients tend to be noncompliant with TCADs.

Metabolism of TCADs

Drug interactions with TCADs. TCADs potentiate the effects of drugs that release NE (e.g., amphetamine) or block M receptors (e.g., atropine) and α-receptors (e.g., phentolamine). By blocking NE transporter sites, TCADs prevent accumulation of certain drugs into sympathetic nerves (e.g., antihypertensive effect of guanethidine is blocked).

When administered with an MAO-inhibitor, TCADs may produce fatal hyperthermia, delirium, convulsions, coma and death. An interval of approximately 2 weeks should intervene between treatments with MAOIs and TCADs.

Uses of TCADs
1. Agitated depression is often treated with amitriptyline (Elavil) and doxepin (sinequan), TCADs that are sedating.
2. Retarded depression is often treated with the other TCADs: imipramine (Tofranil), desipramine (Pertrofrane), nortriptyline (aventyl) and protriptyline (Vivactil).
3. Enuresis in older children is sometimes treated with imipramine, because of its prolonged antimuscarinic effect (trigone sphincter in urinary bladder).

B. Irreversible monoamine oxidase inhibitors (MAOIs)

These drugs irreversibly inhibit MAO_A and MAO_B by covalently binding to the FAD cofactor. The antidepressant effect is correlated with inhibition of the MAO_A in NE nerves and 5-HT nerves, an action that elevates intraneuronal levels of NE and 5-HT. Even though substantial MAO-inhibition occurs after the first dose, an antidepressant effect is not observed for 2 weeks. About 80% MAO-inhibition is considered to be optimal.

MAOIs are perhaps slightly more effective for depression than TCADs, but their use is restricted because of potentially serious adverse effects.

Chemistry and activity (Figure 9.31). All of the MAOIs resemble

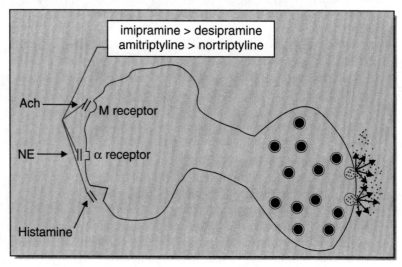

Figure 9.29

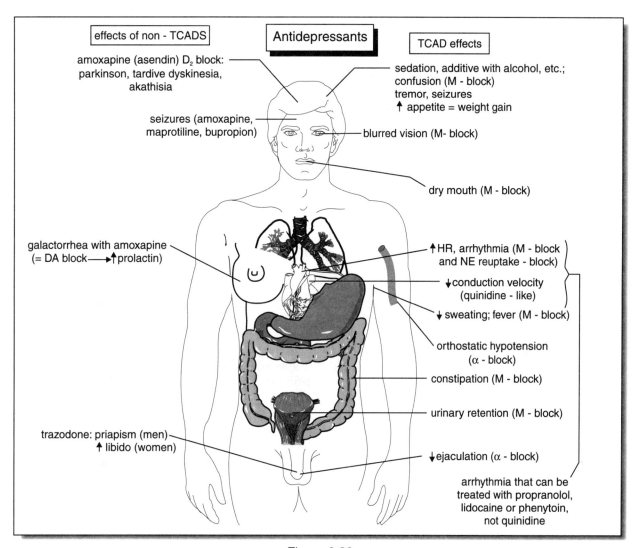

Figure 9.30

amphetamine and acutely produce a sympathomimetic effect like amphetamine. Although the hydrazine group is associated with overdose liver toxicity, this effect is rare (but 20-25% mortality). MAOIs inhibit enzymes other than MAO and accordingly interfere with hepatic metabolism of many drugs.

Pharmacokinetics. MAOIs are well-absorbed. The hydrazide MAOIs are cleaved to hydrazines, the active species. The hydrazine group is inactivated mainly by acetylation. Unless dosage adjustment is made, toxic effects may occur with hydrazine MAOIs in the large percentage of people that are slow acetylators.

Pharmacological actions (Figure 9.32)
1. MAOIs produce amphetamine-like *sympathomimetic* activity.
   Some foods (cheeses, cured meats) and beverages (wines, beers), high in tyramine content, produce a hypertensive crisis in patients taking MAOIs. The patient and family must be given a list of foods to avoid, although this list could possibly be used to aid in suicide.
2. MAOIs also have some antimuscarinic activity and produce a spectrum of activity similar to that of atropine.

3. The antidepressant action of MAOIs is associated with a suppression of REM sleep. MAOIs tend to correct the disordered sleep found in depression, whether that be excessive or insufficient.

Drug interactions. MAOIs potentiate the effects of:
   a. CNS depressants (e.g., anesthetics, sedatives, alcohol, H_1- and M-blockers).
   b. Drugs metabolized by MAOs (e.g., L-DOPA and sympathomimetics amines).
   c. Drugs metabolized by MAOI-inhibited enzymes (e.g., opioids like meperidine cause a reaction resembling combined MAOI and TCAD).
   d. Protein-bound drugs (e.g., oral hypoglycemics, TCADs).

SECOND GENERATION DRUGS

These generally produce less sedation, less M block, less α-block and fewer CDV effects than TCADs. Second generation drugs are also highly protein bound.

A. Atypical antidepressants. These are chemically unrelated to TCADs and MAOIs.

   1. *Amoxapine* (Asendin) is a metabolite of the antipsychotic, loxapine. Amoxapine inhibits reuptake of both NE and 5-HT, but also blocks D_2 receptors and occasionally produces extrapyramidal side effects including parkinsonian symptoms and tardive dyskinesia **(Figure 9.33)**.

   2. Bupropion (Wellbutrin) resembles amphetamine chemically and produces some similar effects (i.e., agitation, insomnia). DA reuptake is partly blocked. There are fewer drug interactions with bupropion than with TCADs, although bupropion interacts with MAOIs. The main risk from bupropion is seizures.

   3. *Trazodone* (Desyrel) is metabolized to products that produce an anxiogenic effect by virtue of their being potent agonists at 5-HT_{2C} receptors. Trazodone is also associated with sedation and priapism. Trazodone, co-administered with yohimbine, is being tried as a treatment for impotence.

   4. Maprotiline (Ludiomil) is chemically similar to TCADs and has a similar activity profile.

B. Selective Serotonin Reuptake Inhibitors (SSRIs)

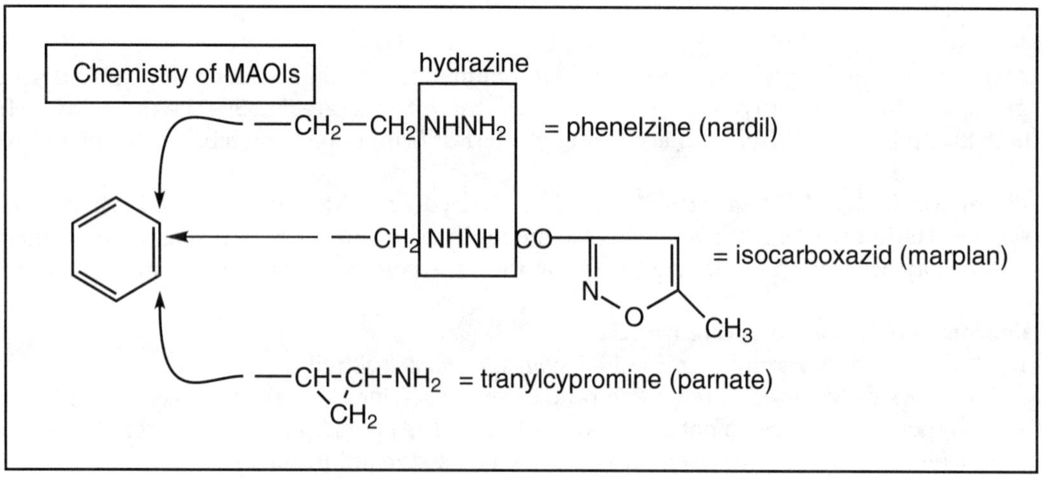

Figure 9.31

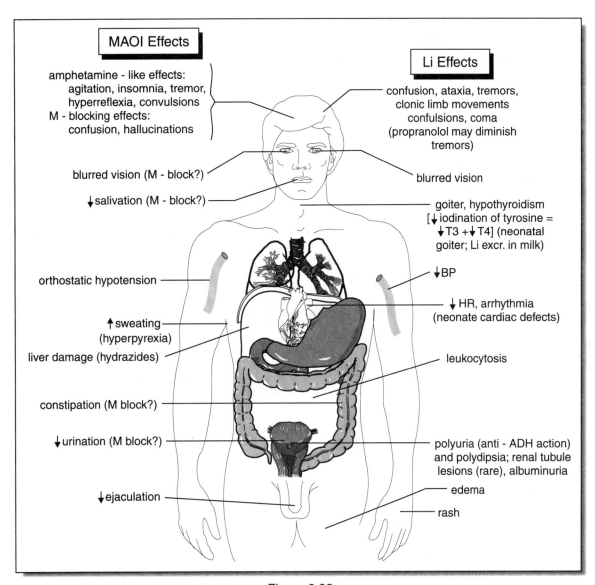

Figure 9.32

These drugs are relatively selective 5-HT uptake blockers. SSRIs are superior to TCADs, MAOIs and "atypical" antidepressants. In contrast to the weight gain often seen with TCADs, weight loss often accompanies SSRI treatment. However, problems with orgasm and ejaculation are as common as with TCADs. SSRIs, if used with MAOIs, could produce a "serotonin syndrome" of hyperthermia, autonomic instability, myoclonus, delirium and coma.

1. *Fluoxetine* (Prozac) has the greatest market share among antidepressants.

    Pharmacokinetics of fluoxetine. Fluoxetine is well absorbed and peak plasma levels occur in 6-8 hours. Its $t_{1/2}$ is approximately 3 days, although the major metabolite, norfluoxetine, has a longer $t_{1/2}$ (about 10 days) and is active. Both substances compete with other drugs for metabolism by cyt P_{450} enzymes. Being highly protein-bound, fluoxetine also displaces many protein-bound drugs. Fluoxetine enhances activity of many drugs by interfering with their metabolism and/or binding.

Pharmacological actions of fluoxetine. Fluoxetine and other SSRIs have a wider safety margin than other antidepressants. Most patients experience few adverse effects and seem to feel better after treatment, so compliance is generally not a problem. However, fluoxetine may produce insomnia, feelings of nervousness and even akathisia and tremors (i.e., excitatory versus sedative actions of TCADs). Bradycardia may occur. About 4% develop a rash with fluoxetine.

2. Paroxetine (Paxil) undergoes extensive first-pass metabolism. Its $t_{1/2}$ is approximately 20 hours.

3. Sertraline (Zoloft) has a $t_{1/2}$ of about 1 day and its major metabolite has little activity. Of the SSRIs sertraline poses the least risk of drug interactions since it induces cyt P_{450} enzymes (versus fluoxetine-inhibition of cyt P_{450}).

4. Venlafaxine (Effexor) is chemically related to bupropion. Venlafaxine inhibits NE reuptake as well as 5-HT uptake. Its $t_{1/2}$ is about 6 hours, while its major metabolite has a $t_{1/2}$ approximately 12 hours.

5. Fluvoxamine and citalopram are other SSRIs. Nefazodone (Serzone) is a new SSRI that also blocks $5-HT_{1A}$ receptors. This effect is thought to enhance synaptic release of 5-HT. Nefazodone reputedly is free of cardiotoxicity and dose not produce insomnia or sexual dysfunction.

C. Reversible MAOIs

Selective and reversible MAOIs, namely moclobemide and bromfaromine, are under study. These pose a lower risk than irreversible MAOIs, since the consequence of food- and drug-interaction is much less.

D. Partial $5-HT_{1A}$ agonists

Partial $5-HT_{1A}$ agonists, gepirone and ipsapirone, are under study as antidepressants. Buspirone, another $5-HT_{1A}$ agonist, was described earlier as an anxiolytic drug.

ELECTROCONVULSIVE THERAPY (ECT)
Electroconvulsive therapy is a non-drug approach to treating depression. ECT is considered by some to be more effective than antidepressant drugs (approximately 80%). An advantage is the often seen immediate alleviation of depressed mood after ECT. Since all antidepressant drugs require a minimum of about 14 days for their effect, ECT is frequently implemented on suicide patients. However, ECT is often held in reserve because of undesired acute effects and memory loss.

Treatment of Obsessive Compulsive Disorder (OCD). The almost uncontrolled urge to perform a repetitive action like hand-washing or hair-plucking (trichotillomania) is a characteristic of OCD. The SSRIs fluoxetine and clomipramine (Anafranil) are the only drugs approved for OCD in the United States, although fluvoxamine may soon be approved.

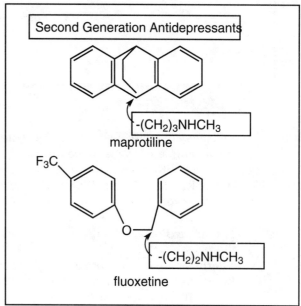

Figure 9.33

Treatment of panic attacks. The benzodiazepine alprazolam is the only drug approved for panic attacks in the United States, but low dose fluoxetine is commonly used instead.

5. TREATMENT OF MANIA

Lithium (Li) is the drug of choice for treating mania. An antipsychotic like haloperidol is often first given to "control the patient" before starting Li. Valproate (Depakene), carbamazepine (Tegretol), clonazepam and alprazolam are alternatives to Li. Li is usually given daily for several months until the patient responds (75% response). ECT is also effective for mania. A lower maintenance dose of Li is then often given for several months.

Pharmacokinetics. Li^+ is completely absorbed in 6-8 hours, reaches peak plasma levels in approximately 1½ hours, distributes with total body water and is excreted in the urine ($t_{1/2}$ of about 20 hours). In the kidneys Li^+ competes with Na^+ for reabsorption. Li elimination can be enhanced, as when toxic levels are present, by increasing Na intake. The excess Na^+ burden impedes Li reabsorption in the nephron, thereby promoting Li excretion. Conversely, in (a) dehydrated patients and in patients (b) on a low Na^+ diet, (c) on diuretics (e.g., furosemide, thiazides) or (d) with renal disease, Li^+ excretion is diminished and toxic effects of Li occur. The optimal Li plasma level is 0.8-1.0 mEq/l. The therapeutic index of Li is narrow, as adverse effects become prominent at a plasma level less than 1.5 mEq/l. The plasma level of Li should be monitored.

Li Mechanism of Action. Li participates in ion transport processes, altering electrolyte balance. In animals Li attenuates the antipsychotic-induced increase in dopamine D_2 receptor number. This effect and a modulation of neurotransmitter release (NE, 5-HT) are postulated as additional actions of Li.

Li Actions. These are shown in **Figure 9.32**. It is important to know that toxicities of Li include tremor, goiter and polyuria.

For acute toxicity, dialysis and/or sodium bicarbonate will help remove Li from the body. An anticonvulsant would also be given.

E. DRUGS OF ABUSE

Drugs of abuse are substances that are taken illicitly to produce euphoria or hallucinations. The Controlled Substances Act in 1970 classified drugs into five "Schedules," with Schedules I and II having the highest abuse potential. The pharmacology of most of these agents is described in a different chapter, according to the class of drug (e.g., heroin is described with opiates).

Schedule I. These substances have no approved medical use in the United States:
    a. Heroin
    b. Some amphetamines (MDA, MDMA; =3,4-methylenedioxy-(meth)-amphetamine)
    c. Hallucinogens (LSD, THC, marijuana, peyote, mescaline)

Schedule II narcotics. Most opioids (morphine, meperidine, methadone, codeine) and cocaine.

Schedule II nonnarcotics. Short acting barbiturates (amo-, pento-, secobarbital), methamphetamine and phencyclidine (PCP).

Schedule III narcotics. Smaller amounts of morphine, codeine, etc.

Schedule III nonnarcotics. Long-acting barbiturates.

Schedule IV narcotics. Dextropropoxyphene (dosage units).

Schedule IV nonnarcotics. Most sedatives-hypnotics (benzodiazepines, phenobarbital).

Schedule V narcotic. Buprenorphine.

Nomenclature
  Addiction: State of physical and psychological dependence on a drug, and characterized by drug-seeking behavior.
  Physical dependence: Neural adaptation to the drug (e.g., receptor number or sensitivity is changed), whereby a withdrawal from the drug triggers undesirable physiological effects: cramps, tremors, etc.
  Psychological dependence: A compulsive desire for the drug of abuse.
  Tolerance: A gradual diminished response to the same amount of drug, possibly reflecting functional tolerance (receptor changes) and/or enhanced metabolism of the drug (metabolic tolerance) and/or increased ability to compensate for the drug effect (behavioral or neuronal tolerance). As a consequence of tolerance more and more drug must be taken to produce the desired drug effect.
  Withdrawal: A series of undesirable physiological effects occurring when the drug is abruptly withdrawn and its blood conc. falls below a threshold level. Reactions consist of epigastric pain, vomiting and diarrhea, headache, syncope, chills, hallucinations, tremor, convulsions, coma and death.

Treatment of drug abuse. The objective is to control the acute adverse withdrawal symptoms. Sometimes (1) an occasional small amount of the abused drug is administered to control symptoms (e.g., ethanol during alcohol withdrawal), (2) a less addicting drug is substituted, to wean the patient through withdrawal (e.g., methadone for heroin; diazepam for alprazolam), or (3) a drug that relieves withdrawal is given (e.g., clonidine). Symptomatic assistance is needed for possible convulsions and in case of respiratory failure. Counseling and group therapy may be required afterwards.

SPECIFIC DRUGS OF ABUSE AND WITHDRAWAL TREATMENTS

A. CNS Stimulants

1. **Nicotine.** This is the primary addictive in cigarette smoke. Tolerance develops rapidly (receptor adaptation), as well as psychological dependence and overt withdrawal symptoms. Adverse effects of smoking include increased risks of cancer, ulcers and cardiovascular disease. These actions are attributable to the smoke as well as nicotine. Nicorette (nicotine resin) chewing gum and nicotine patches (Habitrol, Nicoderm, Prostep) are aids to discontinue smoking. Nicotine is discussed in **Chapter 2**.

2. **Caffeine.** This is sometimes considered an abused substance in heavy coffee drinkers (over 5 cups of coffee daily). Withdrawal signs include slight tremor, irritability and changes in heart rhythm.

3. **Amphetamine and analogs** (Methamphetamine [speed, ice, crystal]; MDA [love pill]; MDMA [speed, ecstasy]; methcathinone [cat]; methylphenidate. Amphetamines are discussed in **Chapter 2** and earlier in this chapter as CNS Stimulants. These are indirect-acting sympathomimetic amines that (1) release NE from sympathetic nerves, thereby affecting blood pressure, cardiac dynamics, etc. and (2) release DA, NE and 5-HT in the brain, thereby producing CNS excitation and euphoria, but also paranoia with delusions. (3) MDA and MDMA are 5-HT neurotoxins in animals and produce overt destruction of 5-HT nerves. Because of the long half life of amphetamines, the "high" can continue for hours to days, especially if the drug is taken repeatedly in a short time (a "run"). Diazepam or haloperidol may be given acutely to control the drug abuser.

4. **Cocaine.** This drug blocks (1) reuptake of NE at sympathetic nerves, affecting blood pressure and heart rate (sometimes causing fatal ventricular arrhythmias) and (2) reuptake of NE and DA in the brain, thereby producing CNS excitation and euphoria. Cocaine is commonly taken as snuff (coke, snow) or smoked (crack, free base). Effects are similar to the amphetamines, including seizures and hemorrhagic/occlusive stroke. Diazepam may control seizures. Cocaine is discussed in **Chapter 2** and earlier in this chapter as CNS Stimulants.

B. CNS Depressants
1. **Ethanol** (alcohol). This drug is thought to act on membrane fluidity and as an agonist at the GABA receptor. Development of tolerance is related to neural adaptations, not so much to minor changes in metabolism. However, tolerance does not develop to the lethal effect. Withdrawal is known as the DTs (i.e., delirium tremens) and this is characterized by delirium, confusion, visual hallucinations, hyperreflexia, tremor and convulsions. Treatment of withdrawal is symptomatic: diazepam for convulsions, ethanol to reduce withdrawal symptoms, supplemental K^+, Mg^{++}, thiamine. For acute alcohol intoxication, dialysis, gastric lavage and/or respiratory assistance may be required; bicarbonate, for the ensuing metabolic acidosis. Adverse effects of ethanol are discussed in **Chapter 8**. Naltrexone (ReVia) blunts the craving for alcohol and is approved for treatment of alcoholism. Liver toxicity is a complication of naltrexone overdose.

Disulfiram (Antabuse) is a drug that inhibits metabolism of the first product of ethanol metabolism, namely acetaldehyde. Disulfiram does this by competing with NAD^+ for aldehyde dehydrogenase. This action is of importance only if alcohol is consumed. It is acetaldehyde that produces most of the adverse effects of alcohol. In the presence of disulfiram, acetaldehyde accumulates to excess behind the block of aldehyde deH_2ase, producing an aldehyde syndrome characterized by:

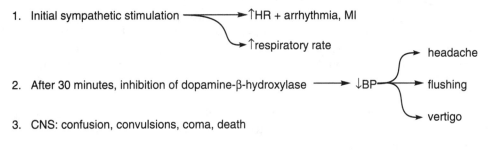

Disulfiram is only given to compliant patients wanting to give up alcohol. Because of its long $t_{1/2}$ (approximately 3 days), effective amounts of disulfiram will remain in the body for as long as 2 weeks after the last dose. Disguised forms of alcohol (cough syrups, wine sauces, etc.) cause problems with patients on disulfiram.

2. **Sedative-Hypnotics**. These primarily include short-acting barbiturates and benzodiazepines. Oral benzodiazepines are rarely lethal, unless taken with ethanol or barbiturates. A benzodiazepine antagonist, flumazenil, is useful for benzodiazepine overdose.

Barbiturate intoxication resembles alcohol intoxication. Tolerance is related to neural adaptations and metabolic tolerance (induction of cytochrome P_{450} enzymes). As with ethanol, there is little tolerance to the lethal effect of barbiturates. There is cross-tolerance between barbiturates, benzodiazepines and ethanol, but not with opiates.

Withdrawal from barbiturates and benzodiazepines is similar to withdrawal from alcohol: delirium, hyperreflexia, hallucinations, tremor, convulsions, death. Withdrawal from short-acting sedatives and hypnotics can be treated with longer acting drugs, like phenobarbital or diazepam. Long-acting drugs can be withdrawn by gradually prolonging the interval between doses of the abuse drug. These drugs were discussed earlier in this chapter.

3. **Opiates (Narcotics)**. These produce euphoria by acting as agonists at mu receptors for enkephalins and endorphin. Heroin (dope) is the most common opiate drug of abuse, because a simple chemical addition to the morphine molecule (i.e., diacetylation) increases pharmacological activity and euphoric potential tenfold. Tolerance (receptor adaptation) develops rapidly. Methadone is a one-a-day orally active opiate that is available in drug clinics in the United States. It has been used to wean addicts off of heroin. L-α-acetyl-methadol (LAMM; Orlamm) is a long-acting meperidine analog that is distributed only at authorized drug clinics, as per methadone, to wean addicts off opiates.

Acute or suspected opiate intoxication can be treated with an i.v. dose of the opiate receptor antagonist, naloxone (Narcan). Instant recovery is diagnostic for opiate abuse. Naloxone treatments must be repeated at intervals because of the short half life ($t_{\frac{1}{2}}$ less than 4 hours). Naltrexone (Trexan) ($t_{\frac{1}{2}}$ is 10 hours) is sometimes used instead, often being administered on alternate days. While these antagonists are life saving, they precipitate withdrawal in a narcotic addict. Withdrawal from opiates is often signaled by yawning and sweating with rhinorrhea. Later there may be mydriasis, hyperthermia and diarrhea—effects opposite to that produced by opiates in non-addicts. Muscle spasms ("kicking the habit") and cardiovascular collapse can occur during withdrawal from opiates, but seizures do not occur. Generally, withdrawal from opiates is not life-threatening.

In the 1980s a synthetic meperidine analog ("synthetic heroin") briefly became popular, until one batch inadvertently became contaminated with the by-product known as MPTP. This substance produced Parkinsonism in users, by actual destruction of dopamine cells in the substantia nigra. This is described earlier in this chapter in the section on Antiparkinsonian Drugs.

C. Hallucinogens (Psychedelics)

1. Lysergic acid diethylamide (LSD; acid) group (mescaline, psilocybin). These drugs are thought to produce hallucinations by inhibiting 5-HT activity, either by blocking 5-HT receptors or by acting as agonists at presynaptic receptors. These drugs have sympathomimetic activity and produce increased blood pressure, changes in heart rate, tremor, etc. In high doses convulsions can occur. However, the major effect consists of visual hallucination with a distorted sense of perspective. Tolerance can develop if daily doses are taken, but physical and psychological dependence do not occur, nor does withdrawal.

Ergot alkaloids, which include useful drugs like bromocriptine and methysergide, also act at 5-HT receptors and can produce hallucinations.

2. Phencyclidine (PCP, angel dust, hog) is essentially a dissociative anesthetic with high hallucinogenic activity. It is an anesthetic in veterinary medicine. PCP, an agonist at opiate sigma receptors, also inhibits NMDA receptors, blocks DA reuptake and blocks M receptors. Acutely, PCP produces visual hallucinations and distorted perspective. Overdoses are associated with cardiac failure, hyperthermia and seizures, which can be treated with diazepam. An acute toxic psychosis may occur, resembling paranoid schizophrenia sometimes with catatonia. An antipsychotic may be used for treating this state.

3. Ketamine is a dissociative anesthetic that produces effects similar to PCP.

4. Marijuana (grass, pot, hashish). The active principal is Δ-9-THC (tetrahydrocannabinol). Marijuana is smoked to produce euphoria, heightened visual and auditory perception, altered sense of time and a dream-like state. Effects begin in several minutes, peak in about 30 minutes and last for 2-3 hours. Marijuana is hydroxylated, and metabolites are detectable for days. Tolerance does not occur and withdrawal is mild.

   Adverse effects. These include vasodilatation (red conjunctiva, tachycardia) and tremor. Because of a prominent antiemetic effect, 9-THC is used in medicine to inhibit emesis in cancer patients (Dronabinol [Marinol]).

D. Other. Organic volatile substances used as solvents or in commercial glues, etc. are inhaled to produce lightheadedness and a sense of euphoria or exhilaration. Most such materials are toxic to the liver. Halogenated hydrocarbons, like chloroform, may produce fatal cardiac arrhythmias.

CHAPTER 10
ENDOCRINE PHARMACOLOGY

A. **HYPOTHALAMIC-PITUITARY AXIS**

  A. Posterior Pituitary Hormones

  The two posterior pituitary hormones are vasopressin and oxytocin. These nonapeptides are synthesized from the same prohormone, and differ from each other by only two amino acids **(Figure 10.1)**.

  Vasopressin and oxytocin are co-released **(Figure 10.2)** but vasopressin (also known as anti-diuretic hormone or ADH) is released in far greater amount during dehydration (i.e., reduced blood volume or blood hyperosmolality), while oxytocin is released in far greater amount during breast feeding. Each hormone retains some of the identical or opposite activity of the other hormone. Vasopressin regulates hemodynamics by modulating arterial resistance and renal excretion of water (diuresis). Vasopressin is discussed in the sections on hypertension and diuretics. Oxytocin has important roles in lactation and regulation of uterine tone. Oxytocin is discussed in the section on uterine drugs.

  B. Anterior Pituitary Hormones **(Figure 10.3)**

  A variety of hypothalamic hormones regulate the release of the anterior pituitary hormones.

  1. Growth hormone (GH) release is stimulated by growth hormone-releasing hormone (GHRH) and is inhibited by somatotropin-release-inhibiting hormone (SRIH is somatostatin).

  2. Prolactin (PRL) release is inhibited by dopamine (DA), which is released from nerve endings in the median eminence.

  3. Thyrotropin-stimulating hormone (TSH is thyrotropin) release is stimulated by thyrotropin-releasing hormone (TRH).

  4. Follicle stimulating hormone (FSH) and luteinizing hormone (LH) release are both stimulated by gonadotropin-releasing hormone (GnRH).

  5. Adrenocorticotropin (ACTH) release is stimulated by corticotropin-releasing hormone (CRH).

  The actions of these substances are summarized in the following chapters. Other substances are produced by the hypothalamus and anterior pituitary, but their hormonal

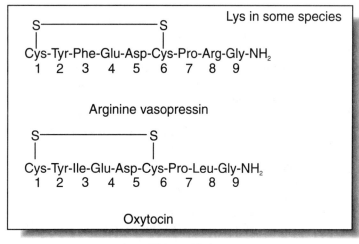

Figure 10.1

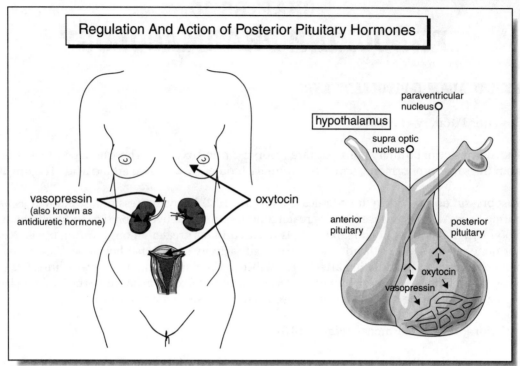

Figure 10.2

role is not as dramatic. Also, the hypothalamic releasing hormones and inhibiting hormones may affect the release of several kinds of pituitary hormones. As yet, these are not considered to be important factors in medical treatment of endocrine disorders.

Growth hormone-releasing hormone (GHRH; Sermorelin) is a polypeptide chain of 44 amino acids ($GHRH_{1-44}$), with activity residing in the segment $GHRH_{1-29}$. GHRH has a $t_{1/2}$ less than 10 minutes, and is sometimes used to test whether a youth is a candidate for growth hormone therapy. GHRH should stimulate the release of growth hormone (GH) if the pituitary cells (somatotrophs) produce GH **(Figure 10.4)**.

Somatostatin (SRIH) is a polypeptide chain of 14 amino acids. It inhibits release of both TSH and GH from the pituitary. SRIH is also produced in the GI tract and pancreas where it probably has a local

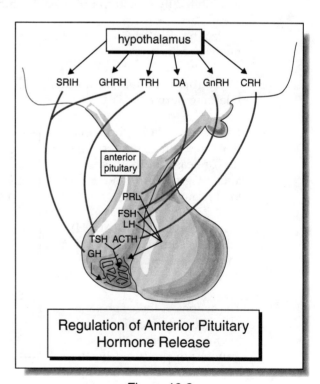

Figure 10.3

action. Actions and uses are shown in **Figure 10.5**. Since SRIH has a $t_{½}$ of less than 5 minutes, the 8 amino acid analog octreotide (Sandostatin) with a $t_{½}$ greater than 1 hour and duration of about 10 hours is used in medicine. This analog is less potent than $SRIH_{1-14}$ in inhibiting insulin release.

Human growth hormone (hGH) is a polypeptide chain of 191 amino acids. The biosynthetic form, Humatrop, has a $t_{½}$ less than 30 minutes but a duration of about 1½ days, reflecting the 5 hour $t_{½}$ of somatomedins that are produced in response to GH. The most important of the somatomedins is insulin-like growth factor (IGF-1), a 70 amino acid peptide, 90% of which is made in the liver. IGF-1, in theory, could effectively induce growth in both pituitary dwarfs (deficiency of GH) and Laron dwarfs (deficiency of GH receptors). This is not yet practical. IGF-1 has a feedback inhibition on GH Secretion, as shown in **Figure 10.4**. Actions of GH are shown in **Figure 10.6**. The analog of GH_{1-191} is somatrem (Protropin). Antibody formation is more frequent with the latter polypeptide.

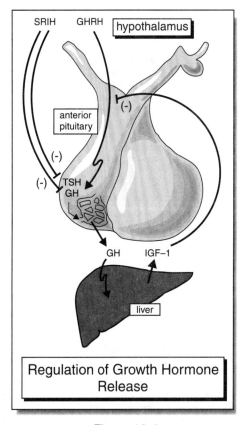

Figure 10.4

PROLACTIN

Prolactin (PRL) is a 198 amino acid glycoprotein, chemically similar to GH. The major action of PRL is stimulation of lactation. There is a progressive increase in PRL production during pregnancy, which acts with a placental lactogen to stimulate development of the mammary glands. Lactation *per se* is inhibited by high estrogen and progesterone levels in pregnancy. Lactation is induced by PRL when the placenta is delivered at birth and estrogen and progesterone levels drop precipitously. Suckling is the biological stimulus for continued PRL secretion.

In excess amount, as with prolactinomas and idiopathic hyperprolactinemia, PRL produces
      a. galactorrhea and
      b. hypogonadism and amenorrhea.

Bromocriptine (Parlodel) is a dopamine (DA) D_2 receptor agonist that acts at D_2 receptors in the median eminence to inhibit prolactin secretion, an action like that of endogenous DA. Bromocriptine also acts directly at D_2 receptors on lactotrophs to inhibit lactation. Bromocriptine is used for hyperprolactinemia and to inhibit lactation in women that don't opt to breast feed newborn infants.

B. OXYTOCIN AND UTERINE DRUGS

Several classes of drugs and hormones increase or decrease uterine tone and are used to modify labor and postpartum bleeding. These substances are described in this chapter.

UTERINE STIMULANTS
1. *Oxytocin* is a posterior pituitary peptide hormone **(see Section A)** that:
      a. Contracts myoepithelial cells which surround alveoli in the mammary glands, thereby promoting milk ejection (i.e., milk letdown). Suckling enhances oxytocin release. The oxytocin effect on

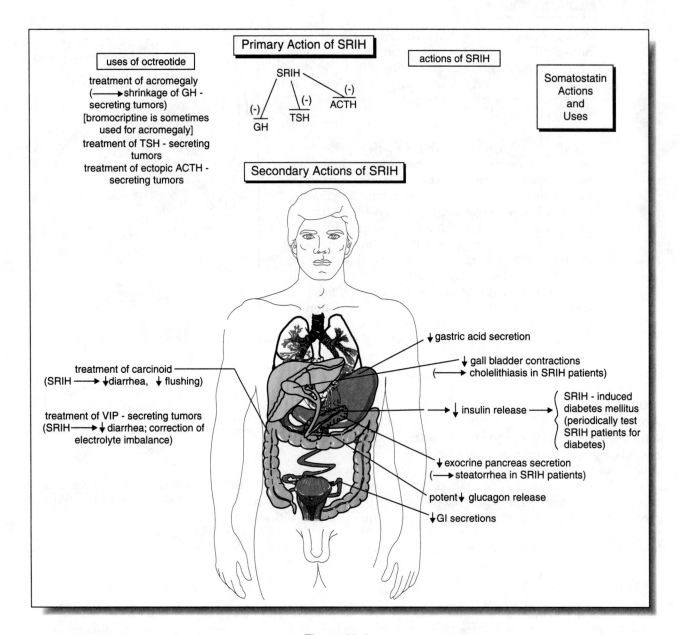

Figure 10.5

lactation may be its major physiologic role. Oxytocin nasal spray is sometimes used for this purpose.

b. Contracts the estrogen-primed uterus and relaxes the cervix. There is increased sensitivity of the uterus during the last trimester of pregnancy. Oxytocin (Pitocin, Syntocinon) is sometimes infused to induce labor and maintain labor. Oxytocin should not be used if vaginal delivery is not intended, if there is danger of uterine rupture, or if there is fetal distress. Because of its rapid metabolism in liver and kidneys ($t_{1/2}$ about 5 minutes), the oxytocin effects can be terminated in minutes by stopping the infusion. After childbirth oxytocin-induced uterine contractions promote delivery of the placenta, while diminishing postpartum bleeding. Oxytocin has little effect on the nonpregnant uterus.

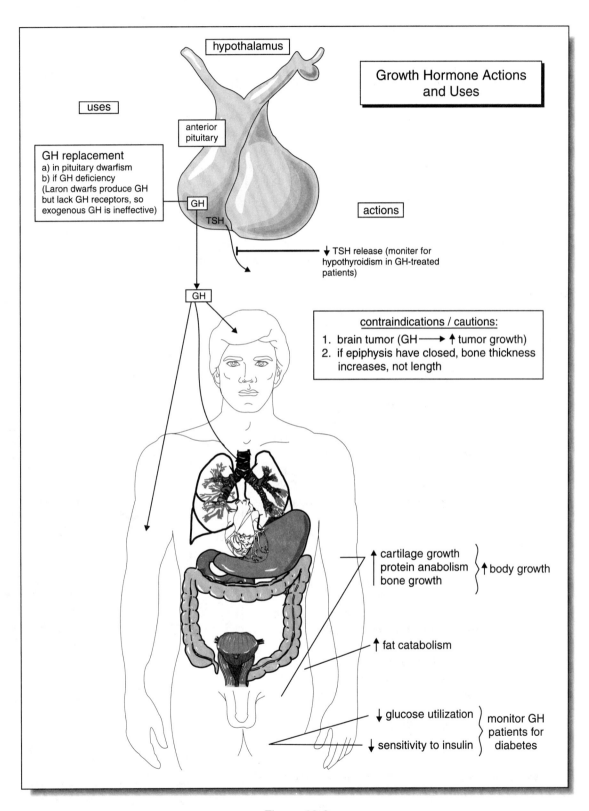

Figure 10.6

Side effects and toxic effects of oxytocin are largely related to residual vasopressin activity. This mainly includes water retention and hypertension or hypotension. Sympathomimetics should not be used with oxytocin since β agonists relax the uterus, thereby countering oxytocin's effect.

2. *Ergot Alkaloids* are potent stimulants of sustained uterine contractions. The pregnant uterus is far more sensitive to ergots than the nonpregnant uterus. Ergonovine (Ergotrate) and methylergonovine (Methergine) are used to expel the placenta and to control postpartum bleeding. These drugs are not used for inducing labor. The sustained uterine contraction poses a risk of uterine rupture. This same action cuts off the blood supply and leads to fetal anoxia. These two ergots have minimal direct vasoconstrictor activity.

3. *Dinoprostone* (Prostin E_2) is a prostaglandin (PG) that induces contractions of both a pregnant and nonpregnant uterus. This drug is used to induce labor. Side effects include a dysmenorrhea-like pain, and excess intestinal action (vomiting, diarrhea). Dinoprostone will produce abortions.

UTERINE RELAXANTS

1. *$β_2$ agonists* act at uterine β adrenoceptors, to reduce uterine tone and inhibit premature labor. Ritodrine (Bricanyl) is a preferred drug of this class, although terbutaline and others have a similar action.

2. *Ethanol* inhibits release of oxytocin and vasopression (i.e., antidiuretic hormone). This action, as well as its direct inhibitory action on uterine muscle, made ethanol a useful agent for inhibiting premature labor. Because ethanol crosses the placenta and can produce CNS depression in the unborn child, other drugs are preferred.

3. *Magnesium* (Mg) salts depress uterine muscle and thereby inhibit their contraction. Mg has been used to inhibit premature labor. Complications are CNS depression including the respiratory center and myocardial depression. β agonists are preferred over magnesium.

4. *PG synthesis-inhibitors*, like ibuprofen (Motrin), relax the uterus. These drugs are used to treat dysmenorrhea.

C. THYROID AND ANTITHYROID DRUGS

The thyroid gland, through production of thyroid hormones, plays a major role in regulating metabolic activity of virtually all cells throughout the body.

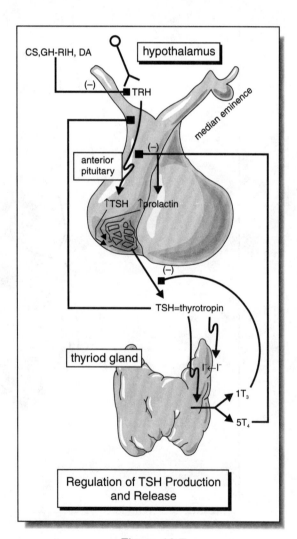

Figure 10.7

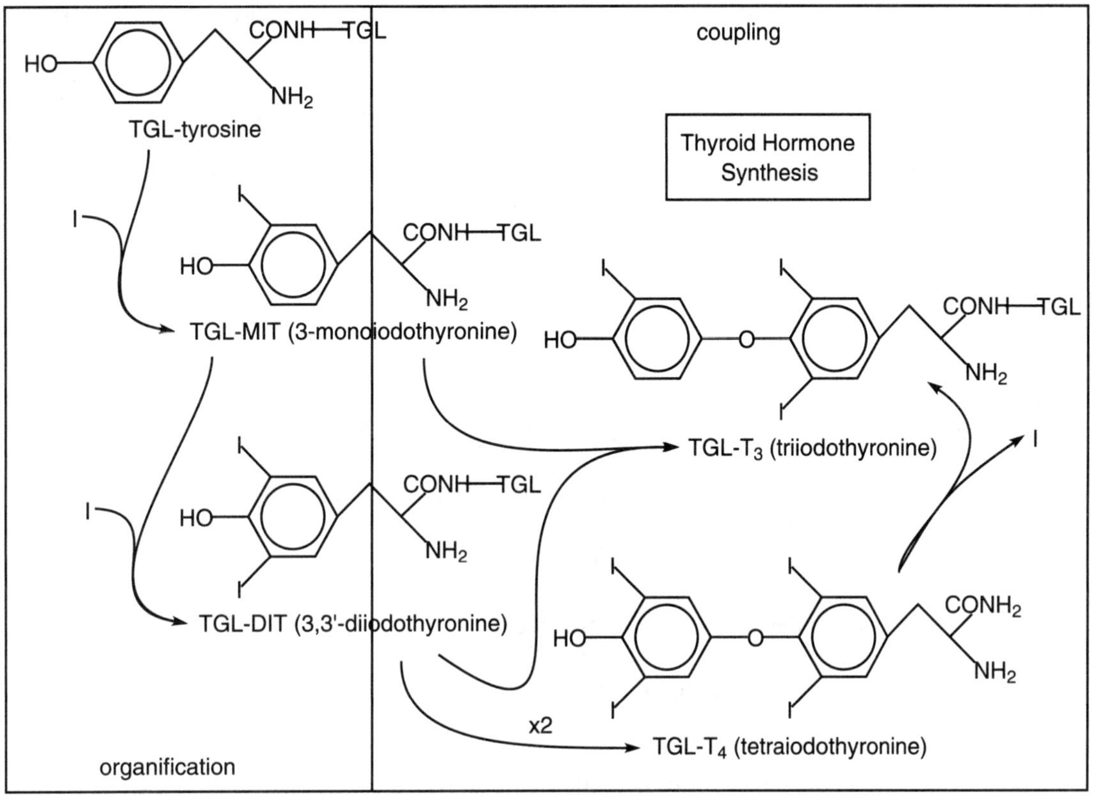

Figure 10.8

Regulation of thyroid hormone production. *TRH* is a tripeptide ([pyro] Glu-His-Pro-NH$_2$) released by nerves into the median eminence and carried in the portal circulation to the anterior pituitary, where TRH induces *thyrotropin* (TSH) synthesis and release. TSH is carried in the blood and induces the production of T_3 (triiodothyronine) and T_4 (tetraiodothyronine) in the thyroid gland. These two thyroid hormones exert effects throughout the body. Circulating T_3 has a negative feedback on TSH levels. Circulating T_3, T_4 and TSH have a negative feedback on TRH **(Figure 10.7)**.

TRH also tends to release prolactin from the pituitary. Several hormones (corticosterones, GH-RIH and DA) tend to inhibit release of TRH.

Thyroid hormone synthesis. This occurs in follicular cells and follicular colloid of this gland, as shown in **Figure 10.8**.
1. Iodide (I⁻) is actively accumulated by follicular cells, to a concentration of about 30 to 200 times that of plasma. This is known as *Iodine trapping*.
2. Peroxidase converts I⁻ to I°.
3. I° iodinates tyrosyl moieties complexed with thyroglobulin (TGL) in the colloid. This is known as *Organification*.
4. Iodotyrosine residues complex with one another. This is known as *Coupling*.

Thyroid hormone release. TGL-T$_3$ and TGL-T$_4$ are sequestered into colloid droplets which traverse the follicular cell. During this migration a lysosome fuses with the droplet and cleaves TGL-T$_3$ and TGL-T$_4$ complexes, releasing active T$_3$ and T$_4$ into the blood. There is a ratio of about 1 T$_3$ to 5 T$_4$, with T$_3$ being about 3 times as active as T$_4$. Most T$_3$ is formed outside the thyroid by deiodination of T$_4$. About 99.9% of T$_3$ and 99.5% of T$_4$ are bound to a thyroxine-binding globulin (TBG) in blood. Some drugs, like salicylates, displace T_3 and T_4 from their binding sites (activation).

Thyroid hormone action. This occurs at target cells via T_3 and T_4 binding to receptors
- a. on outer cell membranes causing increased uptake of glucose and amino acids;
- b. on mitochondrial membranes causing increased O_2 consumption and
- c. on nuclear membranes causing increased protein synthesis **(Figure 10.9)**.

Catabolism of T_3 and T_4 occurs mainly by deiodination, but also decarboxylation, deamidation, glucuronidation and sulfation, primarily in liver **(Figure 10.10)**. T_3 has a $t_{1/2}$ about 1 day; T_4 has a $t_{1/2}$ about 1 week.

HYPOTHYROIDISM
Types of hypothyroidism
1. Primary (low T_3 and T_4 synthesis in the thyroid gland)
    a. Idiopathic
    b. Hashimoto's thyroiditis (autoimmune disorder, with IgGs and IgMs directed against the colloid or thyroglobulin.)
    c. Overtreatment of hyperthyroidism

2. Secondary (low TSH synthesis in the pituitary)

3. Tertiary (low TRH synthesis in the hypothalamus)

Treatment of hypothyroidism
1. *Levothyroxine* sodium (Synthroid) is T_4. Being of high purity, and with T_4 having a long $t_{1/2}$, this is the drug of choice.

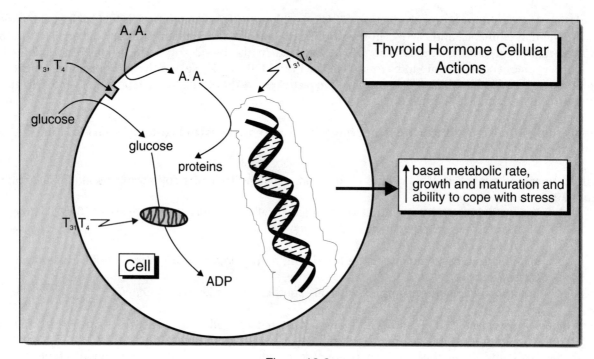

Figure 10.9

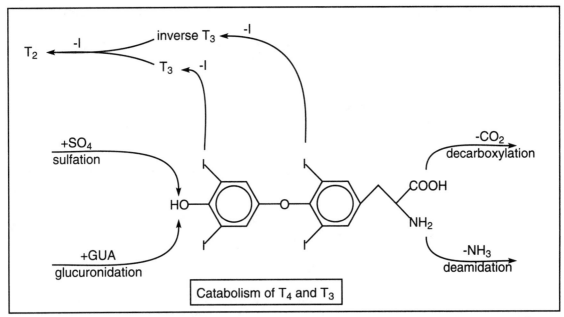

Figure 10.10

2. *Liotrix* (Euthroid) is $4T_4:1T_3$. Although this drug approximates the $T_3:T_4$ ratio in blood, this is no advantage, since T_4 is normally and adequately converted in the body to T_3.
3. *Levothyronine* sodium (Cytomel) is T_3. Because of a short $t_{1/2}$ and high potency of T_3 (causing cardiac effects), this drug is impractical for maintenance therapy.
4. Thyroid (USP) and Thyroglobulin (Prolid) is combined MIT, DIT, T_3 and T_4. This is a desiccated, defatted powder of beef or pork thyroids containing variable ratios of MIT, DIT, T_3 and T_4. Consequently, standardization of biological activity is necessary. This drug is not recommended.

Replacement therapy. Drugs for thyroid hormone replacement are given in low amounts initially, to avoid cardiac effects (arrhythmia, myocardial infarct, etc.). As the body adapts, the maintenance dose is gradually increased over several weeks. If there is adrenal insufficiency this must be corrected with adrenal cortical hormones before instituting thyroid hormone replacement, to avoid a life-threatening crisis. Replacement therapy is life-long. Abrupt withdrawal from therapy would cause reappearance of the hypothyroid state in about 2 weeks.

Aggressive thyroid replacement would be undertaken only for (a) neonatal hypothyroidism (cretinism), to prevent mental retardation and (b) myxedema coma. Levothyronine (T_3) is preferred for myxedema coma, because of its high potency.

HYPERTHYROIDISM (HIGH T_3 AND T_4 LEVELS)
Types of hyperthyroidism
1. *Grave's disease* is an autoimmune disorder in which IgG antibodies to the TSH receptor also have inherent thyroid stimulating activity. There is no negative feedback on these long-acting "thyroid stimulating immunoglobulins (LATS)," so excess T_3 and T_4 is produced. This is the major cause of hyperthyroidism.

2. Toxic adenoma is a disorder in which the thyroid tissue is overactive, producing excess T_3 and T_4 (e.g., thyroid carcinoma).

Treatment of hyperthyroidism. Antithyroid drugs are preferred for patients under 35 years of age.
1. *Thionamides* act by inhibiting (a) peroxidation of I⁻, a necessary step for MIT and DIT formation (organification) and (b) coupling of MIT and DIT. Improvement is seen after 3-4 weeks.
   a. *Propylthiouracil* (Propacil; PTU) additionally inhibits formation of T_3 from T_4.
   b. *Methimazole* (Tapazole) has 10 times the potency of PTU.

   Major adverse effects. The major adverse effects of thionamides include skin rash (about 5%) and blood dyscrasias (rarely). There is often cross-sensitivity to these two drugs. Because PTU does not easily cross the placenta, it is safer during pregnancy.

2. *Iodides* (KI, NaI; Lugol's solution is 5% I_2 and 10% KI), in high amounts, temporarily inhibit both synthesis and secretion of T_3 and T_4. Iodides are used for a week or less, to reduce the size of the thyroid prior to surgery. Iodides cannot be used if ^{131}I is to be administered, since thyroid uptake of ^{131}I would be inhibited.

   Adverse effects. The adverse effects of iodides include prolonged suppression of the thyroid (hypothyroidism) and induction of hyperthyroidism in a euthyroid but iodide-deficient patient.

3. ^{131}I (Iodotope-131) is actively accumulated in the thyroid. Gamma and beta rays produced by this isotope destroy thyroid cells. The objective is destruction of excess thyroid tissue, with consequent production of a euthyroid state. Unfortunately, permanent hypothyroidism is frequently produced.

4. Subtotal thyroidectomy is also used for hyperthyroidism.

Medical issues related to thyroid status. Hypothyroid patients are more sensitive to CNS depressants and digoxin. The opposite is true for hyperthyroid patients. Because anticoagulants are less effective in hypothyroid patients, there is increased bleeding tendency as a euthyroid state is produced. Lithium tends to produce goiter, specifically through inhibition of T_4 release by the thyroid. Many other drugs interact with thyroid hormones. The cabbage family has antithyroid action because of high content of thiocyanate (SCN⁻), one of several anions that inhibit thyroid uptake of I⁻.

Supplemental therapy of thyrotoxicosis. β-adrenergic blockers (e.g., propranolol) are often included as adjunct therapy in thyrotoxicosis, to block the enhanced adrenergic effects of elevated thyroid hormone levels.

D. ADRENOCORTICAL HORMONES: CORTISOL, ALDOSTERONE AND ANDROGENS

GLUCOCORTICOIDS
Glucocorticoids are steroid hormones from the adrenal cortex that tend to increase blood glucose concentration. Hence, the name "gluco" "corticoid." These hormones, of which cortisol is the most prominent in humans, are regulated by the anterior pituitary hormone, adrenocorticotropic hormone (ACTH), which is released by the hypothalamic peptide, corticotropic releasing hormone (CRH). ACTH stimulates the production and release of cortisol and androgens from the zona fasciculata (zf) and zona reticularis (zr). ACTH has a lesser and unsustained influence on aldosterone release from the zona glomerulosa (zg). There is a negative feedback inhibition of CRH and ACTH by circulating levels of cortisol, not androgens or aldosterone **(Figure 10.11)**.

CRH is a 41 amino acid peptide used only for diagnostic purposes. If Cushing's disease (excess cortisol release) is of pituitary origin, CRH will elevate ACTH levels which would have been extremely low because of negative feedback inhibition by cortisol. If Cushing's disease is due to ectopic ACTH production, CRH will not increase ACTH levels further.

ACTH is a 39 amino acid peptide, derived from pro-opiomelanocortin (i.e., a protein from which α-endorphin, a "natural opioid" and melanocyte stimulating hormone [MSH] are derived). ACTH is released in highest amount in the morning (5:00-7:00 AM) and in lowest amount in the afternoon (3:00-6:00 PM). Cortisol levels parallel the changes in ACTH. Stress is the single greatest stimulus for ACTH secretion.

Exogenous CS produces adrenocortical suppression (i.e., long-lived suppression of CRH and ACTH release from the hypothalamus and pituitary, respectively). To avert this, CS is often administered (a) as alternative day therapy and/or (b) in the A.M. to coincide with the natural peak level of CS. As an alternate to oral administration, CS is used as an inhalant (asthma), topically as a cream, or as an intra-articular injection (joint inflammation in arthritis). CS can be injected i.m. and i.v. to attain a rapid effect.

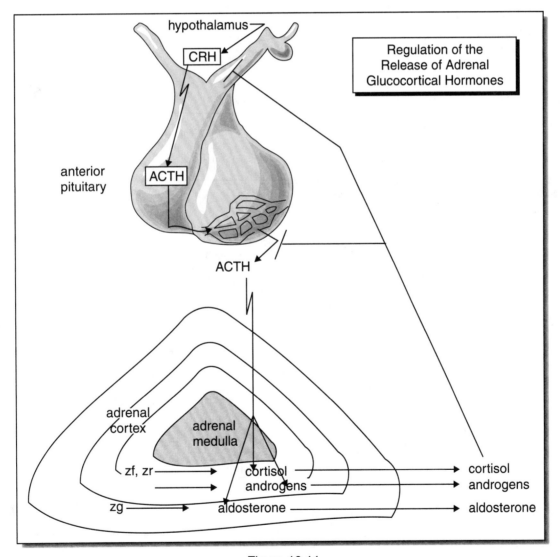

Figure 10.11

PHARMACOLOGY

Synthesis of cortisol (Figure 10.12). In the sequence for cortisol synthesis, desmolase is the rate-limiting enzyme. Its activity is largely under regulation by ACTH. Cholesterol is the starting substrate, typical for steroid synthesis.

Glucocorticoid disorders: diagnosis and treatment

1. *Addison's disease* is a syndrome resulting from inadequate cortisol production. Exogenous $ACTH_{1-39}$ (natural ACTH) or the fully active analog $ACTH_{1-24}$ (Cosyntropin) is sometimes used for diagnosis: Addisonism is likely if ACTH fails to induce a rise in cortisol levels. $ACTH_{1-24}$ is less antigenic so immuno-inactivation is less of a problem with this analog.

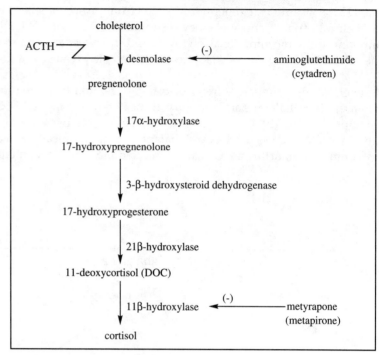

Figure 10.12

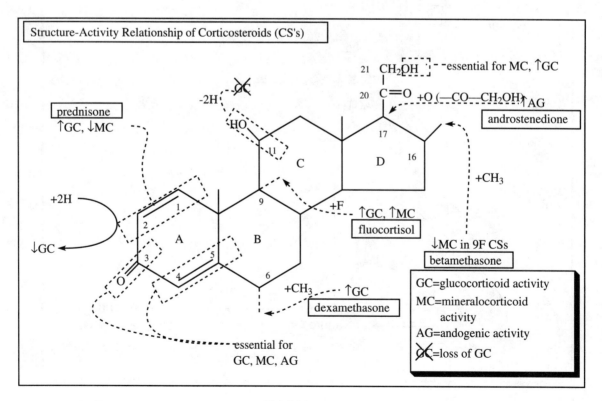

Figure 10.13

Addison's disease is usually treated with gluco-plus mineralocorticoids.

2. *Cushing's disease* is a syndrome resulting from excess cortisol production. Several approaches are used to treat this disease.
   a. *Mitotane* (Lysodren) produces overt destruction of cells in the zf and zr, thereby reducing levels of cortisol.

   b. *Aminoglutethimide* (Cytadren) is an inhibitor of desmolase **(Figure 10.12)**. This agent reduces plasma levels of cortisol. It effectively treats ectopic ACTH synthesis, with associated excess cortisol (e.g., adrenocortical carcinoma). Aldosterone synthesis is impaired with this drug.

   c. *Metyrapone* is an inhibitor of 11 β-hydroxylase **(Figure 10.12)**. This agent reduces plasma levels of cortisol, but increases plasma levels of 11-deoxycortisol (inactive) and ACTH. It is also used to treat ectopic ACTH synthesis and adrenocortical carcinoma. Aldosterone synthesis is inhibited, but its precursor DOC, is still present and has some mineralocorticoid activity. Excess androgen production persists.

   Metyrapone has been used to test for hypopituitary hypoadrenocorticism (low ACTH). If this condition exists, 11-deoxycortisol levels do not rise.

   d. *RU-486* is an antagonist to both the glucocorticoid and estrogen receptor. RU-486 is used in Europe to treat Cushing's syndrome. This drug is discussed in the chapter on Estrogen and Progestins.

Pharmacokinetics. Glucocorticoids, but not aldosterone, are orally absorbed. About 80% of cortisol is bound to corticosteroid binding globulin (CBG is transcortin). Some analogs, like dexamethasone, exist in the body only in the free state, as they do not bind to this protein. GCs are metabolized primarily in the liver by reduction of the double bonds of the A ring, reduction of the ketones at position 3 on the A ring and position 20 **(Figure 10.13)**. Partially metabolized GCs are sulfated and glucuronidated at position 3, for urinary excretion.

Mechanisms and actions. Cortisol binds to cytoplasmic receptors and is transported into the cell nucleus, where new mRNAs are transcribed, for ultimate translation of a host of proteins that evoke a response. The effects of GCs are to elevate glucose levels (anti-insulin action) at the expense of protein, fat and glycogen stores. Specific actions of glucocorticoids are shown in **Figure 10.16**.

Aldosterone (Figure 10.14). Aldosterone secretion is primarily regulated by angiotensin II (AII) and circulating K^+ levels.

Angiotensinogen is the protein substrate for renin, an enzyme formed by cells in the juxtaglomerulosa and released when blood pressure or blood volume declines. The decapeptide product angiotensin I (AI), in turn, is substrate for angiotensin converting enzyme (*ACE*), which is abundant in lungs and blood. The octapeptide product, AII, is the principal regulator of aldosterone. AII has inherent vasoconstrictor activity and is cleaved to the heptapeptide, AIII, which has less vasoconstrictor activity but greater aldosterone-releasing activity than AII **(see Chapter 5)**.

Figure 10.14

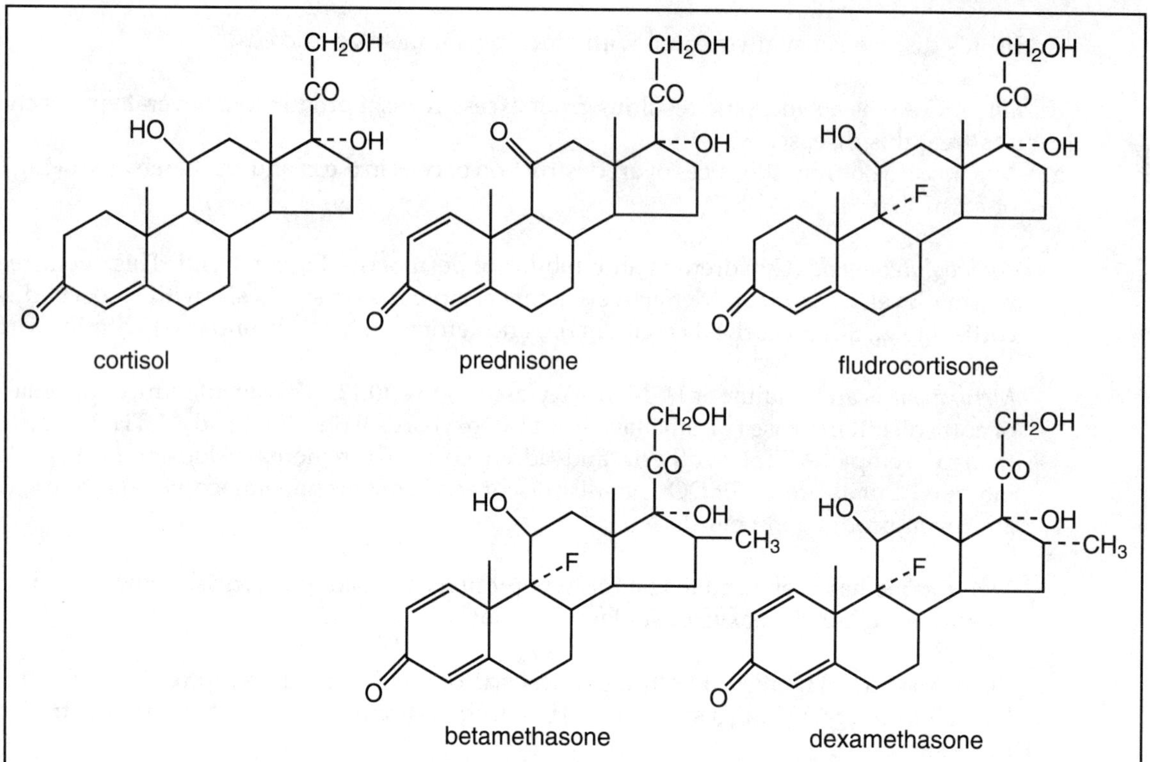

PHARMACOLOGY

Figure 10.15

Mechanisms and actions of mineralocorticoids. Aldosterone binds to cytoplasmic receptors and is transported into the cell nucleus, where mRNAs are transcribed, for ultimate translation of proteins that evoke a response. This mechanism is analogous to that of glucocorticoids. The major action of aldosterone is to increase the renal tubular exchange of Na^+ for K^+, so that Na^+ is conserved. Actions of AII and aldosterone are discussed in more detail in **Chapters 5, on Hypertension**.

Pharmacokinetics. Aldosterone has a short $t_{1/2}$ and cannot be administered orally. About 60% of aldosterone is bound to protein. Metabolism occurs largely in the liver and excretion is primarily through the kidneys. Fludrocortisone is an orally active analog.

Actions of Cortisol Analogs

	Relative	Potency
CS	GC	MC
Cortisol	1	1
Prednisone	4	0.5
Fludrocortisone	5	125
Betamethasone	25	0
Declamethasone	25	0
Aldosterone	0	3000

Preperations

Short acting:	Intermediate:	Long acting:
8-12 hours cortisol	12-36 hours prednisolone	36-54 hours dexamethasone

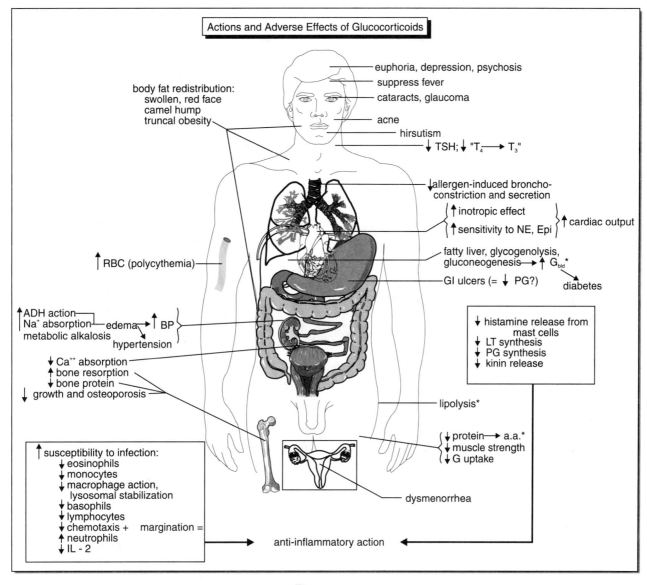

Figure 10.16

Use. Hypoaldosteronism is treated with mineralocorticoids. For long-term treatment, fludrocortisone (**Figure 10.15**) is the agent of choice.

E.   ANDROGENS AND ANABOLIC STEROIDS

Hypothalamic-pituitary axis in the regulation of testosterone. Hypothalamic GnRH enters a portal system to the anterior pituitary and stimulates the release of FSH and LH which enter the circulatory system. LH stimulates testosterone production in Leydig cells of the testes, primarily through induction of enzymes involved in the rate-limiting step in testosterone synthesis—conversion of cholesterol to pregnenolone. Other androgenic products of lesser potency are also formed, including inhibin, which together with testosterone exerts a negative

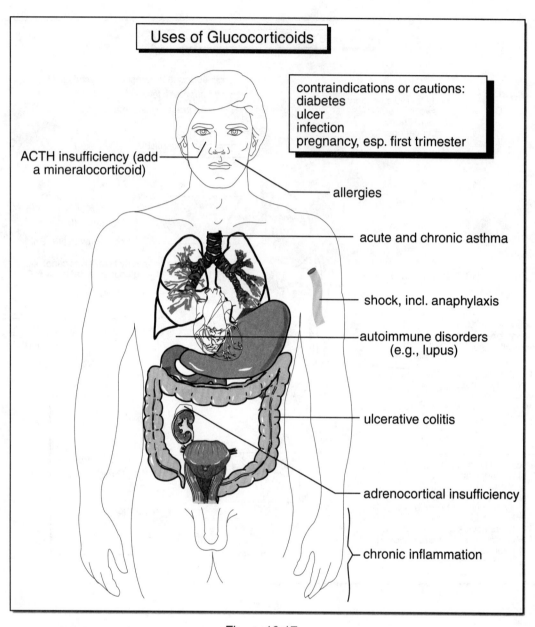

Figure 10.17

feedback inhibitory control on GnRH production in the hypothalamus and on FSH and LH production in the pituitary **(Figure 10.18)**.

FSH, co-released with LH from the pituitary, primarily stimulates spermatogenesis in the seminiferous tubules of the testes. This process occurs only if testosterone levels are sufficiently elevated.

Testosterone (T) activity. Physiological actions of T and clinical uses of androgens are illustrated in **Figure 10.19**. *Androgenic* actions reflect masculinizing effects (primary and secondary sex characteristics), while *anabolic* effects are related to positive nitrogen balance (protein synthesis). Adverse and undesirable effects are shown with the female illustration in **Figure 10.20**. After synthesis, primarily by testicular Leydig cells, T is transported in the blood mostly bound to sex hormone-binding globulin (SHBG) or albumin. The approximate 3% free T and albumin-bound T are active forms.

Androgen receptor antagonists. At the target cell T acts at androgen receptors to exert its effects on protein synthesis, etc. *Flutamide* (Eulexin) and *cyproterone* are competitive antagonists for the androgen receptor. These drugs are useful in precocious puberty, prostate cancer and in reducing the sex-drive. Part of the activity of cyproterone is due to progesterone-like activity. Gonadotropins (FSH, LH) (continuous) may be used alone or combined with flutamide and cyproterone to produce androgen suppression in the treatment of advanced prostate cancer. The diuretic, spironolactone (Aldactone), owing to its competitive antagonist activity at androgen receptors, is sometimes used to treat hirsutism. This drug has an additional inhibitory effect on the 17 α-hydroxylase involved in T synthesis **(Figure 10.21)**.

T synthesis inhibitor. In some tissues T is first metabolized by a 5-α-reductase to 5-α-dihydrotestosterone, a more active species. Finasteride is a selective 5-α-reductase inhibitor that may be useful in reducing androgen-dependent growths, like prostate size in benign prostatic hypertrophy or in prostatic cancer **(Figure 10.21)**.

Androgenic activity (Figure 10.21). The ratio of anabolic/androgenic activity is increased by some

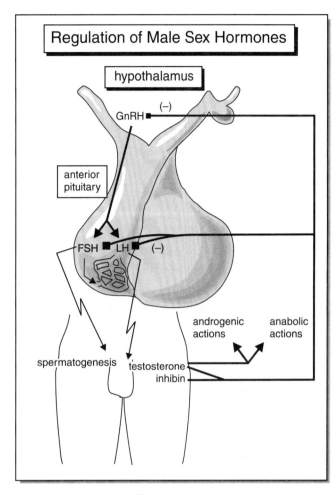

Figure 10.18

C17-alkyl analogs. This modification also makes the steroids resistant to metabolism and orally active. The most active of the anabolic analogs are methandrostenolone (Dianabol) and ethylestrenol (Maxibolin). In general, the 17 α-alkylated agents have more of an adverse effect on liver cells (increased SGOT, jaundice) and LDL content.

ANDROGENS
1. Testosterone would be given as a pellet for long-term use, since more than 75% of oral testosterone is inactivated by first pass metabolism.
2. Testosterone enanthate and cypionate are esters with a long duration of action (i.m.).
3. Fluoxymesterone
4. Methyltestosterone—17-α-alkyl analogs, orally active.
5. Danazol

Danazol (Danocrine) is a long-lived ($t_{1/2}$ about 15 hours) analog that effectively prevents FSH and LH release, even a midcycle rise in women and a postcastration elevation in men. By suppressing FSH and LH, danazol can be used for endometriosis (abnormal gonadotropin-induced growth of endometrial tissue). Danazol finds similar usefulness in fibrocystic breast disease. Gonadotropin suppression occurs despite weak androgenic activity. Danazol also inhibits synthesis of testosterone, as shown in **Figure 10.21**.

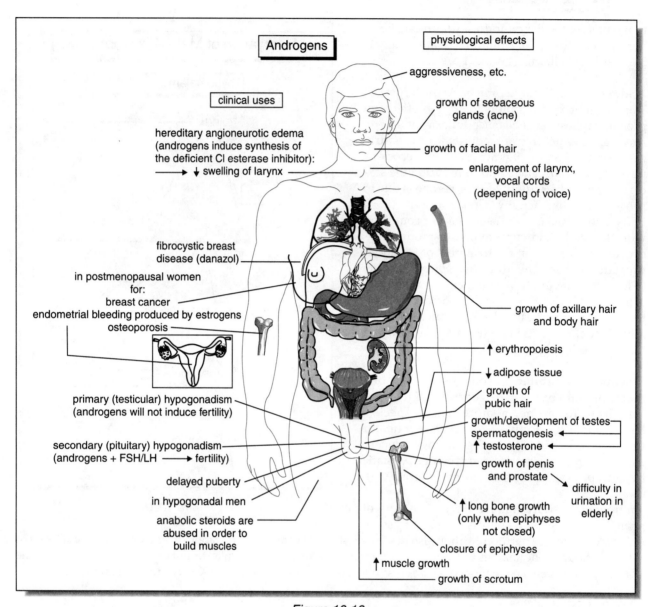

Figure 10.19

Adverse effects. These include deepening of the voice (weak androgenic activity), hot flashes (decreased estrogen level), decreased breast size (decreased FSH/LH), edema and weight gain (steroid effect).

F. ESTROGENS AND PROGESTINS

These female reproductive hormones are widely used in contraception; less often for replacement therapy.

Regulation of female hormone production. *Gonadotropin-releasing hormone* (GnRH), a decapeptide released in pulses (about every hour) from the hypothalamus into the portal circulation to the anterior pituitary, stimulates the release of follicle stimulating hormone (FSH) and leutinizing hormone (LH) by pituitary gonadotrophs into the same portal circulation. FSH stimulates development of the ovarian follicle, which produces estrogen (E).

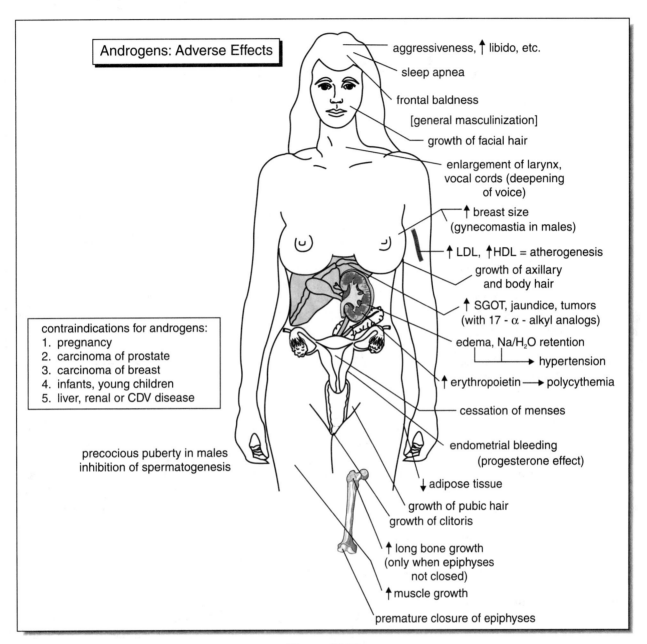

Figure 10.20

After the follicle has matured, LH stimulates the corpus luteum to produce both E and progesterone (P). Both E and P have a negative feedback regulation on GnRH production in the hypothalamus and on FSH and LH production in the pituitary. In the late follicular phase when E levels are elevated, GnRH induces an LH surge which stimulates ovulation **(Figure 10.22)**.

Chemistry of GnRH (Figure 10.23). Natural GnRH is the chemical structure of Gonadorelin; leuprolide and nafarelin are more-potent analogs with a longer $t_{1/2}$ approximately 3 hours versus ½ hour.

Pharmacodynamics. Since peptides are degraded in the GI tract, GnRH must be given i.v., s.c. or i.m.

Actions and uses of GnRH

1. All of the GnRHs, given in pulses, induce FSH and LH release, and ovulation. Consequently, GnRHs are used as fertility drugs and to induce puberty.

2. GnRHs given continuously (depot form), inhibit FSH and LH release. Hence, GnRHs are used to suppress ovulation, to suppress precocious puberty and to treat sex-hormone dependent conditions in women (breast cancer, endometriosis, uterine fibroids, polycystic ovarian disease). Leuprolide is used in this way in men, also, to treat prostate cancer when orchiectomy or estrogen therapy is unacceptable.

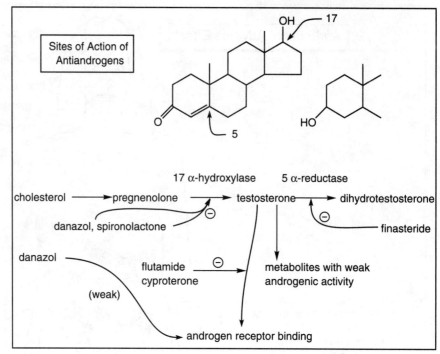

Figure 10.21

3. All of the GnRHs can be used to diagnose anterior pituitary gonadotroph function in males and females. Following GnRH injection (100 µg), LH is measured at intervals during the next 2 hours. Failure of an adequate LH rise is indicative of hypothalamic or pituitary hypogonadism.

Adverse effects. Except for local irritation, there are few adverse effects. Anaphylaxis can occur.

Gonadotropins—FSH, LH, human chorionic gonadotropin (hCG)

Chemistry of gonadotropins

FSH: α-89 amino acid chain: virtually identical with the α-chain of LH, hCG and TSH. β-112 amino acid chain: FSH biological activity resides in the β-chain. Glycosylation (carbohydrate complexation) of a and β-chains is essential for activity.

LH: α-89 amino acid chain. β-115 amino acid chain: LH-like biological activity resides in the β-chain. Glycosylation of α and β chains is essential for activity.

hCG: α-92 amino acid chain. β-145a.a. chain, with hCG_{1-115} being virtually identical with the LH β-chain. Glycosylation is essential for activity.

Actions

FSH, in the presence of E, stimulates growth and development of the ovarian follicle.

LH triggers rupture of the mature follicule and release of an ovum (ovulation), followed by stimulation of the corpus luteum to produce P.

hCG mimics LH, and stimulates the corpus luteum to produce P during pregnancy—until the placenta is able to synthesize adequate amounts.

Preperation
Menotropins (Pergonal) is an FSH and LH isolate from urine of postmenopausal women.
Urofollitropin (Metrodin) is an FSH isolate from urine of postmenopausal women.
hCG (Pregnyl) is an hCG isolate from urine of pregnant women.

Uses
1. Infertility
   a. In anovulatory hypoganotropic females menotropins are administered for 1-2 weeks (to induce follicular development), followed by a single high dose of hCG (to induce ovulation). This mimics hormonal changes in the ovarian cycle.

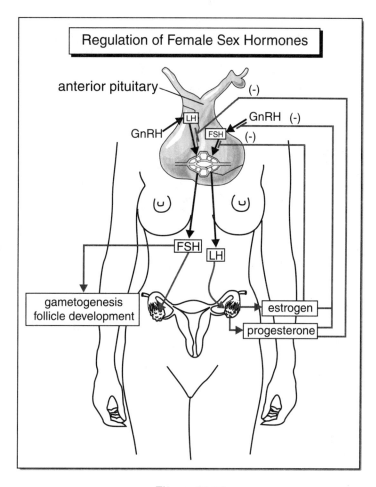

Figure 10.22

   b. In hypogonadotropic hypogonadic males, hCG is first administered for several months (to increase serum T) before addition of menotropin (for several additional months) to induce spermatogenesis.

   c. Urofollitropin (FSH) is used in fertility clinics to induce multiple ova.

2. hCG is used to promote descent of the testes in *cryptorchidism*. hCG should be discontinued if precocious puberty begins to occur.

ESTROGEN
Estrogen (E) is the hormone responsible for maturation and development of the female reproductive system. E also has the major role in development of secondary sex characteristics of women. In the ovarian cycle E governs the thickening and vascularization of the endometrium.

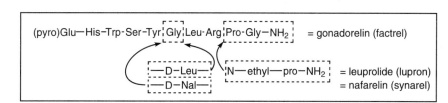

Figure 10.23

Estrogen synthesis and metabolism (Figure 10.24). Estradiol (E_2) is the major estrogen produced by the ovaries. The primary sequence for E_2 synthesis is different during the follicular phase (FSH-dependent) and luteal phase (FSH/LH-dependent) of the ovarian (menstrual) cycle. However, pregnenolone is the initial precursor in each phase, and testosterone (T) is the final precursor in each phase. E_1 and E_3 are made in largest amount in liver (from E_2) and other tissues (from androstenedione), not by the ovaries.

Plasma levels of E_2 are lowest at the start of the ovarian cycle and increase about tenfold just prior to ovulation. Only a small amount of E_2 is transported in blood in free form (2%). Most is bound to sex hormone binding globulin (SHBG) and albumin. There is extensive first pass metabolism of estrogens, so natural estrogens are not given orally, unless micronized, conjugated or esterified. Non-steroidal estrogens are also resistant to metabolism and are usually given orally. Intravaginal estrogens are not subject to first pass metabolism. Estrogens, by inducing cytochrome P-450, facilitate the metabolism of many drugs.

Actions and uses of estrogen (Figure 10.25). Estrogens act at receptors within the cell nucleus and induce protein synthesis and/or DNA replication. These effects highlight estrogen's anabolic actions, particularly the role in growth and development of the reproductive system. E is used in replacement or substitution therapy as follows.

1. E induces puberty in females with hypopituitary (low FSH) or ovarian (low E_2) function.

2. E is used in postmenopausal women to control "hot flashes," palpitations, osteoporosis **(see Section G)**, etc.

3. E is used in men to suppress androgen-dependent tumors, like prostate cancer.

There is much controversy about the benefit and risk of long-term E therapy. In postmenopausal women, the major risks relate to the possibility that E induces endometrial cancer, breast cancer and cancers in E-dependent tissues. It has not been conclusively determined if E does this at the low doses now used, particularly if P is added with E treatment.

Adverse effects of estrogens (Figure 10.26). When used as an oral contraceptive the regimen for E is altered in ways that tend to minimize the undesirable effects. The doses of E and P may be low for the first 12 days to mimic the menstrual cycle.

Preparations
1. Natural estrogens
   a. *Estradiol* (micronized [Estrace], dermal patch [Estraderm], vaginal cream or pellet [Progynon]).
   b. *Conjugated estradiols* include Estrone (E_1), Premarin and equilin (equine means horse; from mares' urine). These are orally active.
   c. *Esterified Es* are also taken orally (Menest).

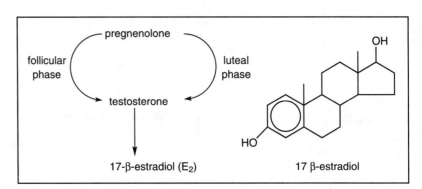

Figure 10.24

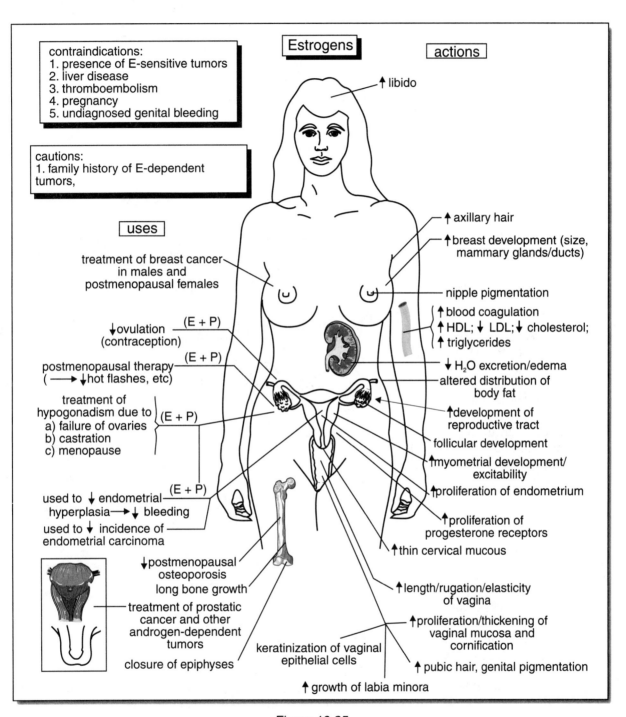

Figure 10.25

2. Semisynthetic estrogens
   a. E_2 esters are long-acting preps for parenteral use (E_2 benzoate, E_2 cypionate).

   b. Ethinyl E_2 (Estinyl) and mestranol are orally-active preps that are common in oral contraceptives.

3. Nonsteroidal estrogens
   a. Diethylstilbestrol (DES), quinistrol (Estrovis) and chlorotrianisene (Tace) are orally active. Ill-fated use of DES in pregnant women was found to produce vaginal adenocarcinoma in offspring.

PROGESTINS

Progesterone (P), as the name implies, is pro-gestational which means it maintains pregnancy. This hormone facilitates implantation of the fertilized ovum in the endometrium and then provides continuous induction of a vascularized, glandular endometrium throughout pregnancy.

Synthesis and regulation. P is synthesized from the precursor, pregnane. P is made in the ovarian follicle prior to ovulation, and by the corpus luteum after ovulation. Plasma P level

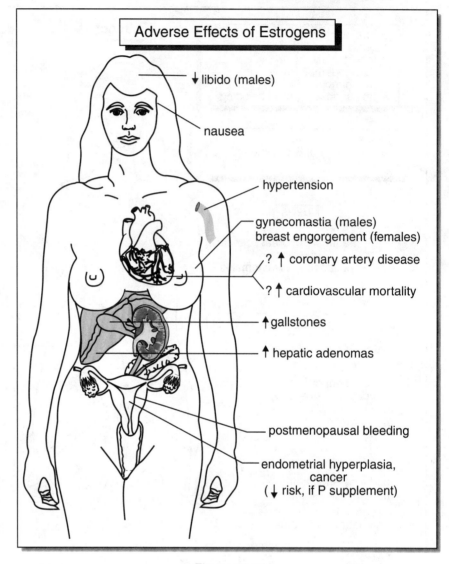

Figure 10.26

is highest during the second half of the ovarian cycle, and this high level is maintained if pregnancy occurs. LH is the principle gonadotropin that regulates P production during the ovarian cycle. hCG replaces LH as the regulator, once the placenta develops.

Analogs of progesterone (Figure 10.27). There are basically three types of P analogs, according to whether they are derived from pregnane (e.g., hydroxyprogesterone), estrane (e.g., norethynodrel) or gonane (e.g., norgestrel). These have high P activity and low testosterone activity.

Pharmacodynamics. Plasma P is largely bound to corticosteroid binding globulin (CBG is transcortin), in contrast to estrane analogs which are largely bound to steroid hormone binding globulin (SHBG); and pregnane analogs which have little affinity for CBG and SHBG. There is extensive first-pass metabolism of P, so natural P cannot be given orally unless micronized. The analogs of P are resistant to metabolism, so they are given orally.

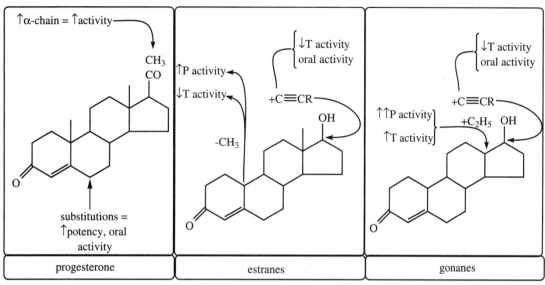

Figure 10.27

Actions and uses of progestins (Figure 10.28). P acts at receptors within the cell nucleus and induces protein synthesis. In the endometrium, E induces an increase in the number of P receptors, while P tends to reduce the number of both E and P receptors.

CONTRACEPTIVES

There are a variety of contraceptives containing E and P or P alone. These are usually taken for 21 consecutive days, followed by a 7 day drug abstinence, to allow menses to occur. There are no preps with E alone, because of the potential for induction of cancer and adverse cardiovascular effects. The dose of E is always low in oral contraceptives.

The objective is to reduce the basal level and surges in FSH and LH, so that ovulation does not occur. Also, the endometrial changes are less favorable to reception of an ovum than naturally-occurring changes.

1. *Monophasic oral preps* are those with a fixed dose of an estrogen (e.g., 20-50 ug/d, ethinyl estradiol or 75-100 ug/d, mestranol) used in combination with the fixed dose of a progestin (e.g., norgestrel, desogestrel, norethindrone or ethynodiol). These are usually taken for 21 consecutive days (e.g., Ovral). Desogestrel is a new less-androgenic progestin that is expected to produce less acne, less hirsutism and less weight gain.

2. *Biphasic oral preps* are those with a fixed dose of an estrogen used in combination with a low dose of progestin for 10 days, then a high dose of progestin for 11 days (e.g., Ortho-Novum 10/11).

3. *Triphasic oral preps* are those with an estrogen (fixed or variable dose) taken (a) for the first week with a low dose of progestin, then (b) for the next 5-9 days with a moderate dose of progestin, then (c) for the third period with a high dose of progestin. The triphasic period is always 21 days total (e.g., Triphasil, Ortho-Novum 7/7/7).

4. The "mini-pill" contains only a progestin (e.g., norethindrone or norgestrel), taken for 21 consecutive days. Thus, E-induced changes do not occur, so the endometrium remains atropic and ill-receptive for the ovum. Ovulation is not as consistently inhibited as when E is present in the contraceptive. *Breakthrough bleeding* is a problem (e.g., Micronor, Ovrette).

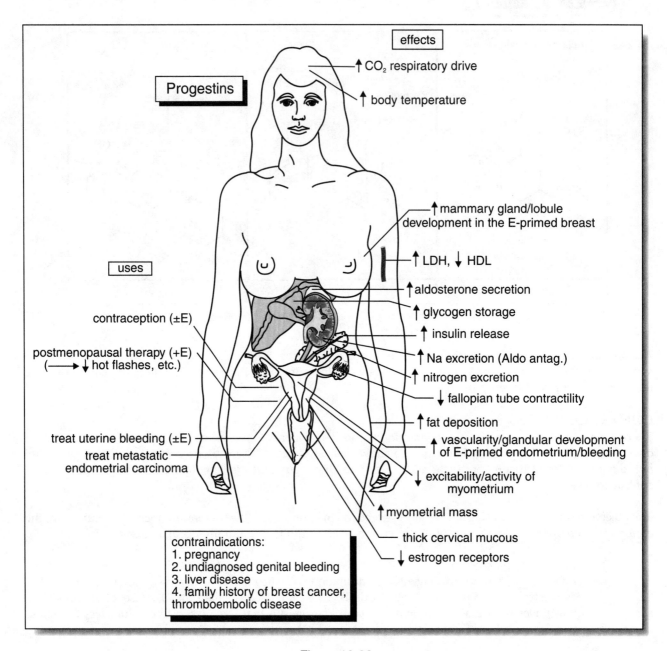

Figure 10.28

5. Progestins are sometimes given as a depot (e.g., Norplant) or with an intrauterine device.

6. The "morning-after pill" is a high dose of estrogen (e.g., diethylstilbestrol or ethinyl estradiol) taken within 48 hours of intercourse and for another 4 days, in order to prevent implantation of the ovum within the endometrium.

Adverse effects of oral contraceptives. These are mainly due to estrogen effects outlined in **Figures 10.25 and 10.26.** and include the following:

1. **Cardiovascular.** There is an increased incidence of myocardial infarction, thrombosis, stroke and hypertension. Tryglycerides are elevated.

2. Cancers of the ovary and endometrium occurred at a higher incidence with older higher dose E preps. The low dose E preps today, combined with a progestin, are thought to reduce the incidence of ovarian and endometrial cancers.

3. Gallstones occur more frequently.

ESTROGEN AND PROGESTERONE ANTAGONISTS

1. *Antiestrogens* are E-receptor antagonists, which prevent E from producing its effect in the hypothalamic-pituitary axis and at target tissues. The antiestrogens have some agonist activity, which is associated with side effects.

    Specific antiestrogens and uses
    a. *Tamoxifen* (Nolvadex) is an orally active nonsteroidal drug, used to treat E-dependent breast cancer in postmenopausal women.
    b. *Clomiphene* (Clomid) is also an orally active nonsteroidal drug. It is used to block the feedback inhibition of FSH and LH by E, resulting in excess FSH, LH and E production. At midcycle, there is a surge in FSH and LH causing ovulation. About 80% of anovulatory women will ovulate after one or more courses of clomiphene treatment. Multiple births sometimes occur.

    Adverse effects. Tamoxifen and clomiphene produce menopausal symptoms, including hot flashes. Tamoxifen may produce hepatotoxicity and endometrial tumors. Clomiphene tends to induce temporary ovarian enlargement.

2. *Antiprogestins* are P-receptor antagonists, which prevent P from producing its effects.

Specific antiprogestin. Mifepristone (RU 486) is used in Europe as an abortifacient. It is not approved in the United States.

1. By blocking P receptors, the hyperplastic endometrium degenerates and hCG production declines. This reduces P secretion by the corpus-luteum and triggers (a) degeneration of the endometrium, (b) increased uterine contractility and (c) increased cervical dilitation. These effects bring about the expulsion of an embryo **(Figure 10.29)**.

2. By blocking P receptors, this drug has the potential to also be used for endometriosis and P-dependent neoplasms like breast cancer.

3. This drug also blocks glucocorticoid receptors, confering a potential for use in Cushing's disease and in glucocorticoid-dependent tumors.

G. PARATHYROID HORMONE (PTH) AND CA++ REGULATION

PTH_{1-84} is derived from a 115 amino acid prohormone in the parathyroid glands. PTH_{1-34} has full activity. PTH increases serum levels of Ca^{++} and PO_4^{-3}.

1. In bone, PTH stimulates osteoclastic more than osteoblastic activity, causing bone resorption and dissolution of hydroxyapatite causing increased Ca^{++} and PO_4^{-3} in blood.

2. In kidneys, PTH enhances reabsorption of Ca^{++} and excretion of PO_4^{-3} (causing increased Ca^{++} and decreased PO_4^{-3} in blood); and increased synthesis of calcitriol, $1,25(OH)_2D_3$.

3. In gut, the PTH-induced calcitriol enhances absorption of Ca^{++} and PO_4^{-3}.

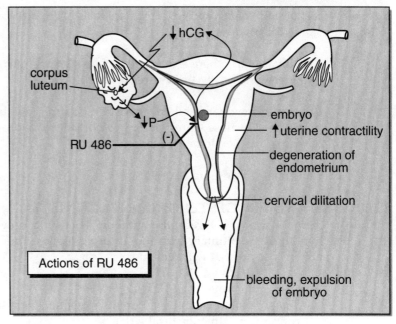

Figure 10.29

The net effect is increased Ca^{++} and decreased PO_4^{-3} in blood **(Figure 10.30)**. Low Ca^{++} concentration in the blood is a powerful inducer of PTH secretion.

VITAMIN D AND Ca^{++} REGULATION

Vitamin D_2 = Ergocalciferol (plant origin) [Calciferol].
Vitamin D_3 = Cholecalciferol (animal/human origin) [vitamin D_3].
These are prohormones.

$25(OH)D_3$ = 25-hydroxyvitamin D_3 = Calcifediol [Calderol].
$1,25(OH)_2D_3$ = 1,25-dihydroxyvitamin D_3 = Calcitriol [Rocaltrol].
$24,25(OH)_2D_3$ = 24,25-dihydroxyvitamin D_3.

These last three substances are hormones.
Vitamin D is a sterol hormone that has actions similar to PTH, except that vitamin D increases renal reabsorption of PO_4^{-3} **(Figure 10.31)**.

1. Vitamins D_2 and D_3 are consumed in the diet, but vitamin D_3 is also formed in the skin upon exposure to UV light.

2. Vitamins D_2 and D_3 are converted primarily in the liver to calcifediol, which
   a. increases Ca^{++} and PO_4^{-3} reabsorption in the kidneys (stimulated by PTH).

   b. is converted primarily in kidney by 1-α-hydroxylase to the most active product, calcitriol which enhances GI absorption of Ca^{++}, Ca^{++} binding protein (CBP) and PO_4^{-3}; and promotes bone resorption of Ca^{++} and PO_4^{-3}. Calcitriol also directly inhibits PTH release, as does the serum Ca^{++} produced by it. The 1-α-hydroxylase is increased by PTH, estrogens and PRL (to mobilize Ca^{++} into breast milk).

   c. is converted in liver also to $24,25(OH)_2D_3$, which promotes bone formation.

3. Excess vitamin D is stored in adipose tissue.

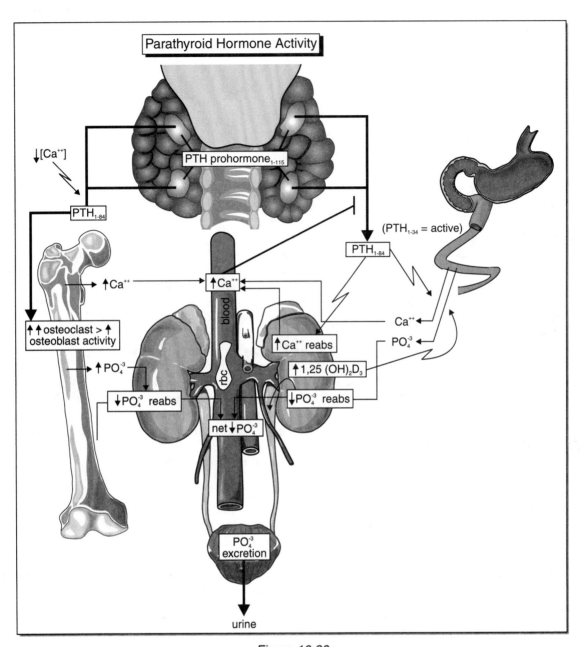

Figure 10.30

Vitamin Ds are administered orally. All are transported in blood by a vitamin D binding protein (i.e., an α_2-globulin). All have long half lives (approximately 1 day for vitamin D_3 to approximately 3 weeks for calciferol).

Therapeutic uses of vitamin Ds. These are used to treat hypocalcemia due to:
1. Vitamin D deficiency
    a. Hypovitaminosis D: give vitamin D.
    b. Vitamin D-dependent rickets
       Type I (low calcitriol production): give vitamin D or calcitriol

Type II (defective calcitriol receptor): give high dose calcitriol, not vitamin D.
   c. Vitamin D-resistant rickets (low renal PO_4^{-3} reabsorption): add PO_4^{-3} to high dose vitamin D.

2. Malabsorption

3. Hypoparathyroidism

4. Osteoporosis

5. Renal disease

Hypervitaminosis D would cause hypercalcemia.

OTHER SUBSTANCES AFFECTING Ca^{++} ACTIVITY
1. $Calcitonin_{1-32}$ is secreted by perifollicular thyroid cells in response to elevated serum Ca^{++} levels. The monomeric form has a $t_{1/2}$ less than 30 minutes, with inactivation occurring mainly in the kidneys. Calcitonin inhibits bone resorption and promotes renal excretion of Ca^{++} and PO_4^{-3}, resulting in reduced Ca^{++} + PO_4^{-3} levels in blood—a net action opposite that of PTH **(Figure 10.32)**.

Uses
   1. Calcitonin deficiency (as with thyroidectomy) or excess (as with medullary carcinoma of thyroid) does not produce known medical problems.
   2. Calcitonin (Calcimar) is used to treat Paget's disease, a disorder associated with (a) excessive osteoclastic activity (followed by disorganized new bone formation), (b) elevated serum alkaline phosphatase (from bone), (c) increased urinary hydroxyproline (from bone), (d) bone deformities and bone pain. Etidronate, a diphosphonate, is sometimes used with calcitonin.
   3. Calcitonin ± etidronate may be useful in osteoporosis.
   4. Calcitonin may inhibit nephrolithiasis, etc. in hypercalcemia.

2. Diphosphonates are chemical analogs of pyrophosphate which reduce bone turnover, inhibit $1,25(OH)_2D_3$ production and inhibit Ca^{++} absorption in the GI tract. Taken orally, they have a $t_{1/2}$ of several weeks in bone. Etidronate (Didronel) is being used for hypercalcemia (Paget's disease, hypervitaminosis D and osteoporosis).

3. Thiazides **(see Chapter 4)** decrease Na^+/Ca^{++} exchange in the distal tubules, thereby increasing Ca^{++} reabsorption in the kidneys. For this reason thiazides are useful in preventing renal stones in patients with hypercalciuria. Thiazides can produce hypercalcemia.

4. Plicamycin (Mithramycin) is an antibiotic that binds to DNA, inhibiting mRNA and protein synthesis. In a low dose this drug inhibits bone resorption and is thereby useful in treating hypercalcemia including Paget's disease).

5. Gallium nitrate inhibits bone resorption and is used for treating hypercalcemia (e.g., Paget's disease).

6. Saline diuretics increase excretion of all ions, including Ca^{++}. Furosemide facilitates Ca^{++} excretion and is used to treat hypercalcemia.

7. Phospate (i.v.) is a drastic measure to reduce serum Ca^{++} when other measures have failed. The risk is precipitation of $Ca_3(PO_4)_2$ in the circulation and organs, leading to renal failure.

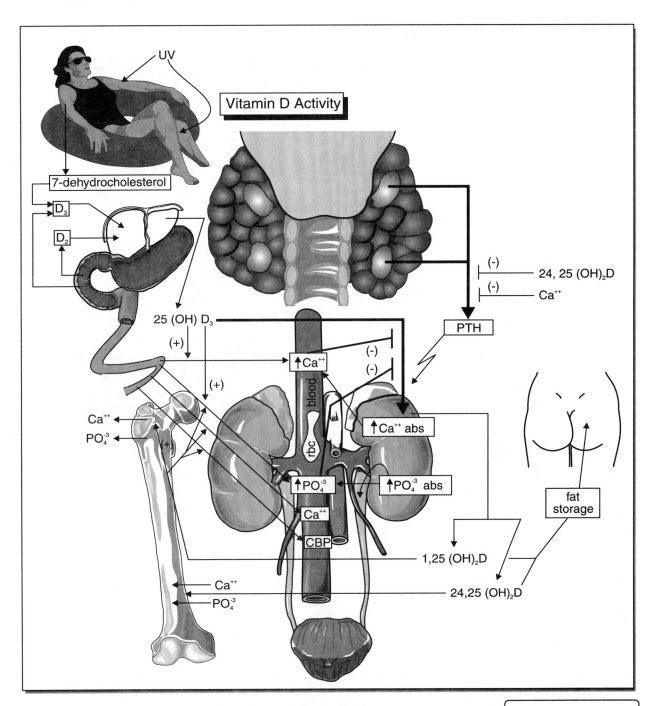

Figure 10.31

8. Glucocorticoids are a non-selective approach towards hypercalcemia, owing in part to anti-vitamin D activity. These drugs are useful in hypervitaminosis A and sarcoidosis, which leads to elevated levels of 1,25(OH)$_2$D.

Lymphatic cancers (e.g., multiple myeloma) often release an osteoclast activating factor which induces bone resorption and hypercalcemia. Glucocorticoids commonly reduce tumor size and correct the hypercalcemia.

9. Calcium salts, oral or i.v., will correct hypocalcemia.

10. Fluoride (1 ppm) in drinking water, or topically applied to teeth, hardens the enamel of developing teeth (i.e., in children), thereby preventing caries. Excess fluoride, because of strong binding to enamel, mottles the teeth permanently. Fluoride is being tried for osteoporosis.

Altered Calcium Homeostasis. *Hypercalcemia* is associated with a variety of symptoms and can be produced by a variety of physio/pathological conditions and/or drugs (**Figure 10.33**). Treatments are shown in **Figure 10.34**.

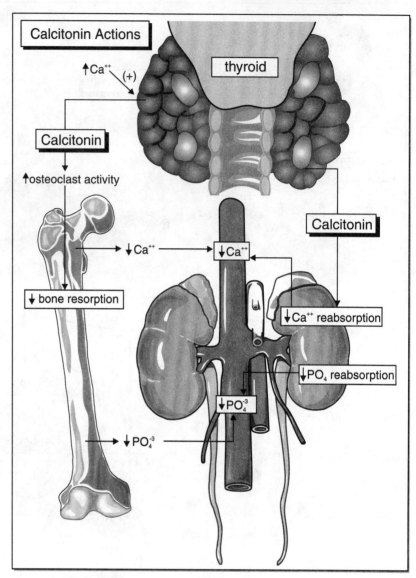

Figure 10.32

Hypocalcemia is associated with symptoms that are generally opposite those in hypercalcemia (**Figure 10.35**). Treatments are shown in **Figure 10.36**. In true hypoparathyroidism there is a relative absence of PTH. In pseudoparathyroidism PTH is present, but receptors for PTH are defective. To differentiate, measure PTH blood level (absent in true hypoparathyroidism) and measure PO_4^{-3} excretion in urine after exogenous PTH injection (no increase in PO_4^{-3} excretion in pseudoparathyroidism). Either form of hypoparathyroidism is treated with vitamin D.

Prevention/treatment of postmenopausal osteoporosis. Bone mass is greatest around 25 years of age and remains constant until around 40 years. From this time Ca^{++} in bone begins to be resorbed progressively, and is greatest in women during the first few years after menopause. The period just before and just after menopause is considered to be the most important time for drug intervention. Possible strategies are:
1. Supplemental Ca^{++}. This seems to be more helpful in recent rather than older menopausal women.

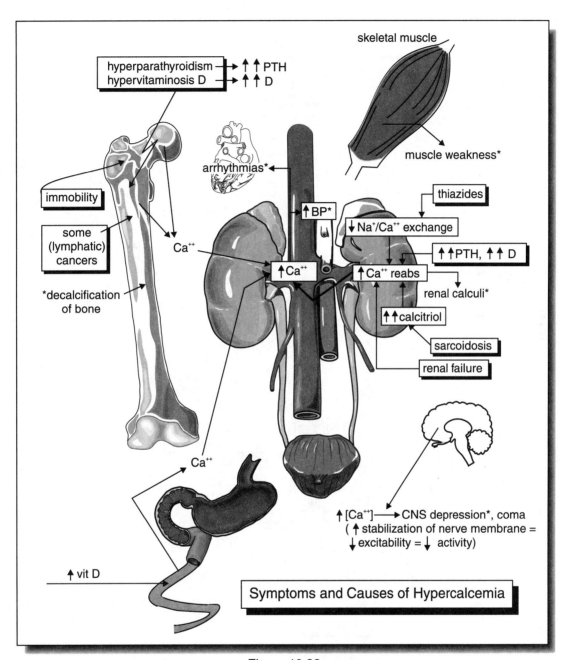

Figure 10.33

2. Antiresorptive drugs such as:
   a. Estrogen replacement, but a progestin must be added to prevent development of endometrial cancer. Progestins produce an unfavorable blood lipid profile and may increase the risk of breast cancer.
   b. Salmon calcitonin (Calcimar), but this must be given s.c. every 1 or 2 days. Human calcitonin is not approved for this use. Calcitonin does not reduce the loss of Ca^{++} from bone in the extremities.
   c. Biphosphonates (etidronate), but these are not approved for this use.

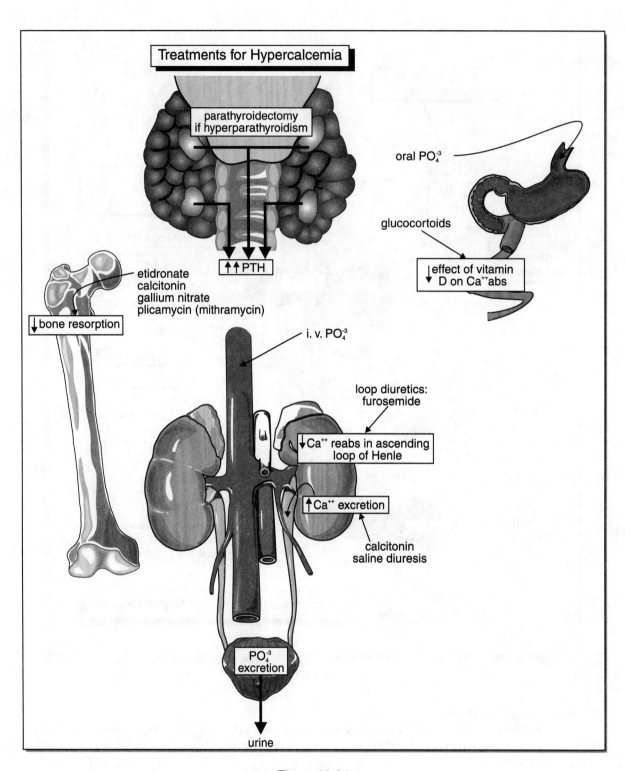

Figure 10.34

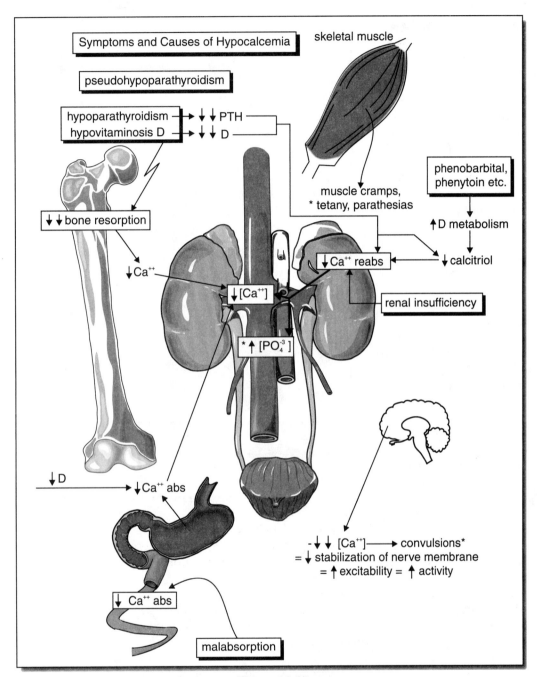

Figure 10.35

3. Vitamin D, which is known to be helpful in mild to moderate osteoporosis.

4. Fluorides, although it seems that this may make bones brittle and more prone to fracture.

276 PHARMACOLOGY

H. INSULIN, GLUCAGON AND ORAL HYPOGLYCEMIC AGENTS

INSULIN

Synthesis, storage and breakdown of insulin (Figure 10.37). Insulin is formed in *pancreatic β-cells* in the islet of Langerhans, when a C-peptide fragment is cleaved from the single chain *precursor proinsulin* protein molecule. The 30 amino acid (aa) B chain fragment of insulin is linked by 2 disulfide bonds to the 21 aa A chain. There are species differences in insulins, with beef insulin differing from human insulin by 3 aa's and pork insulin differing by 1 aa. Insulin is also stored in the β-cells, as a complex of 6 insulins with 2 zinc (Zn) atoms. Degradation of insulin occurs through cleavage of the disulfide bonds by insulinases in the liver and kidney **(Figure 10.37)**. Plasma $t_{1/2}$ of insulin is about 5 minutes.

Release and actions of insulin. The primary signal for insulin release is an elevation in the blood glucose (G) level. Insulin acts at specific insulin receptors on cell surfaces. Glucagon, corticosterones and epinephrine exert anti-insulin effects on carbohydrate metabolism. Insulin receptors are down-regulated in obesity.

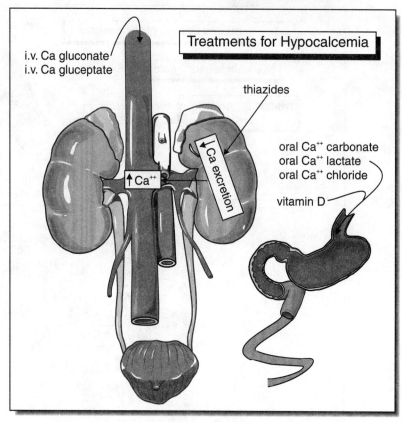

Figure 10.36

The major action of insulin is enhanced energy storage in liver, muscle and fat **(Figure 10.38)**. Insulin facilitates G and aa uptake in skeletal muscle and fat (liver uptake of G and aa's does not require insulin) and induces the enzymes involved in storage of these substrates as either glycogen, fat or protein.

Diabetes. Type I diabetes is known also as insulin-dependent diabetes mellitus ("juvenile diabetes") and is characterized by destruction and loss of pancreatic β-cells and low circulating levels of insulin. Type I diabetes is an autoimmune disorder. Treatment requires insulin replacement. However, cardiovascular and renal complications still progress in insulin-treated diabetics.

Type II diabetes is known as non-insulin-dependent diabetes mellitus ("maturity-onset diabetes") and is associated with insulin receptor defects. If weight reduction alone is not corrective, oral hypoglycemics are usually an effective treatment. Mostly in the case of severe illness or pregnancy, insulin might be added temporarily to control blood glucose. In the United States 80-90% of diabetics have Type II diabetes; about two-thirds of them are obese.

EXOGENOUS INSULIN

Pharmacokinetics. Because insulin is destroyed in the GI tract, parenteral administration is necessary. Usually insulin is injected s.c.; in emergencies, i.m. or i.v. The liver degrades about half the insulin in the portal circulation.

ENDOCRINE PHARMACOLOGY

The kidneys eliminate or degrade about half the insulin filtered at the glomerulus. Dosage adjustment is indicated for patients with liver or kidney disease.

Types of preps. *Beef* and *pork* insulins are obtained from the animals' pancreas and are highly purified (less than 10 ppm of proinsulin). Immunogenicity is low.

Human insulin is semisynthetic from pork insulin or obtained from recombinant DNA methods. There is even less immunogenicity of this form.

Standardization. Biological potency of insulins is expressed as 'units': U-40, U-100, U-500 (for 40, 100 or 500 units per cc).

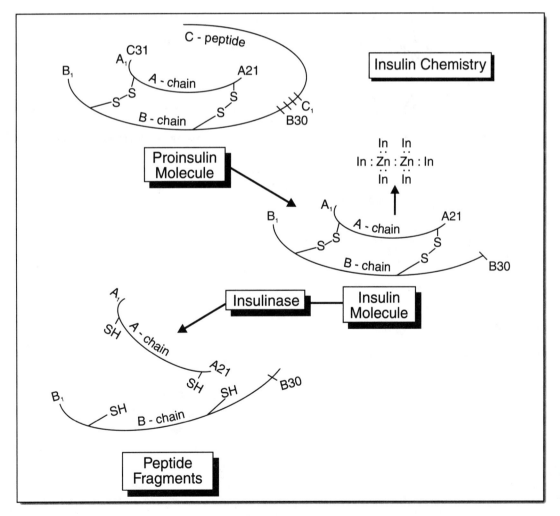

Figure 10.37

Specific insulin preps
1. *Short-acting insulins* are Zn-insulin forms (beef, pork or human insulin). The onset of effect is 15-30 minutes; duration is about 6-8 hours.

    Regular standard (Regular insulin).

    Regular purified (Velosulin).

    Regular human (Humulin).

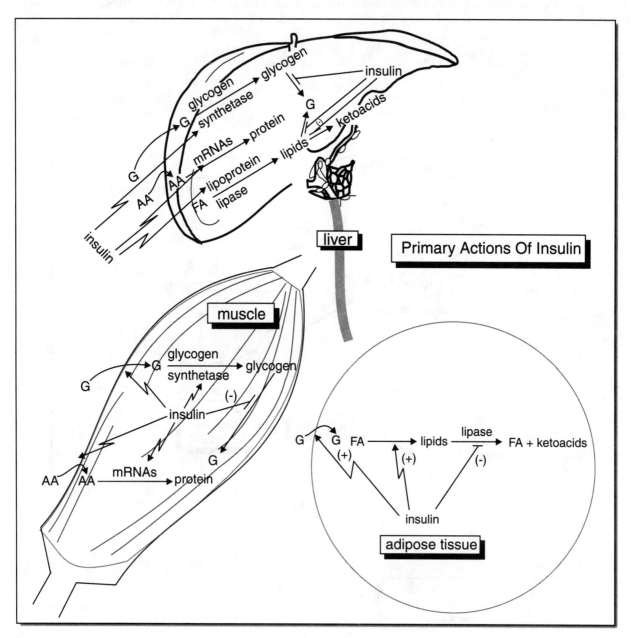

Figure 10.38

Regular forms are soluble and are the only preps that can be given i.v.

Semilente standard (Semilente insulin): Semilentes are fine suspensions with a duration of about 14 hours.

2. *Intermediate-acting insulins* are turbid suspensions of insulin, complexed with either *protamine* or Zn. The onset of effect is 1-2 hours; duration is 6-15 hours.

Lente (Lente, Iletin I).

NPH standard (NPH Iletin I).

3. *Long-acting insulins* are formulated like the intermediate-acting insulins. The onset of effect is 4-6 hr; duration is about 32-36 hours.

Ultralente standard (Ultralente Iletin I).

Combinations of short-, intermediate- and long-acting preps are used to maintain reasonable insulin and carbohydrate levels throughout the day.

Adverse effects of exogenous insulin
1. The most common adverse effect is hypoglycemia, accompanied by increased heart rate, sweating and sometimes even coma. This can be minimized with oral carbohydrates; in emergencies, with glucagon.

2. All patients develop at least a low titer of IgG anti-insulin antibodies. This is normally not a problem. If the titer becomes high, insulin-resistance develops and the diabetic must be switched to a different purer form of insulin.

3. IgE-mediated allergic sensitivity is rare, although there often is an allergic reaction at the injection site, and lipoatrophy or lipohypertrophy.

4. Hyperglycemic rebound may also occur.

ORAL HYPOGLYCEMIC AGENTS
Oral hypoglycemic agents effectively lower blood glucose levels by releasing insulin from the pancreas and sensitizing tissues to the effects of insulin. These drugs are used to treat Type II diabetes.

Sulfonylureas (Figure 10.39)

Mechanism (Figure 10.40)
1. Sulfonylureas are thought to depolarize the pancreatic β-cells, while blocking ATP-sensitive K^+ channels. A compensatory opening of voltage-dependent Ca^{++} channels leads to increased Ca^{++} intracellularly, facilitating fusion of insulin granules with the outer cell membrane. This enhances insulin release.

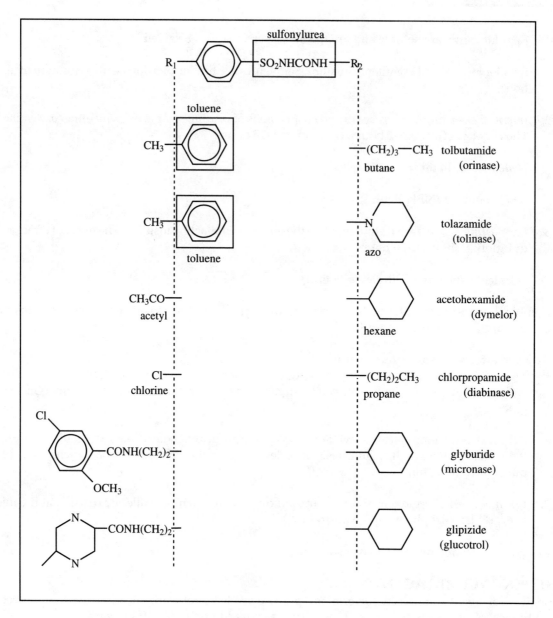

Figure 10.39

2. Sulfonylureas also increase the sensitivity of target tissue to the effects of insulin, possibly by increasing the affinity of insulin receptors.

3. Sulfonylureas inhibit glucagon release, a hormone that generally has an effect opposite to that of insulin.

Pharmacokinetics. The compounds in **Figure 10.39** are clearly non-polar, lipid soluble substances. They are well absorbed from the GI tract, attain peak levels in about 2-4 hours and are highly protein bound (greater than 90%). Tolbutamide has the lowest duration of action (6-8 hours) and therefore is least likely to produce severe hypoglycemia. This makes it preferred in elderly patients. The other sulfonylureas have durations around 12 hours or more.

Being highly protein bound the "first generation" sulfonylureas displace (or are displaced by) other drugs that are highly protein bound (e.g., warfarin, NSAIA's, etc.).

Glyburide and Glipizide are "second generation" drugs, distinguished by their higher potency and the fact that their protein binding is non-ionic, so they do not displace drugs that are protein bound. A second generation drug may act when a first generation drug does not. The second generation drugs are contraindicated if there is hepatic failure or renal insufficiency. All of these drugs are metabolized in the liver; some metabolites are active.

Adverse effects
1. Hypoglycemia is the primary adverse effect, being more dangerous with those drugs having long durations of action. Hypoglycemia is more common when there is renal or hepatic impairment. The consequences of this side effect can be neurological damage and death.

2. Chlorpropamide enhances ADH (antidiuretic hormone) action, which leads to fluid retention and water intoxication.

3. Chlorpropamide also has a disulfiram action, which can lead to severe reactions in people that consume alcohol **(see Chapter 9, on Drugs of Abuse)**.

4. Rash and blood dyscrasias rarely occur. Tolbutamide reportedly increases cardiovascular mortality. There is controversy on this issue.

5. Many drugs will influence the activity of sulfonylureas, so blood glucose must be monitored closely and dosage adjustments must be made when adding or subtracting another drug. Propranolol should be avoided since it prevents the hypoglycemic-induced tachycardia—masking a recognizable danger signal in the patient.

Use of sulfonylureas. Only Type II diabetics with mild or moderate elevation in blood glucose levels are candidates for treatment with sulfonylureas.

Biguanides. These are investigational in the U.S. but are used in Europe. They reduce elevated blood glucose levels but do not produce hypoglycemia. The mechanism seems to relate to enhanced glucose utilization in tissues, not release of insulin from the pancreatic β-cell.

GLUCAGON
Synthesis, release and action. *Glucagon* is a 29 amino acid peptide, produced by pancreatic α-cells in the islets of Langerhans. Low sugar levels

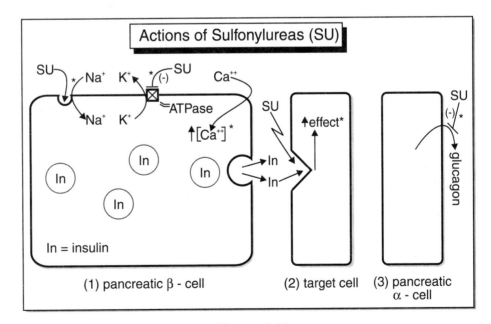

Figure 10.40

induce glucagon release, while high sugar levels inhibit glucagon release. This regulation is the opposite of that for insulin, and indeed glucagon generally produces effects opposite to that of insulin.

Although glucagon is regulated primarily by the glucose concentration in blood, its release is enhanced by amino acids, sympathetic nerve activity and sympathomimetic amines.

Effects of glucagon. After its release, the last 6 amino acids at the N-terminus bind to a receptor on liver cells and initiate a cascade of reactions that result in

    a. glycogenolysis ---> blood glucose
    b. gluconeogenesis ---> blood glucose
    c. lipolysis ---> ketogenesis

Glucagon tends to release insulin, and in Type I diabetics blood glucagon levels are high. In Type I diabetes glucagon has a major role in inducing the catabolism of proteins for energy production.

Use of glucagon. Glucagon (i.m., i.v., s.c.) is used in hypoglycemic emergencies for elevating blood glucose levels. Its $t_{1/2}$ is less than 5 minutes. Adverse effects include nausea and vomiting, release of catecholamines from a pheochromocytoma, inotropy and chronotropy.

CHAPTER 11
ANTIMICROBIAL CHEMOTHERAPY

BASIC PRINCIPLES

Antimicrobial drugs are those that exert relatively selective toxicity on microbes, relying on differences in cellular biochemistry or composition between the microbe and the host.

Use of Antimicrobials. The greatest degree of success with antimicrobials is achieved when (a) the microbe is identified in cell culture and (b) the bacterial sensitivity profile is established in cell cultures. The (c) site of infection sometimes limits the choices of antimicrobials. (d) Health status of the patient determines, in part, how aggressive the therapy must be (e.g., age, immunocompetence, renal/liver status, etc.). Also, the (e) prior experience of a patient with particular antimicrobials should be ascertained (e.g., penicillin-allergy).

Definitions
Antimicrobial—a substance made by one microbe against another.
Antibiotic—an antimicrobial, semi-synthetic form or fully synthetic agent.
Bacteriostatic—a substance that inhibits growth of microorganisms, thereby aiding host immune defenses which kill the bacteria (e.g., sulfonamides, tetracyclines).
Bactericidal—a substance that kills microorganisms (e.g., β-lactams, aminoglycosides).
Broad spectrum—active against a wide variety of microorganisms, frequently both gram-positive and gram-negative bacteria.
Narrow spectrum—active against a small group of bacteria.

Mechanisms of antimicrobial action. There are several major means by which these drugs exert their effects (Figure 11.1).
1. **Inhibition of cell wall synthesis.** β-lactams (penicillins and cephalosporins) inhibit the transpeptidase in bacterial cytoplasmic membrane, thereby inhibiting cross-linking of peptidoglycans. Cell walls become excessively permeable and a bactericidal effect occurs. Vancomycin, bacitracin and

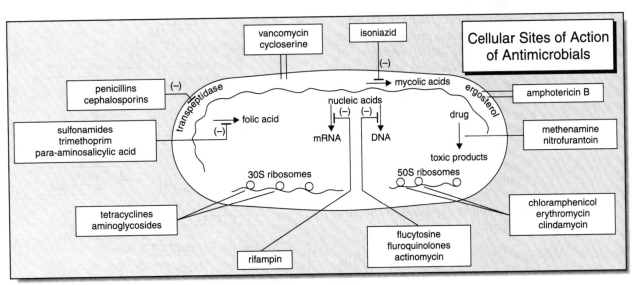

Figure 11.1

cycloserine bind to different sites in bacterial cell membranes and also inhibit cell wall synthesis. Isoniazid inhibits synthesis of mycolic acids in mycobacterial cell walls, exerting a bactericidal effect **(Figure 11.1)**.

2. **Alteration of cell membrane permeability**. Amphotericin B exerts a fungicidal effect by binding to the fungal steroid ergosterol in cell membranes and thereby producing excess permeability **(Figure 11.1)**.

3. **Inhibition of protein synthesis**. Many of the antimicrobials inhibit one of several sites involved in synthesis of microbial protein **(Figure 11.1 and 11.2)**.
    a. Rifampin inhibits RNA polymerase, thus inhibiting mRNA formation.
    b. Aminoglycosides bind to the bacterial 30S subunit of 70S ribosomal protein, preventing formation of the initiation complex (mRNA-tRNA-formylmethionine) involved in protein synthesis. Misreading errors in translation produce non-functional protein.
    c. Tetracyclines bind to 30S ribosomal protein and prevent aminoacyl tRNA binding to the receptor site on 50S ribosomes.
    d. Chloramphenicol binds to 50S ribosomal protein and prevents binding of the amino acid terminus of aminoacyl tRNA to the acceptor site on the 50S ribosome. Consequently, peptidyl transferase is unable to interact with its amino acid substrate.
    e. Erythromycin and clindamycin bind to bacterial 50S ribosomal protein to inhibit translocation of newly-synthesized peptidyl tRNA from acceptor- to donor-sites on the ribosome. Protein synthesis is thereby inhibited.

Selective inhibition of microbial versus mammalian protein inhibition is related to differences in microbial 70S ribosomal protein versus mammalian 80S

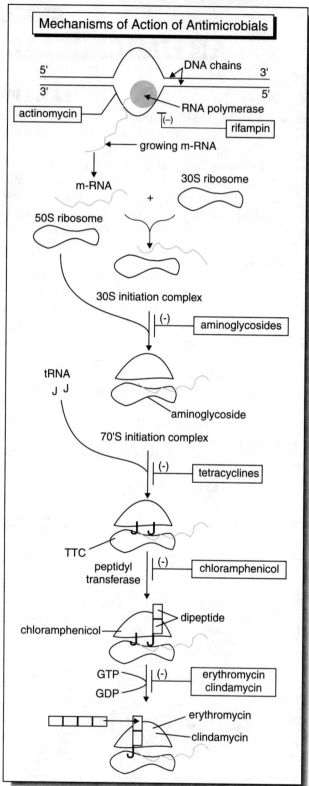

Figure 11.2

ribosomal protein. There is a reasonably close similarity between microbial 70S- and mammalian mitochondrial 70S-proteins. Some adverse effects of chloramphenicol are associated with chloramphenicol binding to mammalian 70S mitochondrial protein.

4. **Antimetabolite effect.** Some antimicrobials prevent microbial synthesis of essential products. For example, sulfonamides and trimethoprim compete with bacterial substrates involved in folic acid synthesis. Inhibition of folic acid synthesis by either class usually results in a bacteriostatic effect; when a sulfonamide is co-administered with trimethoprim, a bactericidal effect often ensues. Para-aminosalicylic acid is a false substrate for folic acid synthesis in mycobacteria **(see Figure 11.7)**.

5. **Inhibition of nucleic acid synthesis.** The antifungal drug, flucytosine, is metabolized to 5-fluorouracil, which inhibits thymidylate synthetase and production of DNA **(Figure 11.29)**. Fluoroquinolones (Norfloxacin, Ciprofloxacin) bind to DNA, inhibiting DNA replication **(Figure 11.1)**. Actinomycin binds to DNA while rifampin binds specifically to the DNA-dependent RNA polymerase in mycobacteria. Both drugs inhibit the DNA-dependent RNA-polymerase, so that mRNA is unable to be synthesized **(Figure 11.1)**.

6. **Liberation of a toxic product.** Methenamine spontaneously decomposes to formaldehyde (HCHO) which is toxic to bacteria **(Figure 11.11)**. Nitrofurantoin is converted to reactive intermediates that damage bacterial DNA. Both drugs are used as urinary tract antiseptics.

Resistance. Microbes may have intrinsic resistance or be able to acquire resistance to antimicrobials by genetic means, usually plasmid-mediated (e.g., enhanced synthesis of penicillinases) but sometimes chromosomally-mediated (e.g., enhanced synthesis of cephalosporinases). A variety of processes are implicated **(Figure 11.3)**.

1. **Inability of the drug to bind to protein**
   a. Some microbes have intrinsic proteins, unable to bind penicillins. These microbes are intrinsically resistant to penicillins. After exposure to penicillins, some microbes alter the protein structure of the cell wall, so that it does not bind the drug.
   b. The 30S ribosomal protein may be modified so that it is unable to bind aminoglycosides.
   c. The 50S ribosomal protein may be modified so that it is unable to bind erythromycin.

2. **Reduced uptake of the drug into the microbe**
   a. Upon exposure to tetracyclines, transporter protein may be modified in microbes, so that the modified transporter protein is unable to efficiently transport the drug inside the microbe.
   b. The relative impermeability of the cell wall of streptococci to aminoglycosides can be overcome by partially damaging the cell wall with penicillins.

3. **Increased metabolism of the drug**
   a. Some microbes are able to enhance production of β-lactamases (penicillinases) which cleave the β-lactam ring of penicillin, thereby inactivating the drug (e.g., Staphylococci).
   b. Plasmid-mediated production of adenylating, acylating and phosphorylating enzymes may be enhanced in gram-negative bacteria, to inactivate aminoglycosides.

4. **Increased production of substrate to overcome antimetabolite effect.** Upon exposure to sulfonamides which compete with the natural substrate p-aminobenzoic acid (PABA) during folic acid synthesis, bacteria may increase production of PABA and override the effect of a sulfonamide.

5. **Development of an alternate pathway.** Some microorganisms treated with sulfonamides develop pathways that utilize exogenous folic acid.

Genetic mechanisms involved in resistance (Figure 11.3)

1. **Chromosomal mechanisms.** Within the genetic code for microorganisms are specific regions able to confer resistance to select antibiotics. A genetic mutation is frequently involved in alterations of binding proteins for (a) penicillin in the cell wall, (b) aminoglycosides (30S protein) and (c) erythromycin (50S protein).

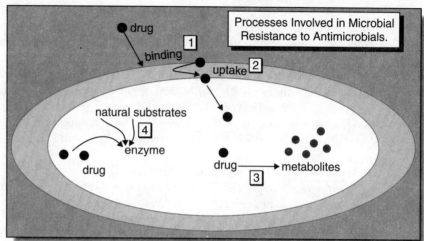

Figure 11.3

2. **Plasmid-mediated resistance.** Plasmids are circular strands of DNA present either within the cytoplasm or attached to the bacterial chromosome. Plasmids carry many codes for production of proteins/enzymes (e.g., β-lactamases) that are involved in the development of antimicrobial resistance. These specialized regions on plasmid DNA are called R-factors. Synthesis of activator protein for the start codon of these genes is often induced by antibiotics.

3. **Resistance can be transmitted between bacteria by the following processes.**
    a. **Conjugation.** Bacteria "mate," so that there is unilateral transfer of plasmid-genes **(Figure 11.4)**.
    b. **Transduction.** A viral-carrier of plasmid genes "infects" another bacterium **(Figure 11.5)**.
    c. **Transposition.** Bacteria "mate" and bilaterally exchange plasmid genes or plasmid-chromosomal genes.

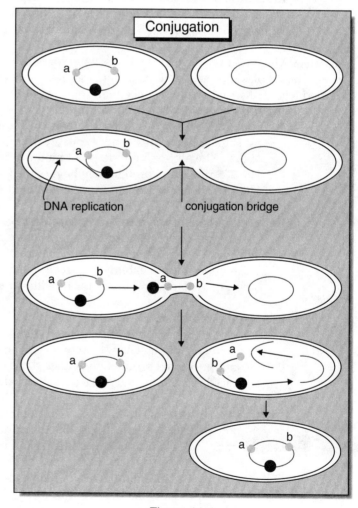

Figure 11.4

Adverse effects of antimicrobials. Dose-related adverse effects may occur with any of the antimicrobials, although these are extremely rare with β-lactams, because of their high therapeutic index. Allergic reactions are among the most frequent adverse effects of antimicrobials. Idiosyncratic reactions may also be seen (e.g., marrow or hepatic effects). Superinfections represent an inadvertent adverse effect, not caused by the antibiotic molecule but caused by the antibiotic effect. That is, when the growth of one organism is inhibited by an antibiotic, the growth of a competing organism goes unchecked. The end result is a superinfection. For example, when bacteria are killed, fungi often proliferate. Superinfections are more common with broad-spectrum than with narrow-spectrum antibiotics (penicillin G less than tetracyclines, chloramphenicol less than third generation cephalosporins).

Combined use of antimicrobials. Antimicrobials are sometimes used in combination for additive or synergistic effects, especially in the following situations:

1. When there is a mixed bacterial infection.
2. When there is severe infection.
3. To prevent emergence of resistant organisms.

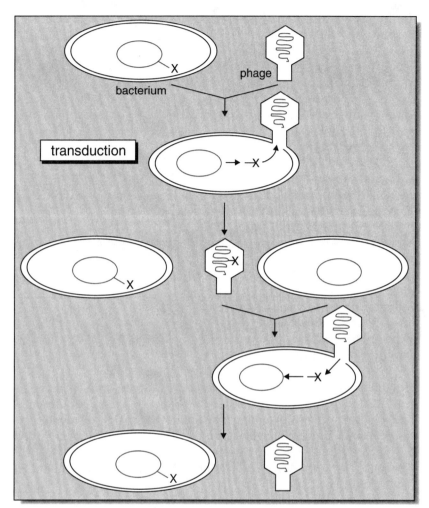

Figure 11.5

It should be appreciated that one antimicrobial can antagonize another. For example, penicillin is cidal only when bacterial are replicating. If a bacteriostatic agent like tetracycline is co-administered, the effect of penicillin is negated and bacterial overgrowth with greater physical debility and death can occur.

1. ANTIFOLATE BACTERIOSTATIC AND BACTERICIDAL DRUGS

A. SULFONAMIDES

These were the first antimicrobial agents. Because of the superiority of newer antibiotics, penicillins in particular, the use of sulfonamides is largely restricted to urinary tract infections.

Chemistry and mechanism. Sulfonamides are chemical analogs of para-aminobenzoic acid (PABA)**(Figure 11.6)**, a substrate used in microbial synthesis of folic acid **(Figure 11.7)**. Sulfonamides compete with PABA for dihydropteroate synthase, the first of a series of enzymes involved in folic acid synthesis. Consequently, de novo synthesis of purines and DNA is impaired. A bacteriostatic effect is produced.

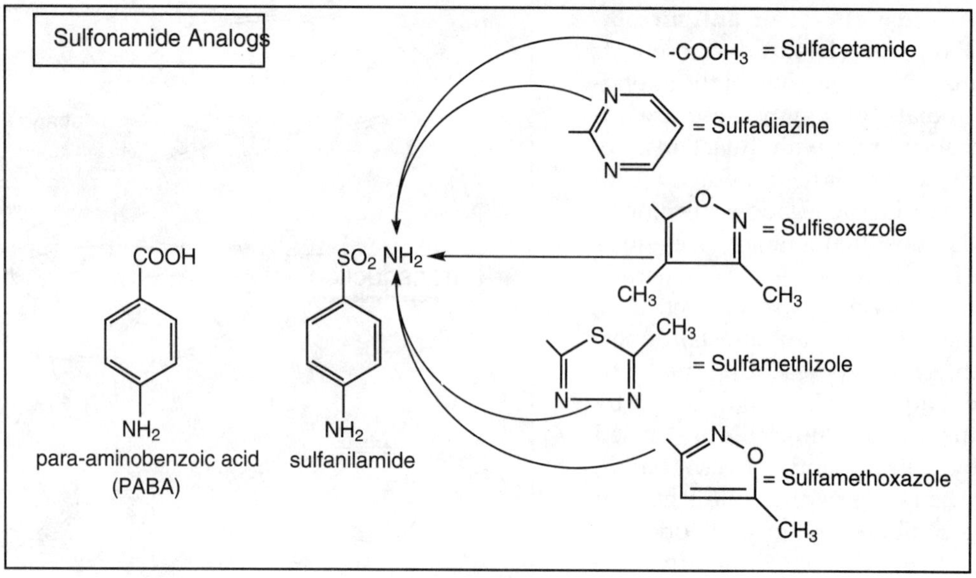

Figure 11.6

Although sulfonamides have much greater affinity for microbial versus mammalian enzymes, some toxicity is seemingly related to reduced folic acid production in humans, especially in people with folic acid deficiency. Since folic acid permeates mammalian but not microbial cell membranes, some toxicities, but not antimicrobial activity, can be prevented by co-administering folic/folinic acid with the sulfonamide.

Substances that are metabolized in humans to PABA (e.g., local anesthetics) could conceivably inhibit the action of sulfonamides.

Pharmacokinetics. Water-soluble sodium salts of sulfonamides can be taken orally and liberate the base form in the alkaline duodenum. The latter, being highly lipid-soluble, are well-absorbed (about 90%) and distribute to all sites including the CNS and across the placenta. The $t_{1/2}$ is approximately 6-12 hours with most sulfonamides, but only 2 hours for sulfamethizole and about 10 days for sulfadoxine.

Most (over 50%) is plasma-protein bound. Bilirubinemia and kernicterus occurs in infants, since sulfonamides displace bilirubin from its binding sites. Accordingly, sulfonamides should not be given to pregnant women, infants or nursing mothers. Sulfonamides displace other protein-bound drugs, enhancing their activity (e.g., phenytoin, sulfonylureas, oral anticoagulants). In turn, some drugs (e.g., salicylates, phenylbutazone) displace sulfonamides, enhancing their activity.

Sulfonamides are metabolized by acetylation and glucuronidation. The acetylated form is inactive but still toxic. Renal nephrolithiasis was a particular hazard with older sulfonamides, but newer agents like sulfisoxazole are quite soluble. The potential for renal precipitation can be reduced by alkalinizing the urine (with bicarbonate, etc.) and increasing water intake. Since sulfonamides are excreted primarily by the kidney, these drugs should not be used when there is renal impairment.

A combination of three sulfonamides (sulfadiazine, sulfamerazine and sulfamethazine) is sometimes used to increase solubility. These have independent solubilities, so a higher absolute concentration of sulfonamides can be achieved.

Adverse effects
1. Hypersensitivity reactions (about 5%) include rash, photosensitivity, erythema, exfoliative dermatitis and Stevens-Johnson syndrome with erythema multiforme and mucosal ulceration.
2. *Hepatotoxicity* and jaundice occurs.
3. *Bone marrow suppression* occurs: aplastic anemia, megaloblastic anemia, eosinophilia, thrombocytopenia, agranulocytosis. Supplemental folinic acid suppresses these effects on the marrow, especially in malnourished or alcoholic patients. Marrow suppression and hemolytic anemia are more likely to occur in patients with a genetic deficiency of glucose-6-phosphate dehydrogenase. The incidence of thrombocytopenia and purpurea is increased with thiazides.

B. TRIMETHOPRIM

Chemistry and mechanism. Trimethoprim, like sulfonamides, inhibits folinic acid synthesis in bacteria. Trimethoprim competes with dihydrofolic acid as a substrate for dihydrofolate reductase **(Figure 11.7)**. When used alone, the effect is bacteriostatic; when co-administered with sulfamethoxazole the effect is often bactericidal.

Pharmacokinetics. The pharmacokinetic spectrum is much like that of sulfonamides. However, in contrast to sulfonamides, trimethoprim is an alkaline drug. Accordingly, trimethoprim is active in alkaline fluids like prostatic secretions and is used for treating prostate infections. Renal elimination is enhanced by acidifying the urine (e.g., ammonium chloride).

Adverse effects. These include hypersensitivity reactions and blood dyscrasias as per sulfonamides.

Special formulation. A preparation of trimethoprim with sul-

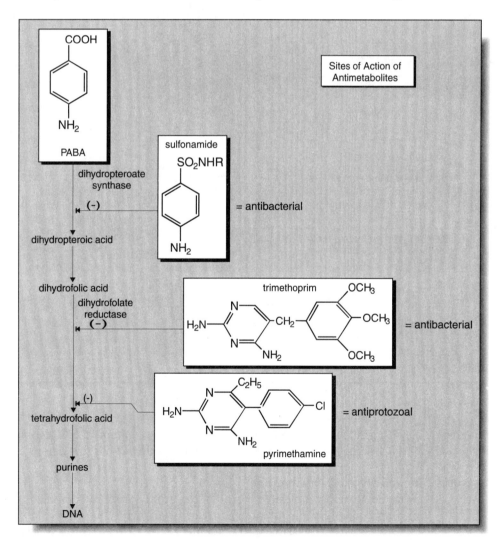

Figure 11.7

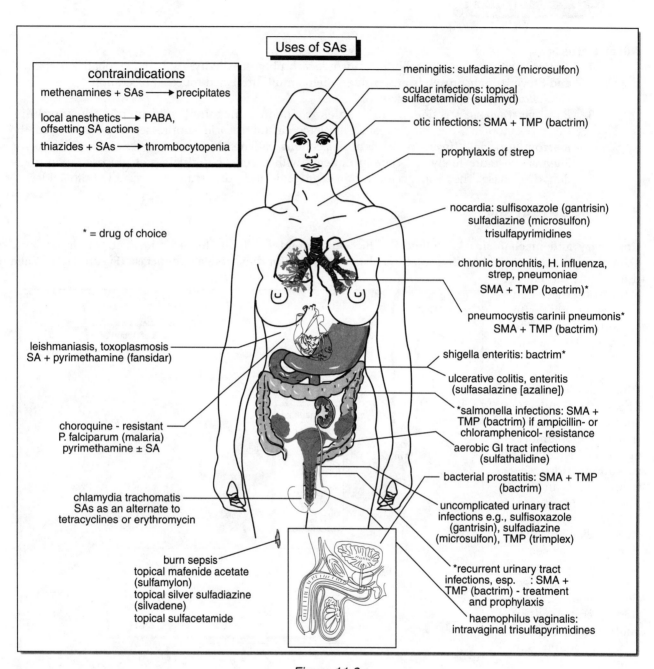

Figure 11.8

famethoxazole (Bactrim) is formulated to provide an in vivo ratio of 1:20, a combination that is most effective for most bacteria. There is a greater incidence of adverse effects with this versus either drug alone.

Uses of sulfonamides and trimethoprim
1. Treatment of uncomplicated, chronic/recurrent urinary tract infections:
    a. Occasionally, sulfisoxazole (Gantrisin) or sulfamethoxazole (Gantanol) alone or with phenazopyridine (Azo Gantrisin; Azo Gantanol) or with trimethoprim (Bactrim). Trimethoprim accumulates in vaginal secretions, increasing antimicrobial effectiveness.

b. Sulfadiazine, sulfacytine (Renoquid) and sulfamethizole (Proklar) are rapidly excreted sulfonamides that concentrate in urine and are sometimes effective as sole treatment of urinary tract infection.

c. Ampicillin or a quinolone represent alternatives.

2. Bacterial prostatitis: Bactrim.

3. Respiratory infections caused by H. influenzae, Strep. pneumoniae or Pneumocystis carinii: Bactrim.

4. Nocardia is treated with bactrim; lymphogranuloma and chancroid, with sulfisoxazole. Sulfonamides can be used for prophylaxis against streptococcal infections, especially in patients resistant to penicillin.

5. Special uses are outlined in **Figure 11.8** and summarized here:
   a. *Burn sepsis* can be treated topically with
      i. *Silver sulfadiazine* (Silvadene), which inhibits bacterial and fungal growth.
      ii. *Mafenide acetate* (Sulfamylon), which inhibits gram-positive and -negative bacteria. If appreciable amounts are absorbed, metabolic acidosis may occur as a consequence of carbonic anhydrase inhibition in renal tubules by mafenide and its metabolites.
   b. *Ophthalmic infection* can be treated topically with *sulfacetamide* sodium (Isopto cetamide), which is nonirritating and able to penetrate into ocular fluid.
   c. Acute *otitis media* in children, caused by H. influenzae or Strep. pneumoniae is treated with *Bactrim*.
   d. *Ulcerative colitis* can be treated with *sulfasalazine*, which is cleaved by intestinal flora to the active non-absorbed species 5-aminosalicylate and the toxic absorbed species sulfapyridine which may produce hemolysis, blood dyscrasias, rash, etc.
   e. *Sulfathalidine*, a poorly-absorbed drug that remains in the lumen of the GI tract, is used to reduce the number of GI flora prior to abdominal surgery.

6. Sulfadiazine and pyrimethamine is the treatment of choice for toxoplasmosis.

7. Sulfadoxine and pyrimethamine (Fansidar) is used as prophylaxis for Malaria, against chloroquine-resistant strains of Plasmodium falciparum. Because Stevens-Johnson syndrome is a possible adverse effect, this treatment is used only in areas with a high risk of resistant malaria.

2. QUINOLONES AND FLUOROQUINOLONES

Chemistry and activity. These drugs have the general structure shown in **(Figure 11.9)**. Quinolones (e.g., nalidixic acid) are uncommonly used. These bind to DNA and inhibit DNA replication **(Figure 11.10)**. Fluoroquinolones (Norfloxacin and Ciprofloxacin) are becoming

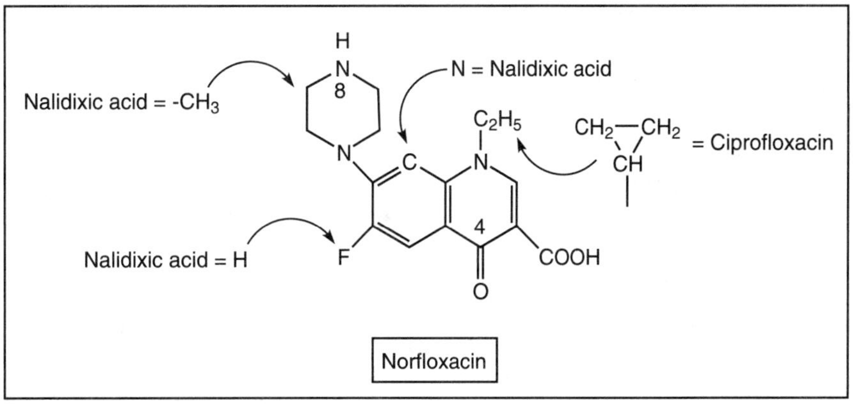

Figure 11.9

292 PHARMACOLOGY

more widely used because of their broader spectrum of antimicrobial activity and relative safety.

Pharmacokinetics. These are well-absorbed and rapidly excreted and concentrated in renal tubules and urine as active species and metabolites. Antacids impair the absorption of norfloxacin and ciprofloxacin.

Adverse effects. Hypersensitivity reactions include hemolytic anemia, marrow suppression, rash and photosensitivity. Seizures, excitement and hallucinations may occur.

Atovaquone (Mepron) is a hydroxynaphthoquinone used for Pneumocystis carinii in patients unable to tolerate bactrim. Atovaquone inhibits the electron transport chain in mitochondria of P. carinii.

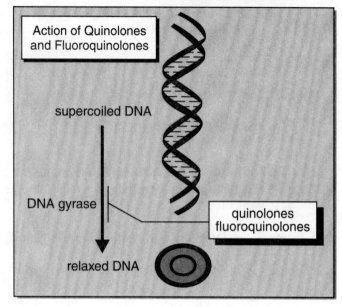

Figure 11.10

URINARY TRACT ANTISEPTICS

These are drugs that have no therapeutic systemic action, but produce urinary tract antisepsis by virtue of their accumulation in renal tubules and urine.

1. *Methenamine* is a hexamethylenamine that spontaneously decomposes in an acid pH to formaldehyde (HCHO) **(Figure 11.11)**. Nearly all bacteria are susceptible to HCHO, so resistance does not develop to methenamine. Hippuric or mandelic acid is used to maintain an acid urine (e.g., methenamine hippurate [Urex] and mandelate [Mandelamine]). Methenamine is used for chronic suppressive treatment of urinary tract infections.

   Sulfamethizole and sulfathiazole react with HCHO to form a precipitate. These sulfonamides should not be used with methenamine.

2. *Nitrofurantoin* (Furadantin)**(Figure 11.12)** is a prodrug that is reduced by bacterial enzymes to the active species. Highly reactive metabolic intermediates in this process damage bacterial DNA. A bacteriostatic effect is produced. In patients with impaired renal function, adequate concentrations of the drug may not be attained in the urinary tract. Nitrofurantoin produces a *brown urine*.

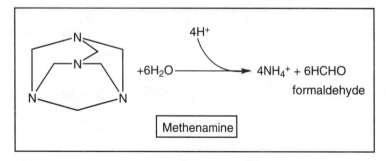

Figure 11.11

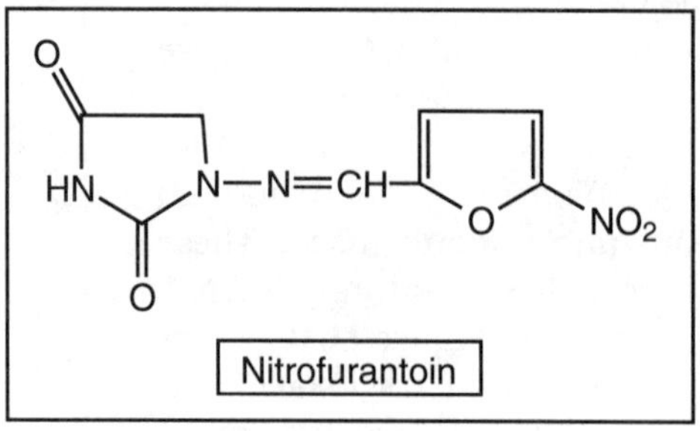

Figure 11.12

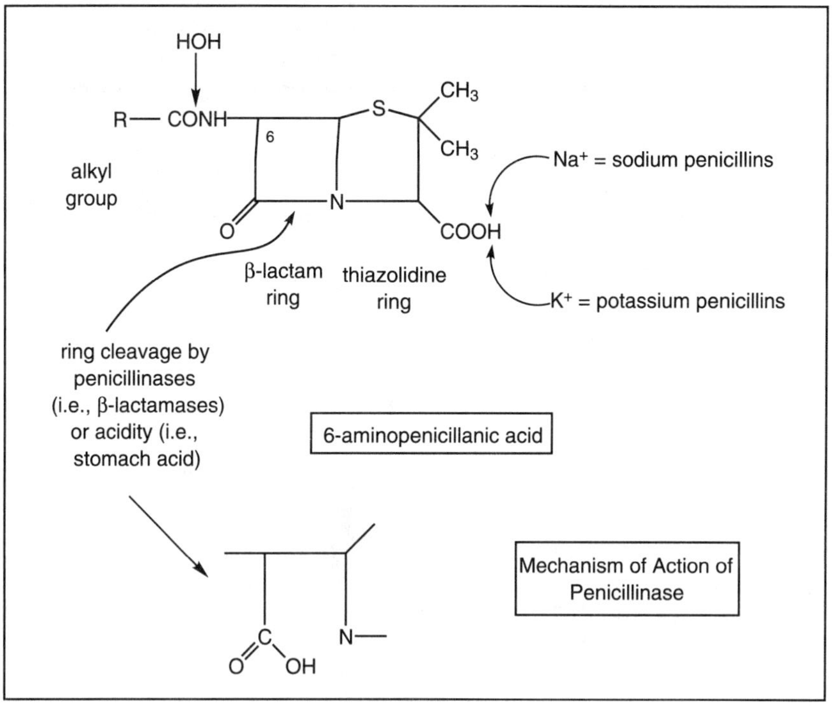

ANTIMICROBIAL CHEMOTHERAPY

Figure 11.13a

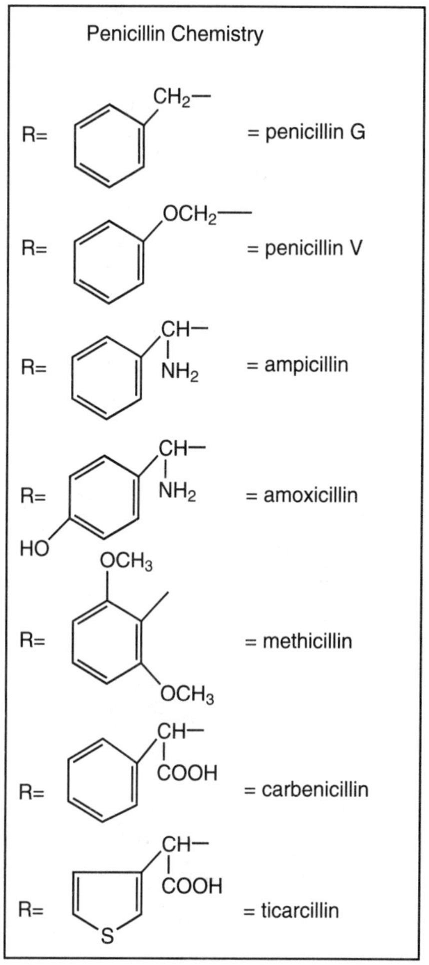

Figure 11.13b

Adverse effects. These include hypersensitivity reactions (granulocytopenia, leukopenia; hemolytic anemia; liver necrosis, pneumonitis) and neurotoxicities (demyelination of sensory and motor nerves with muscle atrophy).

4. BETA-LACTAM ANTIBIOTICS

The principal groups of β-lactam antibiotics are *penicillins* and *cephalosporins*.

A. PENICILLINS

These are a diverse group of compounds, some of which are produced as natural products of fermentation—natural penicillins are Penicillin G and Penicillin V—and others, by adding a side-chain to these fermentation products (semisynthetic penicillins).

Chemistry. The penicillin molecule consists of a thiazolidine ring linked to a β-lactam ring which in turn is attached via an amide bond to an alkyl group **(Figure 11.13a and b)**.

1. The β-lactam-thiazolidine nucleus is necessary for antimicrobial (pharmacodynamic) activity. Hydrolysis of the amide group produces 6-aminopenicillanic acid, the nucleus of all penicillins (the exception is 6-amidinopenicillanic acid, for Amdinocillin) and the starting material for semisynthetic penicillins.

2. The β-lactam ring is associated with acid-instability and penicillin-allergy. Microbial penicillinases (β-lactamases) abolish activity by cleaving the β-lactam ring and producing a penicillanic acid. Most penicillin-resistant bacteria contain penicillinases.
3. The thiazolidine ring is associated with penicillin-allergy.
4. The alkyl group governs many of the pharmacokinetic properties (e.g., duration of action, penicillinase-resistance).

Biological activity. Penicillin G was originally assayed by its ability to impair the growth of a standard culture of Staphylococcus aureus. Penicillin G is now so pure that 1 unit is known to be equivalent to 0.6 µg of penicillin G sodium (i.e., 1 mg equals 1,667 units). Only penicillin G is administered on the basis of units. Others are on the numbers of µg.

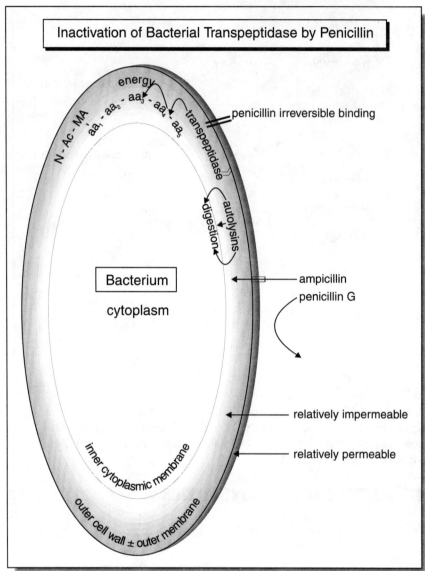

Figure 11.14

Mechanism

1. **Structure of bacterial cell envelope.** The bacterial cell envelope consists of an inner cytoplasmic membrane plus an outer cell wall. In gram-negative bacteria the outer cell wall has an additional outer membrane. It is the inner cytoplasmic membrane that largely regulates permeability. The outer membrane, when present, constitutes a second barrier to permeability **(Figure 11.14)**.

2. **Chemical composition of cell envelope.** This consists of repeating units of N-acetylglucosamine (N-Ac-Gl•NH$_2$) and N-acetylmuramic acid (N-Ac-M=N-Ac-Gl•NH$_2$-LA) (muramic acid is composed of N-acetylglucosamine complexed with a lactic acid). N-Acetylmuramic acid is additionally attached to a small peptide (peptidoglycan) that is conspecific for each strain of bacterium (e.g., E. coli has ala-gly-lys-ala-ala) **(Figure 11.15)**.

3. **Nature of synthesis of cell wall**. This wall is inaccessible to energy sources like ATP which are in cytoplasm. As part of the process of outer cell wall synthesis, transpeptidases cleave the terminal amino acid (aa_5 in **Figure 11.16**; e.g., ala in E. coli) in the peptidoglycan, releasing energy which triggers the final linkage of the penultimate amino acid (e.g., next to last amino acid is aa_4 in **Figure 11.16**; ala for E. coli) to an amino acid on an adjacent peptidoglycan (aa_x in **Figure 11.16**; e.g., for E. coli, ala is linked to a lysine on the adjacent peptidoglycan). This crosslinking of peptidoglycans provides membrane rigidity **(Figure 11.16)**.

inner cytoplasmic membrane:

(-N-Ac-Gl-NH_2---------N-Ac-MA-peptide-)$_n$

MA = N-Ac-Gl-NH_2-LA

Figure 11.15

Site of action of penicillin. The portion of the penicillin molecule shown in **Figure 11.17** resembles two adjacent ends of amino acids. These serve as a substrate for a transpeptidase located on the bacterial cell membrane. When transpeptidase hydrolyzes the penicillin β-lactam ring, an acylated inactive enzyme is formed **(Figure 11.14)**. Cell wall synthesis is thus inhibited.

Murein hydrolase inhibitors normally act in the bacterial cell wall to prevent cleavage of peptidoglycan bonds by murine hydrolases (autolysins). However, penicillins inactivate murein hydrolase inhibitors, thereby allowing autolysins to digest the wall. This effect produces excess membrane permeability, resulting in cell death **(Figure 11.14)**. For these bacteria penicillins are bactericidal; for bacteria lacking autolysins, penicillins are bacteriostatic.

Penicillin is effective only on actively growing cells. Penicillin should not be used with a bacteriostatic agent, which would inhibit cell wall synthesis and thereby inhibit penicillin action. Because Penicillin G is unable to permeate through the outer membrane to access the transpeptidase on the inner cytoplasmic membrane of gram-negative rods, these bacteria are insensitive to penicillin G and some other penicillins **(Figure 11.14)**. These rods are killed by ampicillin and some other penicillins which do permeate through this outer membrane via porins or channels **(Figure 11.14)**.

Bacterial resistance of penicillins. Some bacterial strains have intrinsic resistance to penicillins, which is related to the absence of penicillin binding proteins (PBPs). Acquired resistance is related to synthesis of penicillinases.

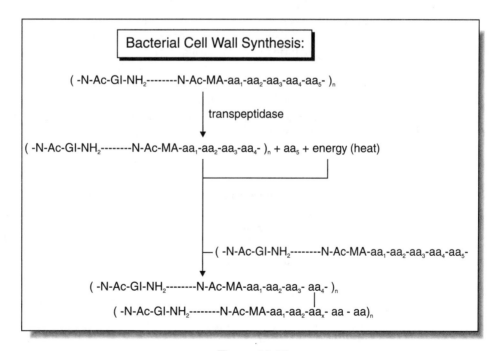

Figure 11.16

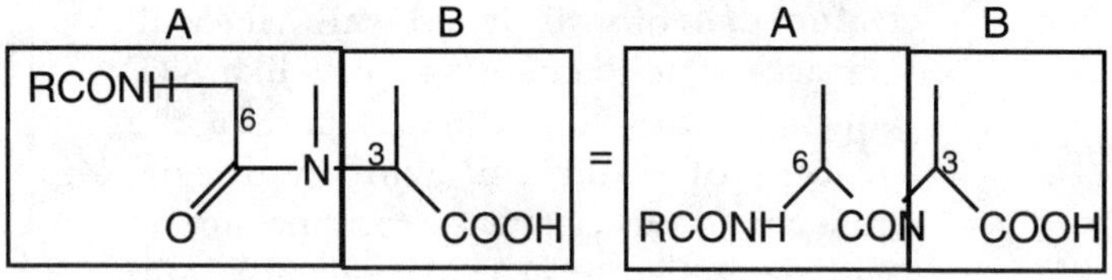

Treatment of β-lactam-resistant bacteria. Clavulanic acid, an irreversible inhibitor of β-lactamase **(Figure 11.18)**, has no bactericidal activity per se but is effective in conferring a "cidal" action to penicillins in an otherwise resistant strain of bacteria. Examples are augmentin (clavulanic acid and amoxicillin) and timentin (clavulanic acid and ticarcillin). Sulbactam is a β-lactamase inhibitor that acts like clavulanic acid. Unasyn is combined sulbactam and ampicillin.

Figure 11.17

Pharmacokinetics. Some are orally absorbed (Penicillin G or V, ampicillin, amoxicillin, others), although food impedes absorption. Most, however, are *acid-labile*, with stomach acid cleaving the β-lactam ring (Penicillin V and amoxicillin are acid-stable). Distribution is throughout the body, partly protein-bound, but not into the cerebral spinal fluid (CSF) unless the meninges are inflamed. Intrathecal penicillin administration is therefore not needed to treat bacterial meningitis. Penicillin does not enter cells.

Penicillin is painful when administered i.m., so a single large dose is often used when this route is employed. Microcrystalline procaine penicillin G (Duracillin) and benzathine penicillin G (Permapen) are administered by this route for prophylaxis.

Penicillins are excreted rapidly and largely intact via renal tubule secretion. Probenecid competes with penicillins for renal tubule secretory sites and accordingly prolongs penicillin action by inhibiting penicillin excretion. Carbenicillin is excreted so rapidly by renal tubules that therapeutic blood levels are not achieved. This drug is used for urinary tract infections. Some highly lipid-soluble penicillins (e.g., nafcillin) are excreted via bile.

Adverse effects. Penicillins comprise one of the safest groups of drugs in medicine. There are few dose-related effects. Some toxicities are listed here:
1. IgE-mediated *hypersensitivity* reactions constitute the most common adverse effect, occurring in about 5% of patients. This is related to binding of the penicilloic acid moiety with lysine residues of protein. Reactions include rash and exfoliative dermatitis; fever, serum sickness or Stevens-Johnson syndrome; bronchospasm, angioedema and anaphylaxis. There is cross-sensitivity among penicillins, but not between penicillins and cephalosporins.
2. Marrow suppression and hepatitis have occurred.
3. Epileptiform seizures may occur after high dose penicillin or after intrathecal administration.
4. Platelet aggregation may be inhibited by penicillins.
5. *Superinfections* are expected.

Uses
1. Gram-positive cocci like Strep pyogenes (e.g., Strep throat), Strep. viridans (e.g., endocarditis) and Strep. pneumoniae are usually treated with penicillin G; alternate, penicillin V.
2. Staph. aureus. is usually resistant to penicillin G. Accordingly, *penicillinase-resistant penicillins* are used:
   Cloxacillin (Cloxapen)
   Methicillin (Staphcillin)
   Nafcillin (Nafcil)
   Dicloxacillin, Oxacillin
   All are orally-active except methicillin, which is acid-labile. In methicillin-resistant staph. infections,

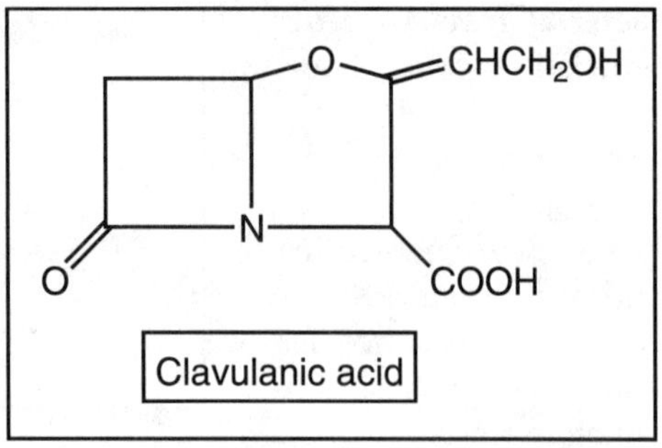

Figure 11.18

vancomycin is the drug of choice.
5. Meningitis is usually treated with penicillin G.
6. Gram-negative bacillary infections (e.g., Bacteroides) can sometimes be treated with penicillin G, but are more likely to be treated with broader spectrum penicillins such as amdinocillin, ampicillin or the carboxypenicillins, carbenicillin and ticarcillin **(Figure 11.13b)**. Piperacillin and the carboxypenicillins are also active against Pseudomonas, Enterobacter and some Proteus or Klebsiella species. Amdinocillin, unlike other broad-spectrum penicillins is penicillinase-resistant.
7. Many other infections are also treated with penicillin G or other penicillins, including clostridia, anthrax and actinomycosis.

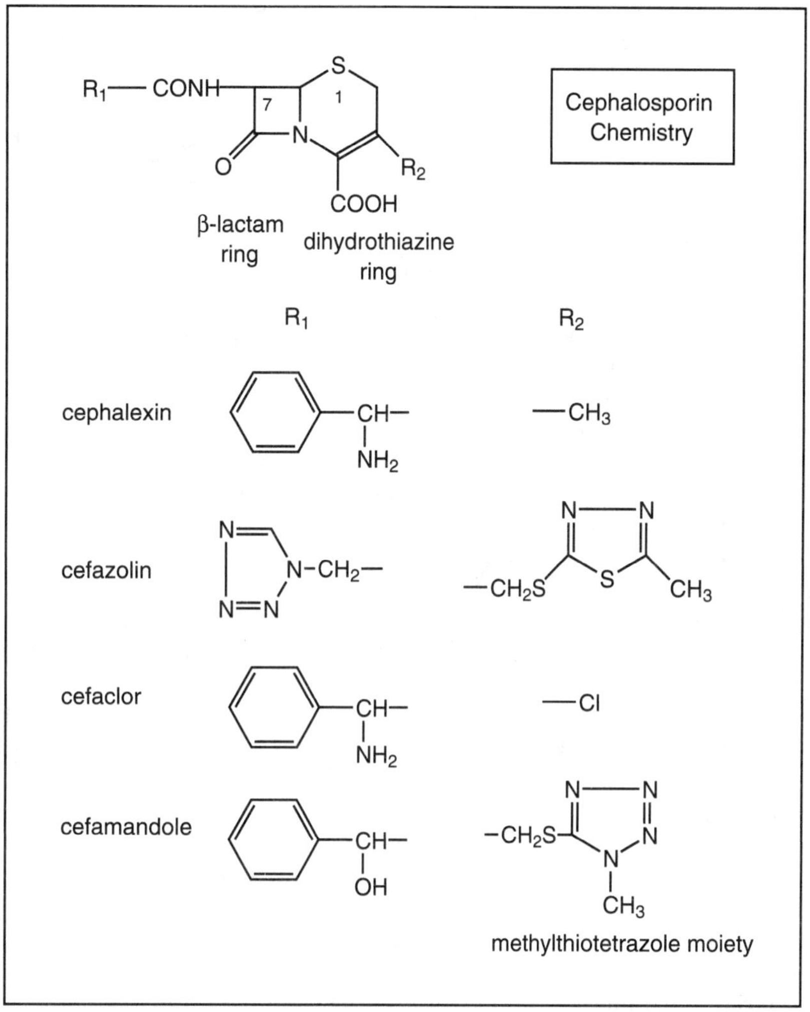

Figure 11.19

B. CEPHALOSPORINS

Cephalosporin was originally isolated from the fungus, Cephalosporium acremonium, which contained three forms: cephalosporin P, N and C. Cephalosporin C is hydrolyzed to 7-aminocephalosporanic acid, the active nucleus. Many semisynthetic derivatives have been produced.

Chemistry (Figure 11.19). The cephalosporin molecule consists of a dihydrothiazine ring linked to a β-lactam ring which in turn is attached via an amide bond to an alkyl group.

1. The β-lactam-thiazolidine nucleus is necessary for antimicrobial activity.
2. Although the β-lactam ring is normally associated with acid-instability, 7-amino-analogs are relatively stable in acid and resistant to penicillinase. The 7-alkyl group also governs antimicrobial activity.
3. The dihydrothiazine ring is associated with allergy.
4. The 3-alkyl group governs pharmacokinetic properties.

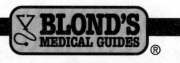

5. The 3-methylthiotetrazoles are associated with a disulfiram-like effect and bleeding diathesis, especially among vitamin K-deficient and elderly patients (e.g., cefoperazone, cefamandole and moxalactam).

Cephamycins are similar to cephalosporins, but have a CH_3O- group at 7.

Activity

First generation: high susceptibility of gram-positive except enterococci and methicillin-resistant Staph. modest susceptibility of gram-negative.

These are used mainly for penicillin- or sulfonamide-resistant urinary tract infections, Klebsiella and as prophylaxis for clean-surgery. The first generations do not enter CSF; some can be given orally (cephalexin, cephradine, cefadroxil).

Second generation: increased susceptibility of gram negative, to include H. Influenza, B. fragilis and Proteus. decreased susceptibility of gram positive.

These are used mainly for infections of the urinary tract, bone and soft tissues; or as prophylaxis for clean surgery. The second generations do not enter CSF, except possibly if meninges are inflamed. Only cefaclor can be given orally.

Third generation: increased susceptibility of gram-negative, to include Enterobacteria and some Pseudomonas decreased susceptibility of gram-positive versus first generation, but still good with cefotaxime.

These are used mainly for serious gram-negative infections and are particularly effective for H. influenzae, N. gonorrhoeae and Enterobacter; moderately effective against anaerobes. Most third generations enter the CSF and are useful for treating meningitis. All must be administered parenterally.

Mechanism. Cephalosporins and cephamycins act identically to penicillins, by
  a. binding to PBPs in the cell wall,
  b. inhibiting the transpeptidase enzyme which is necessary for crosslinking of peptidoglycans and
  c. activating autolysins in the cell wall.
A bactericidal effect is produced in bacteria that contain autolysins.

Bacterial resistance of cephalosporins. This is related to development of cephalosporinases (β-lactamases), which are regulated by chromosomal genes versus plasmid-associated penicillinases. Cefoperazone is particularly susceptible to degradation by cephalosporinases.

Pharmacokinetics. Most cephalosporins are not orally effective. All are partly bound to plasma protein, as shown below. *Ceftriaxone* is unique among the cephalosporins in displacing bilirubin from its protein binding sites and accordingly may cause *kernicterus* in neonates.

Protein binding:

  less than 2/3 bound: cephalothin, cefazolin (first generation)
  cefamandole, cefoxitin, ceforanide, cefonicid, cefotetan (second)
  ceftriaxone, cefoperazone (third)
  more than 1/3 bound: cephalexin, cephradine, cefadroxil (first)
  cefaclor (second)
  ceftazidime (third)
  from 1/3 to 2/3 bound: cephapirin (first)
  cefuroxime (second)
  cefotaxime, ceftizoxime, moxalactam (third)

Distribution into CSF is notable for third generation cephalosporins (e.g., cefotaxime, ceftriaxone, ceftizoxime, moxalactam). However, cefuroxime (second generation) penetrates the meninges when they are inflamed, and therefore can be used for certain types of meningitis.

All cephalosporins have a short $t_{1/2}$ (about 1 hour for first generation drugs; about 2 hours for second and third generation drugs). An exception is ceftriaxone, $t_{1/2}$ about 8 hours. Several are partly metabolized, mainly by deacylation, but most are largely excreted unmetabolized in urine by glomerular filtration and secretion. Except for ceftriaxone and cefoperazone which are largely excreted via bile into the GI tract, dosage adjustments should be made when there is renal insufficiency. Diarrhea and GI superinfections are more prominent with ceftriaxone and cefoperazone.

Adverse effects of cephalosporins
1. Hypersensitivity reactions (less than 5%) with rash, fever and eosinophilia comprise the most common adverse effect. About 5 to 10% of penicillin-allergic patients have a cephalosporin allergy.
2. Nephrotoxicity is uncommon but seems to be increased in the presence of diuretics and particularly when administered with aminoglycosides like gentamicin or tobramycin (e.g., cephalothin).
3. **GI**. Diarrhea and superinfections by Enterococci, Candida and Clostridium difficile are most prominent with cephalosporins that are excreted mainly by this route (over 50%), namely ceftriaxone and cefoperazone.
4. **Bleeding tendency**. This is related to hypoprothrombinemia and reduced platelet aggregation. The effect is caused only by 3-methylthiotetrazole cephalosporins (i.e., cefoperazone, cefamandole and moxalactam). The effect is greatest in elderly and vitamin K-deficient patients and can be prevented with recommended supplemental vitamin K. The 3-methylthiotetrazoles also have disulfiram-like activity, producing adverse effects with alcohol.
5. Marrow suppression and hepatotoxicity are rare.
6. Cefaclor alone produces a serum-sickness-like reaction, characterized by erythema multiforme, fever and arthralgia.

Uses. Most cephalosporins can be used to treat respiratory tract and urinary tract infections, as well as skin, bone and joint infections. Many are also used prophylactically prior to bowel/pelvic/clean surgery (e.g., cefoxitin).

Gram-negative infections, including most nosocomial infections, are usually sensitive to second and third generation cephalosporins, but also some first generation drugs like cefazolin and cephalothin. Several second and third generation cephalosporins are used for anaerobic pleuropulmonary infections and sepsis.

Cefoperazone has unique activity against P. aeruginosa; Ceftazidime, for Pseudomonas. Ceftriaxone is often useful for N. gonorrhoeae infections in women. Cefaclor is used for otitis media.

C. OTHER β-LACTAM ANTIBIOTICS

A. Imipenem

Imipenem is derived from a product produced by Strep. cattleya. Being a β-lactam **(Figure 11.20)**, imipenem acts identically to penicillins and cephalosporins. However, imipenem is extremely resistant to β-lactamases.

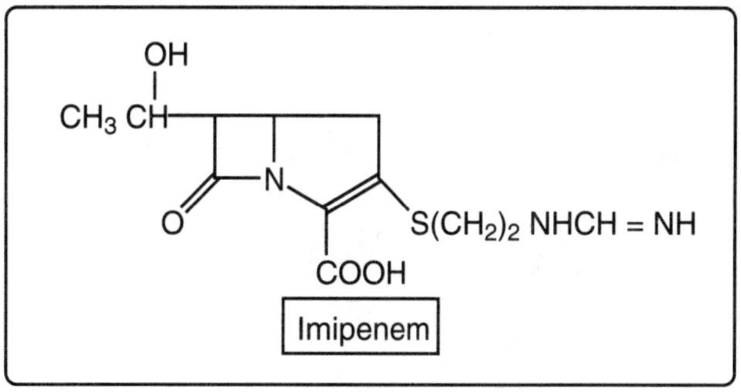

Figure 11.20

Because imipenem is rapidly metabolized by a renal tubule dipeptidase, imipenem is always combined with cilastatin (Primaxin), a drug that inhibits the dipeptidase and extends the $t_{\frac{1}{2}}$ of imipenem to about 1 hour. Imipenem is largely excreted unmetabolized in urine. Primaxin has indications similar to second or third generation cephalosporins. Adverse effects include vomiting, seizures and hypersensitivity reactions.

B. Aztreonam

Aztreonam (Azactam) is a monobactam produced by Chromobacterium violaceum. The antimicrobial spectrum is like that of aminoglycosides, not penicillins or cephalosporins. Enterobacter and Pseud. aeruginosa are particularly susceptible, while gram-positive bacteria and anaerobes are resistant. The $t_{\frac{1}{2}}$ is about 2 hours and most is excreted unmetabolized in urine. Aztreonam is usually safe in patients allergic to penicillins or cephalosporins.

5. AMINOGLYCOSIDES AND AMINOCYCLITOLS

Chemistry. These large molecules are composed of at least one sugar molecule with amino groups ("aminoglycoside"), linked to a hexose-aminocyclitol. The many hydroxyl groups in the sugar/cyclitol moieties confer a high degree of polarity. Some of these microbial antibiotics are analogs of substances naturally produced by Streptomyces strains (i.e., the name "-mycin"). Gentamicin and netilmicin, however, are derived from Micromonospora (i.e., "-micin" rather than "mycin"). Amikacin and netilmicin are semi-synthetic **(Figure 11.21)**.

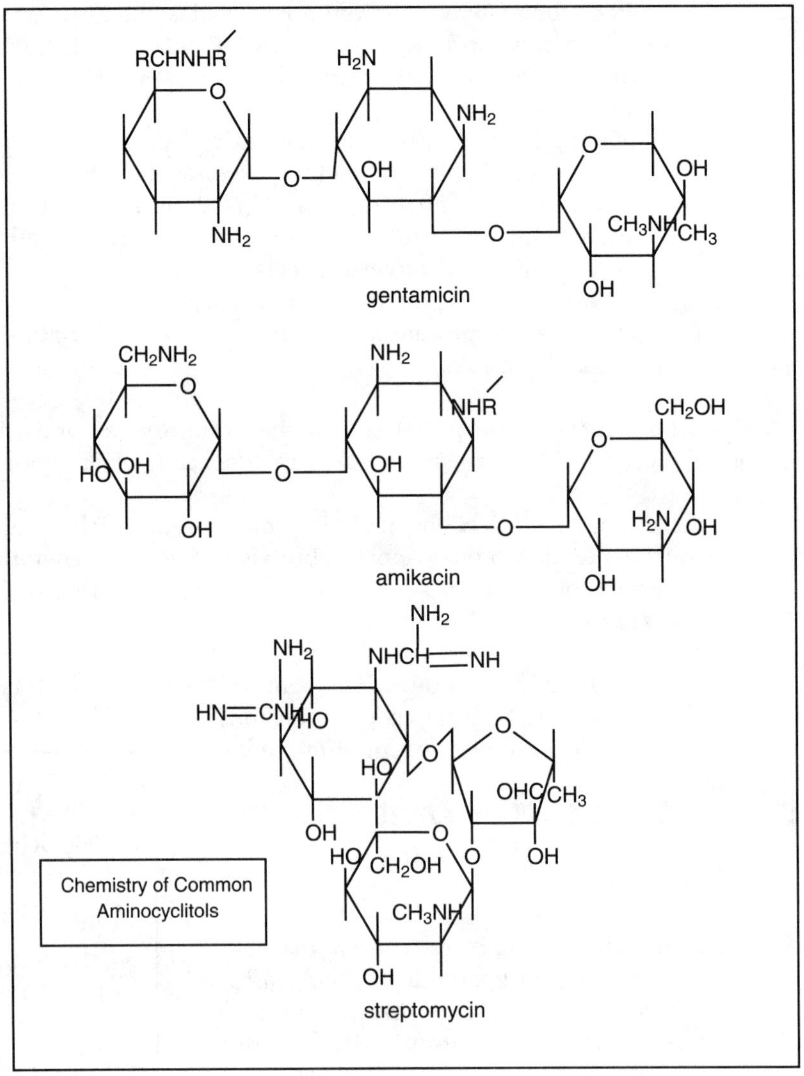

Spectinomycin (Trobicin) has a neutral sugar, not an aminosugar. Hence, it is not an aminoglycoside, but an *aminocyclitol*. Spectinomycin is *bacteriostatic*, not bacteriocidal like the aminoglycosides.

Pharmacokinetics. Being highly polar and bulky molecules, aminoglycosides are not appreciably absorbed. Routinely they are administered i.m. Only 20% is protein-bound. Aminoglycosides tend to accumulate in the renal cortex, as well as the enodlymph and perilymph of the inner ear**(Figure 11.22)**.

Figure 11.21

These are the major sites for toxic effects of aminoglycosides. Although aminoglycosides are excluded from the CNS, they do cross the placental barrier and may cause hearing loss in unborn children. Most of the drug is excreted by glomerular filtration. Dosage adjustment is needed if there is renal impairment and this must be done with precision since the effective and toxic dosages are close.

Mechanism. Aminoglycosides produce a bactericidal effect by binding irreversibly to S12 protein on bacterial 30S ribosomes and permanently inhibiting the initiation of protein synthesis **(Figure 11.2)**. In higher concentrations cell membranes are disrupted by these drugs. This latter effect may account for toxicity at sites (ears, kidneys) where aminoglycosides accumulate **(Figure 11.22)**.

Susceptibility and Resistance. Most aerobic gram-negative bacteria are susceptible to the bactericidal action of aminoglycosides. Anaerobes are not susceptible, since the transport and binding of aminoglycosides to 30S ribosomal proteins requires O_2.

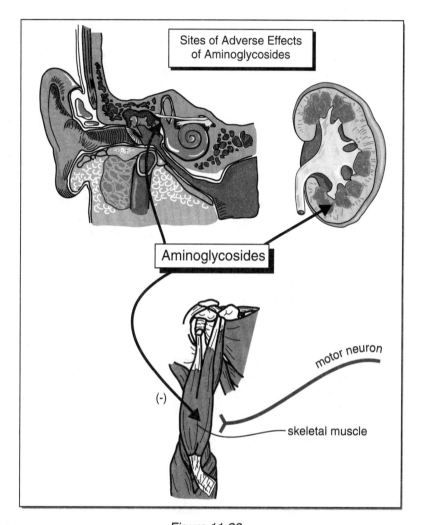

Figure 11.22

Resistant bacterial strains have developed to streptomycin and gentamicin via plasmid-mediated production of aminoglycoside-inactivating enzymes (acetylase, adenylase or phosphorylase). Amikacin and netilmicin are not metabolized by these enzymes.

Adverse effects
1. *Ototoxicity* occurs subsequent to accumulation of aminoglycosides in endolymph and perilymph. Cochlear damage is signaled by a high-pitched tinnitus that lasts for 1 or 2 days. Damage progresses from hair cells in the base (processing of high-pitched sound) to the apex of the cochlea (low-pitched sound). Older people, having fewer cochlear hair cells, are more susceptible to damage. Ethacrynic acid and furosemide potentiate the ototoxic effects. Amikacin, kanamycin and neomycin are more likely to first produce ototoxicity—versus vestibular toxicity.

    Vestibular damage is signaled by an intense headache lasting 1-2 days, followed by equilibrium imbalance and associated nausea and vomiting. If damage is severe there can be chronic labyrinthitis. Streptomycin and gentamicin are more likely to first produce vestibular toxicity. If the dosage is reduced or the drug is discontinued at the first sign of toxicity, permanent damage can be prevented.

2. *Nephrotoxicity* is the consequence of the accumulation and retention of aminoglycosides by proximal tubule (PT) cells. Since PT cells regenerate, nephrotoxicity is usually reversible. However, the reduction in GFR translates into greater retention of aminoglycosides and greater ototoxic effects. Streptomycin is least ototoxic of the aminoglycosides; neomycin is the most ototoxic and this effect precludes in vivo use of neomycin.

3. *Neuromuscular blockade* with apnea can occur with aminoglycosides, especially in myasthenia gravis patients. Generally, this effect is only seen if an anesthesia or true neuromuscular blocking drug is administered. The effect is related to reduced release of acetylcholine (Ach) and reduced post-synaptic sensitivity to Ach.

Uses. Because of the potential for toxicity and necessity for parenteral administration, aminoglycosides are usually used only in combination with a penicillin or cephalosporin for serious or life-threatening infections. This includes but is not limited to bacterial endocarditis, tularemia, plague and many aerobic gram-negative baccili. Streptomycin and gentamicin (Garamycin) are older and cheaper drugs in this class. Tobramycin (Nebcin), amikacin (amikin) and netilmicin (netromycin) are often used when there is drug-resistance to streptomycin or gentamicin. Neomycin (Mycifradin) is only used topically for burns and abrasions; or orally to kill bacteria in the bowel (e.g., prior to surgery).

Vancomycin is a glycopeptide that has a potential for nephrotoxicity and ototoxicity similar to aminocyclitols. As for the latter drugs, vancomycin is used only for life-threatening infections (e.g., methicillin-resistant Staph. aureus). Vancomycin is inactive against gram-negative bacteria. Vancomycin exerts a bactericidal effect by inhibiting one of the final steps in peptidoglycan synthesis in microbial cell walls. Vancomycin is synergistic with gentamicin and tobramycin. Dosage reduction is necessary when there is either hepatic failure or renal impairment. Vancomycin enters the CSF when meninges are inflamed.

6. TETRACYCLINES

Tetracyclines (TTCs) are naturally produced by Streptomyces. Common TTCs are these natural substances or semi-synthetic derivatives. These are broad-spectrum antibiotics that are bacteriostatic to gram-positive and gram-negative bacteria, as well as some nonbacterial species. However, they are rarely used for gram-positive bacteria. TTCs have largely been supplanted by safer and more-effective antibiotics.

Chemistry. TTCs, as the name implies, are 4-ring compounds (**Figure 11.23**). These are acidic and multi-polar compounds, quite soluble in water, that tend to chelate divalent and trivalent cations. When such chelates are absorbed, they deposit in bone and teeth. Since TTCs are yellow, teeth become discolored. For this reason, TTCs are contraindicated in children, starting from the second trimester of pregnancy. Breakdown products of TTC are toxic to the kidneys and other organs, so outdated TTC should never be used.

Pharmacokinetics. Although TTCs are adequately absorbed, heavy metals (Ca^{++}, Al^{+3}) inactivate TTCs by forming chelates. Consequently, TTCs should be taken at least 2 hours before or after meals. Milk and antacids impair absorption of TTCs. Substantial amounts of TTCs pass through the GI tract intact and kill residual flora. Stools become soft, odorless and yellow-green. Yeast superinfections may arise.

Except for oxy-TTC (33%), 2/3 or more of TTCs are bound to plasma protein. The t½ is long, ranging from 6 hours (TTC and oxy-TTC) to 18 hours (doxycycline and minocycline). Although renal glomerular filtration represents the major means of elimination of TTCs, they tend to accumulate in the liver and are also excreted via bile into the GI tract. Dosage reduction is needed when there is renal impairment. Doxycycline is unique among the TTCs, being eliminated mainly via liver/bile into the GI tract. This drug also has less impact among TTCs on intestinal flora. Dosage adjustment of doxycycline is not required if there is renal impairment.

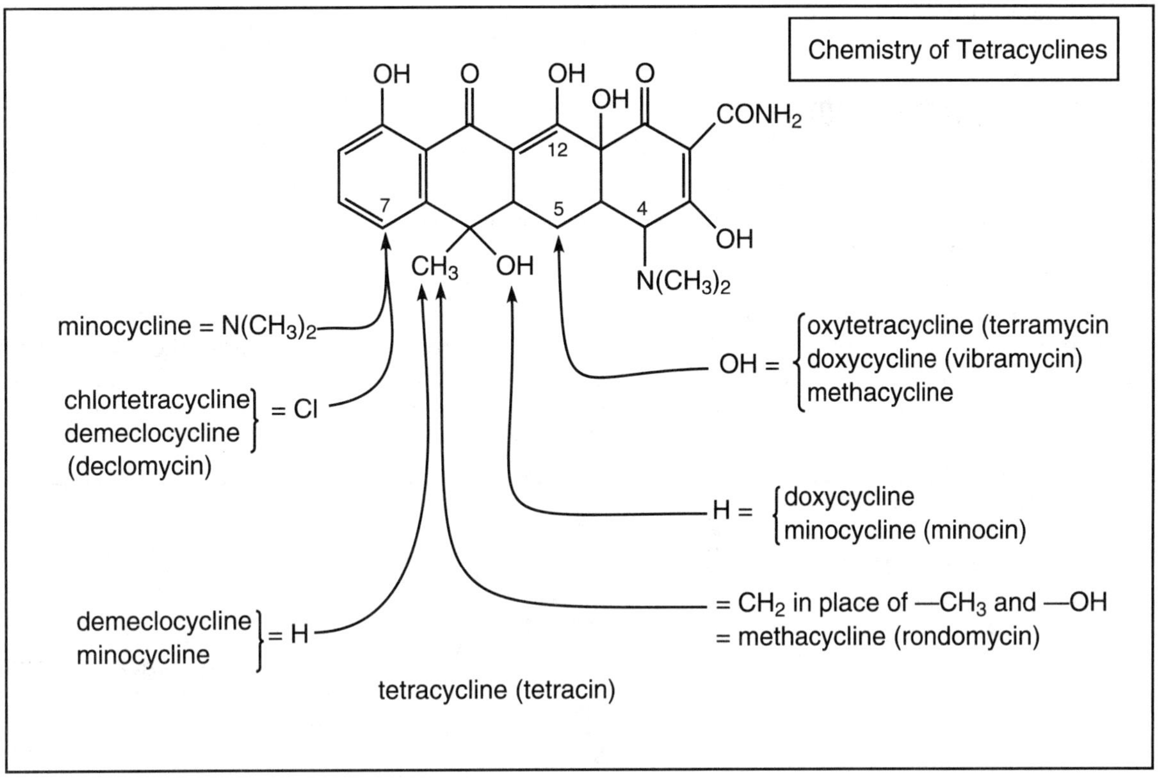

Figure 11.23

Following i.v. but not oral administration, CSF levels of TTCs approximate that of plasma. Minocycline, being very lipid soluble, attains high enough concentrations in tears and saliva to eradicate the meningococcal carrier state.

Mechanism. TTCs bind reversibly to bacterial 30S ribosomal protein and prevent aminoacyl tRNA binding to the receptor site on 50S ribosomes, thereby blocking elongation of a growing peptide chain **(Figure 11.2)**. Microbial protein synthesis is inhibited. TTCs are actively accumulated by microbial but not mammalian cells.

Resistance. Resistance occurs by plasmid-mediated processes, either diminishing TTC uptake (R-factors) or increasing energy-dependent TTC efflux. Because TTC is used as a growth promoter in animal feed, farmers and animal handlers have a greater incidence of TTC-resistant infections.

Adverse effects. There are potentially many adverse effects of TTCs, as illustrated in **Figure 11.24**.

Uses
1. TTCs are drugs of choice for Rickettsia (e.g., Rocky Mountain spotted fever, rickettsiae pox, murine typhus) and Chlamydia infections (e.g., pneumonia, psittacosis, trachoma, lymphogranuloma venereum).
2. Doxycycline is the drug of choice for Mycoplasma pneumoniae.
3. TTCs are alternate drugs for Neisseria gonorrhoea (in combination with ceftriaxone), syphilis, cholera, brucellosis, tularemia, Lyme disease, tetanus, plague, actinomyces, Treponema and urinary tract infection with gram-negative bacteria.
4. TTCs are active against most strains of Staphylococcus aureus, H. Influenzae, and Campylobacter.

5. Low-dose TTC is used for acne, possibly because of its action on propionibacteria that reside in sebaceous follicles. The net effect is less fatty-acid production by sebaceous glands.

7. CHLORAMPHENICOL

Chloramphenicol is naturally-produced by Streptomyces venezuelae, but is commercially prepared by chemical synthesis. This is a broad-spectrum antibiotic that is bacteriostatic to gram-positive and gram-negative bacteria, as well as some nonbacterial species.

Chemistry and Pharmacokinetics (Figure 11.25). Chloramphenicol is a lipid-soluble nitrobenzene analog that is well-absorbed, reaching peak levels in less than 3 hours. A prodrug ester, chloramphenicol palmitate, is sometimes administered orally in place of the active species. Pancreatic lipase readily cleaves the palmitate, liberating free drug. When administered i.m., chloramphenicol succinate is used. About half is protein-bound,

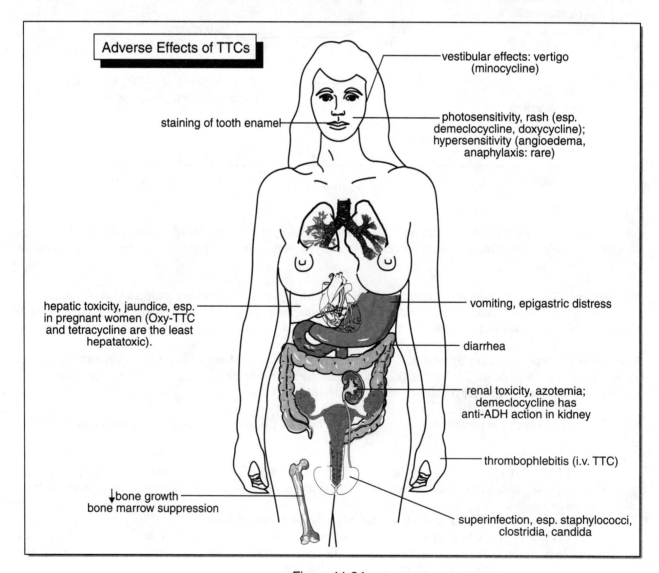

Figure 11.24

so hemodialysis is not so effective for eliminating overdoses. Distribution is to all sites including CNS and across the placenta.

Elimination occurs primarily by hepatic glucuronidation (greater than 75% in 1 day). Dosage adjustment is needed in patients with cirrhosis and in infants, but not in patients with renal impairment.

Mechanism. Chloramphenicol binds reversibly to 50S ribosomal protein and prevents binding of the amino acid terminus of aminoacyl tRNA to the acceptor site on the 50S ribosome. As a consequence, peptidyl transferase is unable to interact with its amino acid substrate, so protein synthesis is inhibited **(Figure 11.2)**.

Chloramphenicol also binds to mitochondrial 70S ribosomal protein and in a similar manner as above, inhibits mitochondrial protein synthesis in mammalian cells. Erythropoietic cells are particularly sensitive, accounting for serious hematologic abnormalities.

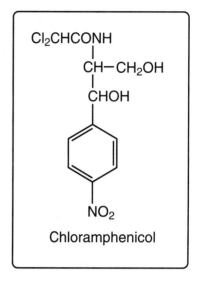

Figure 11.25

Resistance. By a process of conjugation, gram-negative rods and H. influenzae acquire a plasmid that contains the code for an *acetyltransferase* which is capable of acetylating chloramphenicol and thereby eliminating the ability of the drug to bind to ribosomes.

Adverse effects
1. Hematological effects (aplastic anemia more than agranulocytosis and thrombocytopenia) often lead to death. It is because of this action that chloramphenicol is relegated to use only in serious infections.
2. Neurotoxicity is manifested as digital paresthesias and optic neuritis (destruction of retinal ganglion cells and atropy of the optic nerve).
3. GI effects include nausea, vomiting and diarrhea.
4. Hypersensitivity reactions (rash, fever, angioedema) are uncommon.
5. Gray baby syndrome (40% fatality) is related to a slow rate of drug inactivation (glucuronidation) in infants (immature livers) and a slow rate of renal excretion of active drug.

Drug interactions. Chloramphenicol irreversibly inhibits microsomal cyt P_{450} enzymes and displaces other protein-bound drugs, enhancing the free concentration of these drugs as well as their half life. These effects produce toxicities. Chloramphenicol-interactions occur with warfarin, phenytoin, tolbutamide, etc.

Uses. Chloramphenicol (Chloromycetin) is used for
1. Bacterial meningitis caused by Hemophilus influenza. Chloramphenicol is bactericidal for H. influenze and this effect is potentiated by ampicillin. Chloramphenicol is an alternate drug for meningitis caused by N. meningitidis and Strep. pneumoniae in patients allergic to penicillin.
2. Typhoid fever.
3. Anaerobic infections caused by bacteroides.
4. Rickettsial diseases such as Rocky Mountain spotted fever.
5. Brucellosis, in which a tetracycline is contraindicated.

8. MACROLIDE ANTIBIOTICS

These consist of a multi-membered lactone ring that is attached to one or more deoxy sugars. The drugs are highly polar and water soluble.

A. ERYTHROMYCIN

Erythromycin is a natural product of Strep. erythreus.

Pharmacokinetics. Erythromycin is incompletely but adequately absorbed from the upper GI tract. Food decreases the bioavailability of erythromycin. An enteric-coated prep is required, since gastric juice inactivates the drug. Peak levels are attained in less than 4 hours and the t½ is 90 minutes. Most of the drug is protein bound, so hemodialysis is ineffective in treating overdose toxicity. The drug is largely excluded from the CNS, but crosses the placenta. Most of the drug is metabolized (approximately 90%) by liver cyt P_{450} enzymes and excreted in urine. Erythromycin competes with metabolism of many drugs (e.g., warfarin, digoxin, theophylline), thereby enhancing their effects.

Mechanism. Erythromycin and other macrolide antibiotics bind reversibly to 50S ribosomal proteins in microorganisms, inhibiting the translocation of newly-synthesized peptidyl tRNA from acceptor- to donor-sites on the ribosome **(Figure 11.2)**. Mitochondrial protein synthesis is not affected by erythromycin.

Gram-positive bacteria accumulate 100 times more erythromycin than gram-negative bacteria. Plasmid-mediated drug-resistance is related to (a) reduced entry (i.e., altered carrier protein), (b) reduced binding to ribosomes (i.e., modified binding protein) and (c) increased hydrolysis of erythromycin (i.e., synthesis of an esterase) by microorganisms.

Adverse effects. Serious adverse effects are rare.
1. Hypersensitivity-type reactions include rash, fever and eosinophilia; Cholestatic hepatitis with abdominal pain, jaundice, increased SGOT and SGPT.
2. GI effects include pain, vomiting and diarrhea.

Uses
1. Treatment of Legionnaires' diseases, diphtheria and pertussis.
2. Alternative to tetracycline for Mycoplasma pneumoniae, Chlamydia urogenital infections or Chlamydia pneumoniae of infancy.
3. Alternative to penicillin for tetanus, syphilis, Streptococcal and Staphyloccal infections, and prophylaxis in preventing bacterial endocarditis following dental or respiratory tract procedures.
4. Alternative to β-lactam antibiotics for gonorrhea during pregnancy.

B. CLARITHROMYCIN (BIAXIN) AND AZITHROMYCIN (ZITHROMAX)

These are more-stable long-lived analogs that produce fewer adverse GI effects versus erythromycin. Clarithromycin has added activity against Toxoplasma gondii and Mycobacterium intracellulare. Azithromycin is active against gonococcal and chlamidial infections.

9. MISCELLANEOUS ANTIBIOTICS

A. Clindamycin (Cleocin)

Pharmacokinetics. This drug is well-absorbed. Food does not alter drug absorption. Peak levels occur in less than 1 hour, with 90% being protein-bound. This drug does not reach relevant levels in the CSF, but does cross the placenta. About 90% is excreted as an N-demethylated or sulfoxide derivative, mainly in urine but also in bile/feces. Dosage adjustment is needed when there is hepatic failure.

Mechanism. Clindamycin, like chloramphenicol and erythromycin, binds to 50S ribosomes in microorganisms and inhibits protein synthesis **(Figure 11.2)**. These drugs are not used concurrently.

Adverse effects
1. **GI effects.** Diarrhea and *pseudomembranous colitis* are the major adverse effects. Diarrhea is the consequence of a clindamycin-induced secretion of a toxin by Clostridium difficile.
2. Allergic reactions include rash and anaphylaxis; Stevens-Johnson syndrome with blood dyscrasias.
3. Neuromuscular blockade is analogous to that seen after aminoglycosides.

Use. Clindamycin (Cleocin) is used to treat anaerobic infections, particularly *Bacteroides fragilis*.

B. *Polymixins* (Polymyxin B and Colistin) are products of Bacillus polymyxa that act as cation detergents. Being highly nephrotoxic, these drugs are used topically (e.g., Pseudomonas infections of the ear).

C. *Bacitracin* represents a group of polypeptide antibiotics, that inhibits one of the final steps in peptidoglycan synthesis in bacterial cell walls. In this regard its action is similar to that of vancomycin. Because it is so nephrotoxic, bacitracin is only used topically.

10. TREATMENT OF TUBERCULOSIS (TB) (MYCOBACTERIUM TUBERCULOSIS, M. LEPRAE, M. AVIUM)

Treatment approach. Antimycobacterials exert their effect only on growing organisms.

First-line drugs are sometimes used to treat TB, as follows:
1. 9 months: isoniazid and rifampin.
2. 2 months: isoniazid and rifampin and pyrazinamide, plus 4 months: isoniazid and rifampin.
3. When there is primary resistance to isoniazid, isoniazid, rifampin, pyrazinamide and ethambutol or streptomycin.

Second-line drugs may be needed if there is resistance:
1. ethionamide, aminosalicylic acid, cycloserine, amikacin, kanamycin, ciprofloxacin, ofloxacin, capreomycin or clofazimine.

Standard therapy of TB consists in simultaneous administration of at least two drugs, since one in every 10^6-10^8 bacteria are resistant to each drug alone. Since there are as many as 10^9 tubercle bacilli per TB cavity, single drug therapy would promote development of resistant strains.

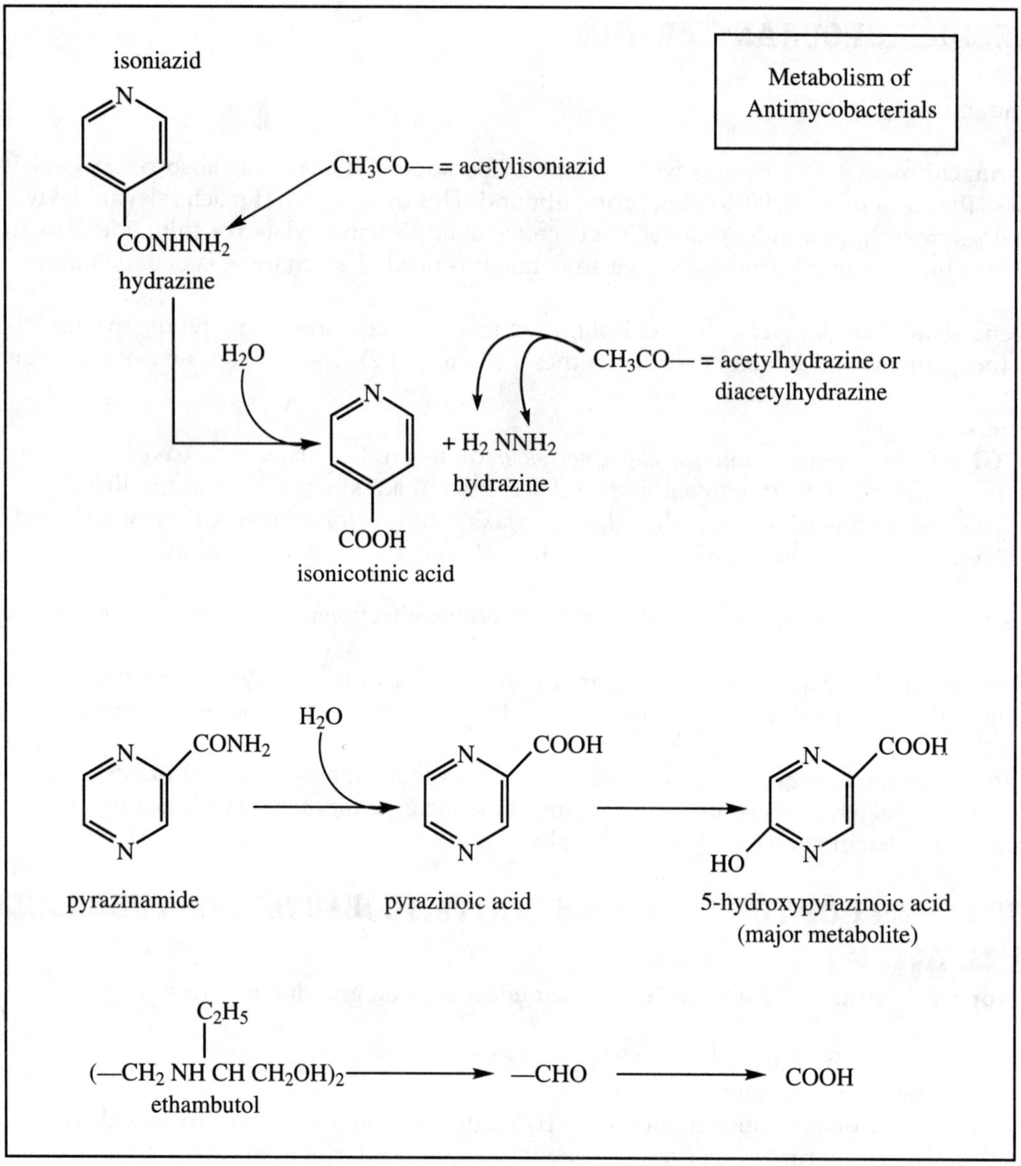

Figure 11.26

First-line drugs

A. Isoniazid (Nydrazid) **(Figure 11.26)**

   Pharmacokinetics. Absorption is rapid but impeded by aluminum antacids. Peak levels occur in less than 2 hours. Isoniazid distributes to all tissues (including the CNS) and compartments (pleura, ascites, TB cavities). About 90% is excreted within 24 hours, mostly in the form of the metabolites, acetylisoniazid and isonicotinic acid.

   A small percentage of people (Japanese, Eskimos) acetylate rapidly; most (Jews, North Africans and Scandinavians) acetylate slowly, so that the $t_{1/2}$ of isoniazid ranges from 1 to 3 hours in these respective groups. Slow acetylators are more at risk of toxic levels of isoniazid when renal function is impaired.

Mechanism. Isoniazid inhibits synthesis of mycolic acids, vital constituents in the mycobacterial cell wall, possibly by (a) preventing elongation of long-chain fatty acid precursors of mycolic acid and (b) inhibiting mycobacterial desaturase, the enzyme that catalyzes the first step in mycolic acid synthesis. Isoniazid is bactericidal.

Neural toxicity is associated with pyridoxine (vitamin B_6) deficiency, subsequent to isoniazid hydrozones that inhibit conversion of pyridoxine to a phosphate. The effect is more likely to occur in slow acetylators, but is prevented by supplemental pyridoxine intake.

Adverse effects are common.
1. Allergic-hypersensitivity-like reactions: rash, fever, jaundice.
2. Hematologic: agranulocytosis, eosinophilia, thrombocytopenia, anemia.
3. Peripheral neuritis is prevented by pyridoxine. Other neuronal effects include seizures (especially in epileptics), optic neuritis and encephalopathy; also euphoria and psychosis.
4. Hepatic toxicity (especially in slow acetylators greater than 35 years of age) (multilobular necrosis) with jaundice and elevated serum SGOT is likely caused by the metabolite, acetylhydrazine. Hepatic effects are largely prevented by pyridoxine. Pyridoxine is also the antidote for overdose isoniazid.
5. Antimuscarinic effects occur: dry mouth, urinary retention.

Interactions. Isoniazid inhibits hydroxylation of phenytoin, predisposing to its toxicity.

Use. Isoniazid is only used in combination with other drugs for treatment of established TB. Isoniazid may be used alone in prophylactic therapy.

B. Rifampin (Rifadin)

Chemistry. Rifampin is a semi-synthetic derivative of one of the large complex antibiotics produced by Streptomyces mediterranei.

Pharmacokinetics. Absorption is rapid but impaired by para-aminosalicylic acid. When co-administered, these drugs should be separated by an interval of greater than 8 hours. Peak levels occur at 3 hours. About 2/3 is excreted in bile as an inactive acetylated metabolite that becomes deacylated and reactivated during enterohepatic circulation. Rifampin induces cyt P_{450} enzymes and thereby enhances its own metabolism. Dosage reduction is required when there is hepatic dysfunction.

About 1/3 of the drug is excreted in urine, partially unmetabolized. Dosage reduction is not required when there is renal impairment. Rifampin imparts a reddish color to urine and other body fluids.

Mechanism. Rifampin inhibits DNA-dependent RNA polymerase in mycobacteria, suppressing a new chain formation in RNA synthesis. Mammalian RNA polymerases in the nucleus and mitochondria are less sensitive to rifampin. Rifampin is bactericidal to mycobacteria, Neisseria, S. aureus and most gram-positive organisms.

Adverse effects
1. Allergic-hypersensitivity-like reactions: fever, rash, blood dyscrasias (anemia, leukopenia, thrombocytopenia), renal failure with hematuria.
2. Hepatitis with jaundice (especially in the elderly and in alcoholics).
3. Flu-like syndrome (fever, chills, myalgia with blood and renal effects), particularly if rifampin is given intermittently.

Uses, in addition to treatment of TB.
1. Rifampin is the drug of choice for meningitis produced by H. influenzae and Neisseria.
2. In combination with a beta-lactam or vancomycin, rifampin is useful in treating Staphylococcal endocarditis and osteomyelitis.
3. In combination with trimethoprim-sulfamethoxazole, rifampin is used to treat methicillin-resistant urinary tract infection caused by Staphylococci or Enterobacteriaceae.

C. Pyrazinamide

Chemistry. Pyrazinamide is an analog of nicotinamide **(Figure 11.26)**.

Pharmacokinetics. Pyrazinamide is well-absorbed, widely distributed and slowly metabolized as shown below ($t_{1/2}$ 12 hours). Elimination is primarily by glomerular filtration.

Mechanism. Pyrazinamide is only active (bactericidal) against actively growing Mycobacteria. An acid environ enhances its activity. However, the precise mechanism of action is not known.

Adverse effects
1. Hepatocellular dysfunction/hepatotoxicity with jaundice is a dose-related effect that is often produced by pyrazinamide. This is generally reversible when the drug is discontinued.
2. Hyperuricemia with gout is related to competition of pyrazinamide acidic metabolites with carrier sites for uric acid in the renal tubules.
3. Arthralgias (mainly fingers, knees, shoulders) occur.
4. Allergic-hypersensitivity-like reactions are also produced.

D. Streptomycin

This antibiotic is described in the section on aminoglycoside antibiotics. For treating TB, streptomycin is *bacteriostatic* and is the least used of the primary drugs. Nearly 10% of patients experience adverse effects, mainly involving auditory or vestibular function.

E. Ethambutol (Myambutol)

Chemistry and Pharmacokinetics. Ethambutol is a simple water-soluble amino alcohol that is well-absorbed (75%), attains peak levels in less than 4 hours, has a $t_{1/2}$ 4 hours, and is largely excreted (2/3) unmetabolized within 24 hours by the kidneys. Dosage adjustment is needed when there is renal impairment. Major metabolites are an aldehyde and di-carboxylic acid **(Figure 11.26)**.

Mechanism. Ethambutol is effective against most strains of mycobacteria, including those resistant to isoniazid and streptomycin. Ethambutol inhibits incorporation of mycolic acids into the cell wall of mycobacteria. It exerts a tuberculostatic effect.

Adverse effects. Ethambutol generally produces few adverse effects. For this reason it has largely replaced para-aminosalicylic acid in the treatment of TB. Ethambutol is used in combination with isoniazid. Major adverse effects are:
1. *Optic neuritis* causes decreased visual acuity and decreased red-green discrimination. This is a dose-related effect that may be unilateral or bilateral. Patients usually recover when the drug is discontinued.
2. Hyperuricemia is an expected effect of a drug that is metabolized to a -COOH which competes with uric acid for secretory sites on the nephron.
3. Hypersensitivity reactions occur infrequently.

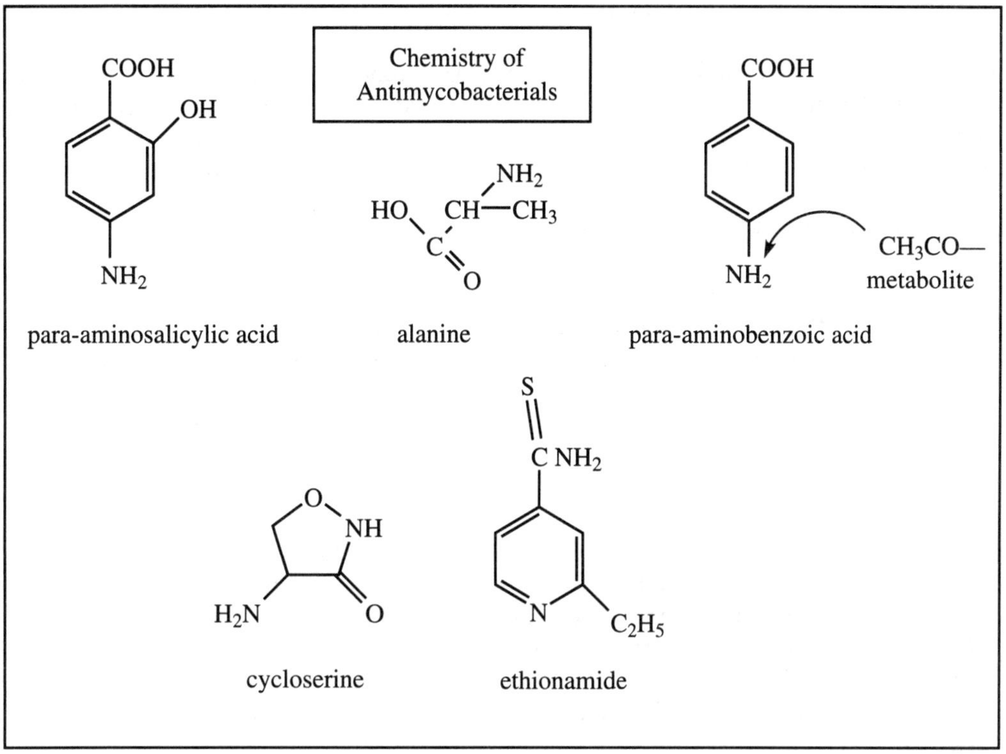

Figure 11.27

Second-line drugs

A. Para-Aminosalicylic Acid (Teebacin)

Chemistry and action. Para-aminosalicylic acid (PAS) acts as a false substrate, being erroneously utilized by folic acid synthesis in place of para-aminobenzoic acid (PABA) during mycobacterial cell growth **(Figure 11.27)**. This action is like that of sulfonamides **(Figure 11.7)**. PAS is bacteriostatic.

Pharmacokinetics. PAS is well-absorbed, attains peak levels at 2 hours, has a $t_{1/2}$ 1 hour and is excreted mainly in urine (greater than 75%) as active and inactivated products. PAS should not be used when there is renal impairment. Concentrations of PAS in the CSF remain low because of outward transport of the PAS.

Adverse effects
1. *GI distress* (pain, vomiting, diarrhea) is so common, that this often limits the usefulness of PAS.
2. Hypersensitivity-like reactions occur:
   Hematologic: leukopenia, lymphocytopenia, agranulocytosis
   Dermal eruptions, fever
   Hemolytic anemia

B. Cycloserine (Seromycin) **(Figure 11.27)**

This drug is a competitive antagonist of mycobacterial enzymes that utilize D-Ala in cell wall synthesis. It is short-lived, being excreted in urine largely unmetabolized. Toxic levels accumulate when there is renal insufficiency. Major toxic reactions are in the CNS: seizures, paranoia, psychosis, catatonia, etc. Since cycloserine lowers the seizure threshold, ethanol should not be consumed.

C. Ethionamide (Trecator-SC) **(Figure 11.27)**

This analog of nicotinic acid acts in a manner analogous to that of isoniazid. Sulfoxide metabolites retain activity. GI side effects often limit the usefulness of ethionamide, as per PAS. Hepatotoxicity and neurotoxicity occur after ethionamide, as per isoniazid; pyridoxine should be co-administered with ethionamide. Hypersensitivity reactions and endocrine disturbances may also be seen. This spectrum of activity of ethionamide relegates it to second-line status in the treatment of TB.

D. Others

The following drugs must be administered parenterally. All have the potential to produce ototoxicity and nephrotoxicity, so they cannot be used simultaneously or with streptomycin.
1. Kanamycin and amikacin are aminoglycosides that were described earlier. Likewise, ciprofloxacin, ofloxacin and clofazimine were described previously.
2. Capreomycin is a bulky molecule first isolated from Streptomyces. Capreomycin consists of 4 active analogs that are bactericidal for mycobacteria. Leukocytosis sometimes is produced as an adverse effect.

11. ANTIFUNGAL DRUGS

Antifungal drugs are generally water-insoluble bulky molecules that are products of modified products of microorganisms. Many are so highly toxic that their use is restricted to topical application. A few are sufficiently safe for systemic use.

A. Systemic Drugs

1. Amphotericin B (Fungizone)
    Chemistry. Amphotericin B is a large water-insoluble molecule with 7 ethylene groups (-CH=CH-) and a large lactone ring, first isolated from a strain of Streptomyces. The name derives from the amphoteric nature of the molecule, being water soluble at extremes in pH, owing to a -COOH group on one end of the molecule and an -NH_2 group on the other end.

    Pharmacokinetics. Orally-administered amphotericin is not appreciably absorbed, so this route is effective only in killing fungi within the lumen of the GI tract. When administered i.v. as a colloidal suspension or liposomal prep, amphotericin is useful for systemic fungal infections. Intrathecal injections are needed for fungal meningitis, since concentrations in the CSF are only 2-3% of that in blood after i.v. injection. More than 90% is protein-bound. Excretion via urine occurs over several days.

    Mechanism. Amphotericin binds to ergosterol in fungal cell membranes, producing pores through which vital intracellular macromolecules are lost. This, plus the altered ionic permeability, results in cell death (fungicidal; **Figure 11.28**). Bacteria, lacking ergosterol, are not affected by amphotericin

B. Toxic effects are thought to be related in part to binding to cholesterol (Ch) in mammalian cell membranes **(Figure 11.28)**, with the consequence being the same as in fungi.

Use. Amphotericin B is used (a) topically for ocular fungal infections, (b) as an irrigant for Candidal infections of the bladder and (c) systemically often in combination with flucytosine for Candidal and Cryptococcal meningitis or systemic candidiasis.

Adverse effects. Amphotericin B, i.v., often produces fever, headache, chills and diarrhea (flu-like symptoms) with azotemia. The major adverse effect is nephrotoxicity, sometimes accompanied by hepatic injury and anemia (decreased erythropoietin production).

2. Flucytosine (Ancobon)/5-Flucytosine (5-FC)

    Pharmacokinetics. Flucytosine is well-absorbed after an oral route, is distributed to all sites including the CNS (75% of serum concentration), is 20%

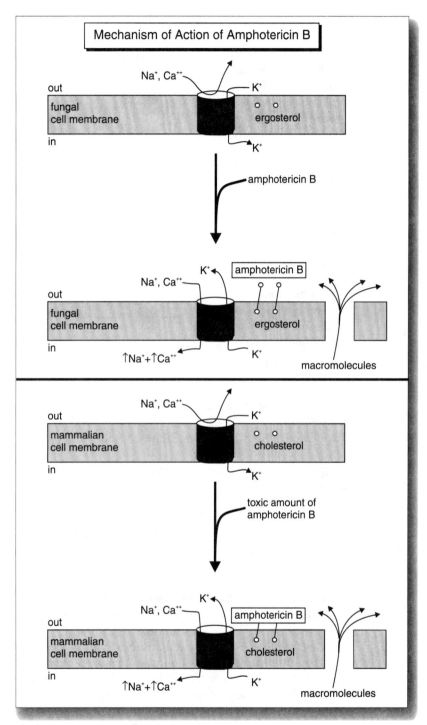

Figure 11.28

protein-bound (so hemodialysis effectively removes overdose flucytosine) and excreted largely unmetabolized in urine (urine has 10 times the serum concentration), so dosage adjustment is needed if there is renal insufficiency. The $t_{½}$ is 4 hours.

PHARMACOLOGY

Mechanism. 5-FC is converted to 5-FU by fungi but not mammalian cells. 5-FU inhibits thymidylate synthetase and the production of DNA by fungi **(Figure 11.29)**. Since resistance develops rapidly, this drug cannot be used alone. Flucytosine is fungistatic and is generally used only with amphotericin B.

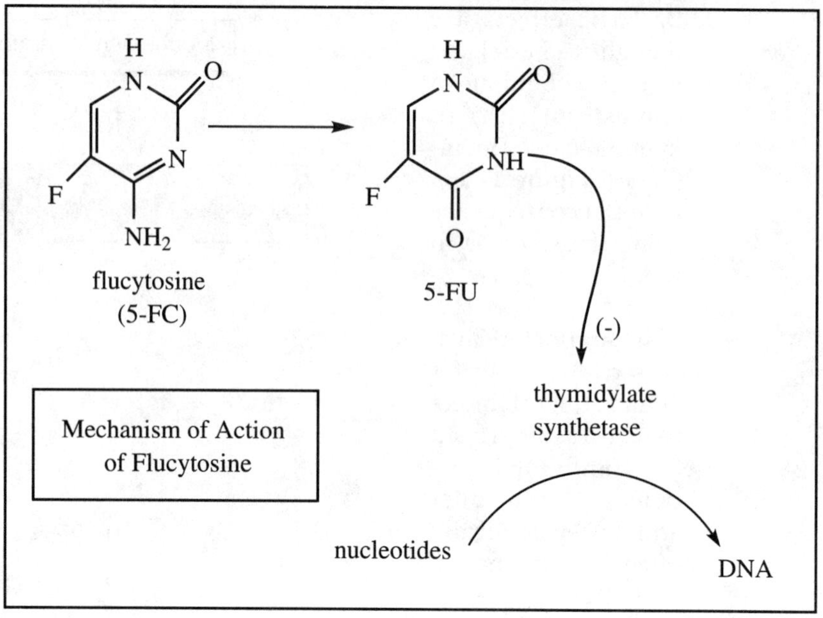

Use. Flucytosine is used for remissions of fungemia, sepsis (systemic candidiasis) or meningitis, especially cryptococcal meningitis.

Figure 11.29

Adverse effects. Flucytosine is relatively non-toxic to humans. However, after prolonged use flucytosine can produce bone marrow suppression (blocked by uracil which does not alter flucytosine activity), diarrhea, alopecia and liver toxicity.

3. **Imidazoles and Triazoles** are synthetic drug classes that share a common mechanism of action and spectrum of activity.

   Mechanism. These drugs inhibit 14-β-demethylase, a cyt P_{450}-dependent enzyme involved in synthesis of ergosterol in fungi. The false metabolites, 14-β-methyl sterols, cannot completely substitute for ergosterol.

4. **Ketoconazole (Nizoral) (Figure 11.30)**
   Pharmacokinetics of ketoconazole
   1. **Absorption.** An acid environment is required for absorption of ketoconazole. Accordingly, antacids and H_2 blockers impair absorption.
   2. Distribution is to all tissues, but less than 1% enters the CNS.
   3. **Protein binding.** 85% is bound to plasma albumin; 14% to RBC; 1% is free.
   4. **Metabolism.** Nearly all of the drug is metabolized prior to excretion, but moderate hepatic impairment does not substantially alter metabolism. However, induction of hepatic microsomal enzymes by rifampin produces a marked increase in ketoconazole metabolism. Conversely, ketoconazole competes for metabolism with warfarin and cyclosporin, potentially pushing these drugs to toxic levels (cyclosporin: nephrotoxicity).
   5. Excretion is via urine and feces.

   Adverse effects of ketoconazole
   1. **GI effects.** 20% with nausea or vomiting (Food reduces the incidence.)
   2. Rash (less than 5%) and pruritus (2%)

3. **Liver**. increased SGOT and SGPT (10%); Hepatitis is rare, but fatal.

Ketoconazole inhibits steroid synthesis in mammals (as in fungi), so adverse endocrine effects are common:
- decreased estrogen and menstrual irregularity (10%).
- decreased testosterone, sperm count, fertility, gynecomastia and libido.
- decreased ACTH, but increased 11-deoxycortisol, water retention and BP.
- Teratogenic effects occur in animals, so ketoconazole should not be taken during pregnancy or when nursing infants. (There is excretion into milk.)

Uses of ketoconazole
1. Drug of choice for most systemic fungal infections (blastomycosis, paracoccidioidomycosis, histoplasmosis, coccidioidomycosis), but not fungal meningitis.
2. Oral, esophageal and mucocutaneous candidiasis.

5. *Itraconazole* (Sporanox) **(Figure 11.30)** has a similar profile as ketoconazole for treatment of systemic fungal infections, but produces fewer adverse effects. Itraconazole has additional activity against invasive Aspergillosis and sporotrichosis.

6. *Fluconazole* **(Figure 11.30)** also has a profile much like that of ketoconazole, except that fluconazole effective enters the CNS and is useful for cryptococcal meningitis. Fluconazole is the drug of choice for oropharyngeal and esophageal candidiasis and has been approved as an oral drug for vaginal candidiasis, equally effective as intravaginal treatment. Fluconazole does not produce the endocrine effects seen with ketoconazole. Since this drug has a long $t_{1/2}$ (1 day) and is excreted in urine largely unmetabolized, toxicity is greatly increased when there is renal insufficiency. Adverse effects include eosinophilia, thrombocytopenia, Stevens-Johnson syndrome and hepatic dysfunction. Fluconazole, like ketoconazole, competes with metabolism of phenytoin, cyclosporine, anticoagulants and other drugs.

7. Griseofulvin (Grifulvin)
Pharmacokinetics. Griseofulvin is poorly absorbed because of its insolubility. Microsize and ultramicrosize particles are used to facilitate absorption. Fatty meals also increase absorption. Most of the drug is metabolized ($t_{1/2}$ of 1 day).

Mechanism and use. Griseofulvin is deposited in keratin precursor cells, which retain the drug and thereby become resistant to fungal infection. A therapeutic effect occurs as old infected skin or nails are gradually replaced by new cells impregnated with the griseofulvin. Consequently, this drug must be administered for a long time, until all of the old infected keratin-containing cells are shed (less than 1 month for scalp or hair infections; 6-9 months for fingernail infections; over 1 year for toenail).

At the cellular level, griseofulvin has an action much like that of colchicine. Griseofulvin disrupts microtubules during spindle formation, producing dysfunctional multinucleate fungal cells. Microsporum, Trichophyton and Epidermophyton infections are effectively treated with griseofulvin.

Adverse effects. Griseofulvin produces a low incidence of adverse effect, which may include headache, peripheral neuritic, hepatotoxicity, leukopenia and serum sickness. Microsomal enzymes are induced, so interactions can occur with many drugs such as warfarin. The drug is teratogenic in animals.

8. Miconazole **(Figure 11.30)** is rarely used systemically for disseminated mycoses.

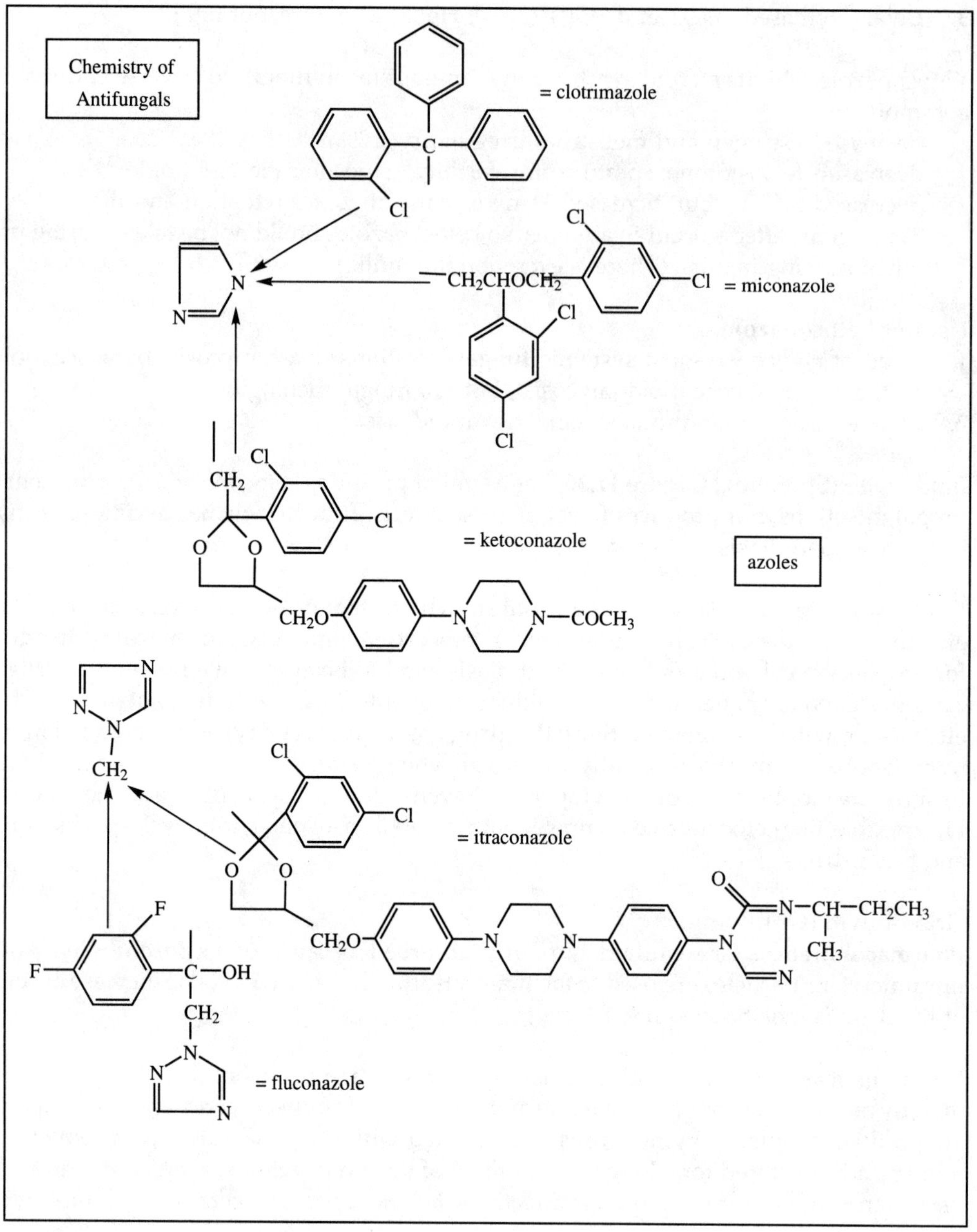

Figure 11.30

B. Topical Drugs

These are used for fungal infections of the
1. skin: tinea corporis, t. pedis, t. cruris, t. versicolor, candidiasis;
2. vagina: candidiasis and
3. mouth: candidiasis.

Since most of the drugs used for vaginal fungal infections are teratogenic, there is concern about their safety during the first trimester of pregnancy.

 **ANTIMICROBIAL CHEMOTHERAPY** **317**

1. Imidazoles and triazoles, individually, are generally useful for cutaneous, vaginal and oral fungal infections. Resistance rarely develops to these drugs.
    a. Clotrimazole (Lotrimin) is not appreciably absorbed through the skin. From the vagina, less than 10% is absorbed and this is largely metabolized in the liver.
    b. Miconazole (Monistat) has a spectrum of activity like that of clotrimazole.
    c. Econazole (Spectazole) is a chemical analog of miconazole.
    d. Butoconazole, terconazole and tioconazole are used only for vulvovaginal candidiasis.

2. Polyenes
    a. Nystatin (Mycostatin) is only used for candidiasis and is fungicidal.
    b. Amphotericin B (topical equals Fungizone) is used for candidiasis.
    c. Natamycin (Natacyn) is the drug of choice for Fusarium solani.

3. Others
    a. Tolnaftate (Tinactin) and naftifine (Naftin) are used only for cutaneous fungal infections. Tolnaftate is ineffective against candidiasis.

12. ANTIVIRAL DRUGS

Viruses are intracellular parasites that instruct host cells to synthesize the nucleic acid embodied in the virus. In concert with this, new viral protein is also synthesized for assembly of the viral DNA or RNA into protein envelopes. This assembly thus constitutes a new virus particle. When released from the infected cell, these new virus particles infect other host cells and replay the entire sequence of intracellular synthesis of viral nucleic acid and protein. Antiviral drugs are those which interrupt the cycle of new virus synthesis. In general, these drugs coincidentally affect host cell biochemistry, so toxicity of antivirals is often high.

The scheme of cellular events involved in virally-infected cells is a multi-step process that is illustrated in **Figure 11.31**. (1a) Viruses must first attach to the host cell. Next, (1b) viruses must penetrate the host cell, (2) with viral nucleic acid becoming uncoated. (3) The information encoded in the viral nucleic acid then directs the host cell to synthesize viral nucleic acids and proteins. Finally, (4) these individual elements of the virus must be assembled for (5) ultimate release of new viruses from the host cell. Drugs may act at any of these different steps in virus synthesis.

A. Drugs that Inhibit Attachment or Penetration/Uncoating of the Virus Particle

1. *Gamma Globulin*, sometimes a hyperimmune globulin prepared from plasma with high antibody titers, is used to attack specific antigens on the protein coat of the virus **(Figure 11.31)**. This has the effect of inactivating proteins that are needed for attachment onto host cells.

    Immunity is conferred for a period of 2-3 weeks. For viruses with prolonged incubation periods, gamma globulin injections must be repeated at an interval of about 3 weeks.

    Use. For hepatitis B, measles, poliomyelitis, rabies and other viruses.

2. *Amantadine* (Symmetrel) is a tricyclic symmetrical amine **(Figure 11.33)** that prevents the uncoating of viruses (first action) and reassembly of viruses (second action) within host cells **(Figure 11.31)**. The net effect is inhibition of viral replication.

    Mechanism. The alkalinity of amantadine prevents acidification of endosomes that encompass viral particles in the host cell, thereby impairing fusion of the virus protein coat with the endosome

(**Figure 11.31**, first action). Consequently, the viral DNA or RNA is unable to be released into the host cell cytoplasm. In addition to this non-specific mechanism, amantadine binds to proteins in the capsid, preventing reassembly of virus particles (**Figure 11.31**, second action).

Pharmacokinetics. Amantadine is well-absorbed, having a $t_{1/2}$ about 12 hours and is excreted largely unmetabolized in urine. Dosage reduction is necessary if there is renal impairment; and in the elderly.

Adverse effects. These are mainly CNS related: agitation, hallucinations and seizures.

Use. Amantadine is useful prophylactically for influenza A (not B), rubella and other viruses. Amantadine is also an antiparkinsonian drug.

3. Rimantadine (Flumadine), a chemical analog of amantadine (**Figure 11.33**), has similar antiviral activity but without the neurotoxicity. Most (about 90%) of this drug is excreted metabolized. Its $t_{1/2}$ is 1 day.

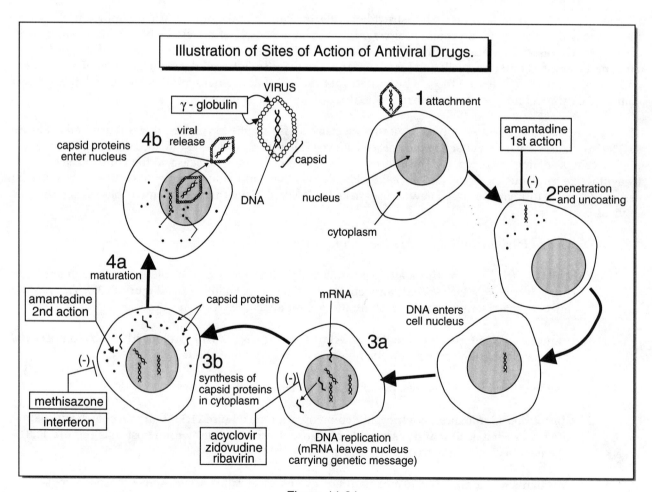

Figure 11.31

B. Drugs that Inhibit Host Cell Synthesis of Viral Nucleic Acid

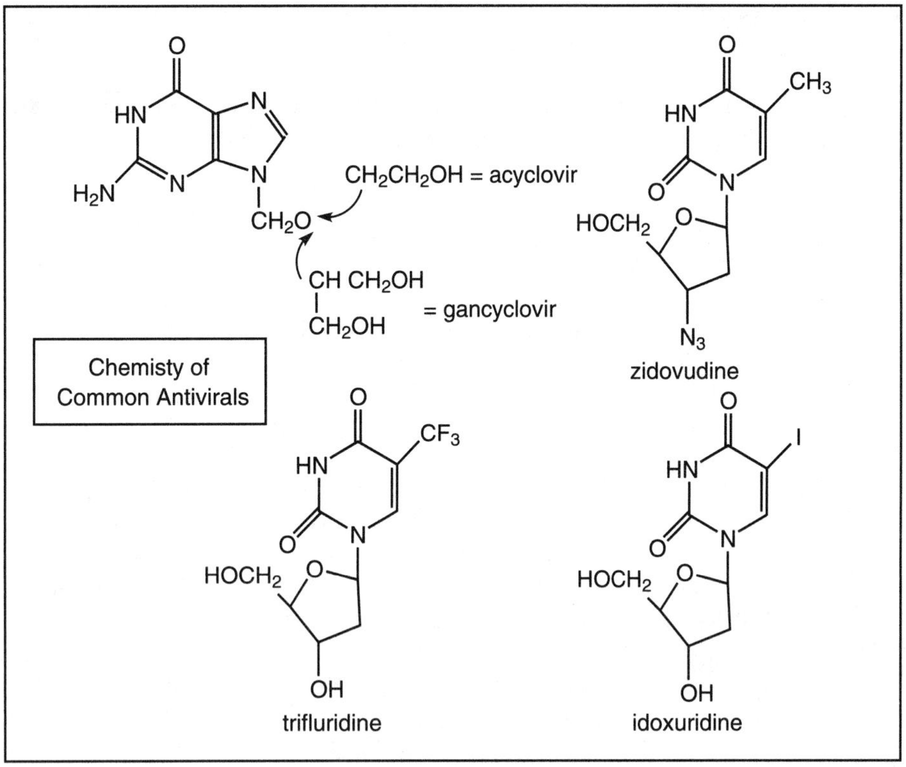

1. *Acyclovir* (Zovirax) is an analog of guanine **(Figure 11.32a)** that acts only on herpesviruses. Acyclovir must be phosphorylated, as shown in **Figures 11.34 a and b**, in order to be active. The thymidine kinase in herpesvirus phosphorylates acyclovir 100 times faster than does host cell thymidine kinase. Consequently, acyclovir has preferential activity in infected cells. The final acyclovir triphosphate inhibits herpes DNA polymerase about 25 times more than host cell DNA polymerase. Acyclovir is also added onto the nucleotide chain. Because nucleotides cannot be added onto acyclovir, this latter process represents chain termination **(Figure 11.34a)**. The defective DNA is unable to maintain vital cell processes **(Figure 11.31)**.

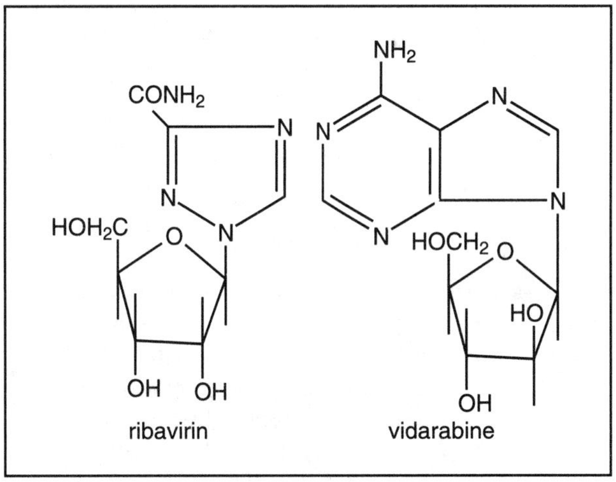

Figure 11.32a

Figure 11.32b

Use. Acyclovir is used for infections with herpes simplex 1, herpes simplex 2 and varicella-zoster viruses. This includes treatment of genital herpes, suppression of herpesvirus infections during immunosuppression therapy (for transplantation procedures) and for treatment of AIDS patients.

Pharmacokinetics. Only about 25% of an oral dose is bioavailable, with about 20% being protein-bound. Approximately 90% is excreted unmetabolized in urine, via glomerular filtration and active secretion into renal tubules. Dosage adjustment is required when there is renal impairment. The $t_{½}$ is less than 4 hours.

Adverse effects. These include local irritation, encephalitis and nephropathy. Acyclovir-resistant, thymidine kinase-deficient herpesviruses may emerge during therapy.

2. *Gancyclovir* (Cytovene) is an analog of acyclovir **(Figure 11.32a)** that acts in a similar manner. Gancyclovir is triphosphorylated in cytomegalovirus-infected cells **(Figure 11.34b)**, inhibiting cytomegalovirus DNA polymerase while being incorporated as a chain terminator into cytomegalovirus DNA.

   Adverse effects. These include bone marrow suppression which is reflected by leukopenia, neutropenia and thrombocytopenia. Also, seizures, hepatic and renal impairment occur. Dosage adjustment is required in patients with renal impairment.

   Use. Because of its high toxicity, gancyclovir is limited to i.v. use in treating life- or sight-threatening cytomegalovirus infections.

3. *Zidovudine* (AZT; Retrovir) is a 3'-azido-substituted analog of thymidine **(Figure 11.32a)** that is converted by cellular enzymes to a triphosphate **(Figure 11.34b)** that inhibits HIV reverse transcriptase of HIV and other retroviruses, about 100 times more so than the mammalian DNA polymerase. Zidovudine is also a chain terminator during DNA synthesis. Both mechanisms are identical to those of acyclovir.

   Pharmacokinetics. After oral administration 65% of zidovudine is absorbed, with peak plasma levels occurring at about 1 hour. About 25% of zidovudine is plasma protein bound. Distribution is to all tissues including brain. Serum $t_{½}$ is 1 hour, with intracellular $t_{½}$ of 3 hours. Most is metabolized by the liver to glucuronides but approximately 20% is excreted unmetabolized in urine.

   Adverse effects. Dose-limiting bone marrow suppression occurs, with associated anemia and granulocytopenia. This effect is related to numbers of CD4 lymphocytes and granulocytes at the start of treatment. Neurotoxicities (Wernicke's encephalitis and delayed poliomyositis-like symptoms) and seizures occur.

   Use. Zidovudine is used to decrease the morbidity and mortality of AIDS. Activity of zidovudine is enhanced by acyclovir and interferon, but reduced by ribavirin which competes for phosphate, thereby reducing formation of zidovudine triphosphate.

4. *Ribavirin* (Virazole) is a synthetic purine nucleoside **(Figure 11.32b)** that is phosphorylated in cells by adenosine kinase to a monophosphate (RMP) that inhibits IMP dehydrogenase and thereby reduces formation of GMP, and ultimately formation of GTP. RMP is further phosphorylated to RTP, a product that inhibits viral RNA polymerase **(Figure 11.34b)**. These actions of ribavirin inhibit viral mRNA formation **(Figure 11.35)**. Ribavirin inhibits both DNA and RNA viruses, including myxoviruses, retroviruses, herpesviruses, adenoviruses and pox-viruses.

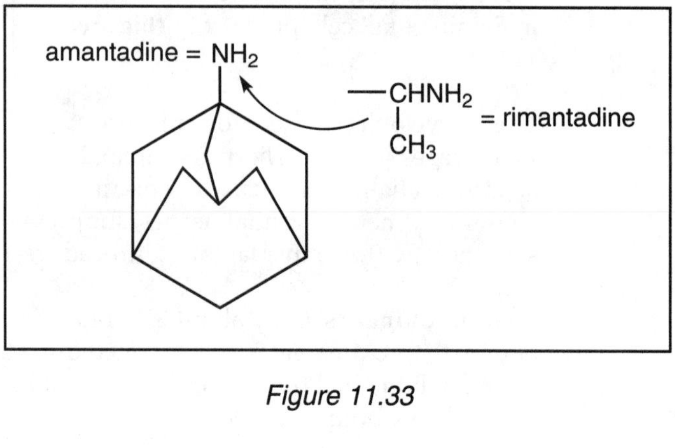

Figure 11.33

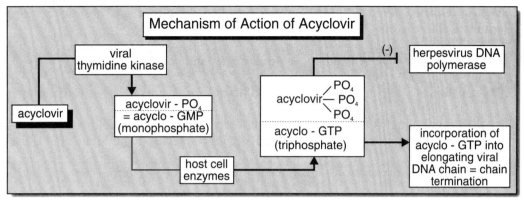

Figure 11.34a

Pharmacokinetics. After oral administration about 50% is bioavailable, with peak plasma levels occurring at 1-2 hours The initial elimination $t_{1/2}$ is about 2 hours, but the prolonged phase of elimination is about 1 1/2 day. RTP accumulates in RBC, with a $t_{1/2}$ greater than 1 month.

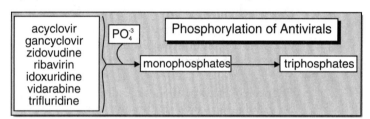

Figure 11.34b

Use. For respiratory syncytial viral infections, influenza A and B, hepatitis and other infections.

Major adverse effects. These include anemia, due to extravascular hemolysis and marrow suppression; and reticulocytosis.

Contraindications. Ribavirin interferes with the action of zidovudine in HIV infections.

5. *Idoxuridine* is the 5-iodo-derivative of 2'-deoxyuridine **(Figure 11.32a)**. This drug is phosphorylated intracellulary like the drugs described above **(Figure 11.34b)**, to a product that is incorporated into DNA. The aberrant DNA synthesizes abnormal viral proteins.

Use. Idoxuridine (Herplex) is used for DNA viruses, primarily herpes and pox viruses. The major use (topical) is for herpes simplex infections of the cornea.

Adverse effects. Corneal irritation or clouding.

6. *Vidarabine* (adenine arabinoside, ara-A) is an adenosine analog **(Figure 11.32b)** that is phosphorylated in host cells, like those drugs

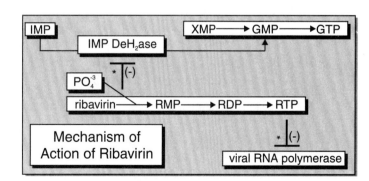

Figure 11.35

already described **(Figure 11.34b)**, to products that inhibit viral DNA polymerase and chain-terminate. There is little toxicity to the marrow, liver and kidneys.

Use. Herpes virus type 1 infections, varicella zoster and others.

Adverse effects. After prolonged use GI and neurological signs may be more prominent.

7. *Trifluridine* (Viroptic) is the 5-trifluoromethyl derivative of 2'-deoxyuridine **(Figure 11.32a)**. This drug is converted to a triphosphate derivative **(Figure 11.34b)** that inhibits viral DNA synthesis.

   Use. Trifluridine is used mainly for keratoconjunctivitis caused by herpes simplex 1 and 2 infections.

8. *Interferons* (IFNs) are glycoproteins having a M.W. of 15-20,000. These potent cytokines are naturally produced by cells (Type I is alpha or beta) and lymphocytes (Type II is gamma) or by recombinant DNA technology. Type I IFNs bind to specific cell receptors and induce host cell ribosomes to synthesize enzymes that powerfully inhibit translation of viral mRNA **(Figure 11.31)**. These enzymes may interfere with any of the other steps in viral production, as well. Type II IFNs stimulate macrophage activity.

   Pharmacokinetics. IFNs are administered i.m. or s.c., have a termination $t_{1/2}$ about 5 hour and accumulate for days at a time in mononuclear cells, enhancing their phagocytotic activity.

   Use. Interferon alpha-2a (Roferon-A) and interferon alpha-2b (Intron A) are used for viral hepatitis B and C, genital warts, hairy-cell leukemia, AIDS-related Kaposi's sarcoma and other neoplastic diseases.

   Adverse effects. These include an influenza-like syndrome (fever, chills, headache, myalgia, diarrhea), bone marrow suppression with granulocytopenia and thrombocytopenia, neurotoxic effects including seizures and EEG changes, renal and hepatic impairments, and cardiac toxicity.

Hookworm and threadworm life cycle. Ova in stools contaminate soil. When ova hatch, larvae penetrate the skin, enter veins and circulate to lungs. Larvae ascend the bronchi and trachea, and are swallowed into the GI tract where they will reside. Threadworms use the contents of the GI tract for nourishment. Since hookworms use blood as nourishment, not GI contents, anemia occurs in the host.

Whipworm life cycle. Whipworm ova in stools contaminate soil and vegetation. When consumed, ova hatch in the lumen of the GI tract and take up residence, using the contents of the GI tract for nourishment.

Pinworm life cycle. In an infected child hookworms deposit ova in the anus, causing irritation and itching. When the child

Drugs Used to Eradicate Intestinal Nematodes (Roundworms) Table 11.1

Species and % global infection	Drug of Choice	Alternate
ROUNDWORM (~25%) Ascaris lumbricoides	Mebendazole	Albendazole, pyrantel, piperazine
HOOKWORMS (~20%) 1. Necator americanus 2. Ancylostoma duodenale	Mebendazole	Albendazole, pyrantel, levamisole
3. A. braziliense (cutaneous larval migrans)	Thiabendazole	Albendazole
WHIPWORM (Trichuriasis) (~15%) Trichuris trichiura	Mebendazole	Albendazole, oxantel, ivermectin
PINWORM (Enterobiasis) (~10%) Enterobius vermicularis	Mebendazole	Albendazole, pyrantel, ivermectin
THREADWORM (1-2%) Strongyloides stercoralis	Thiabendazole	Albendazole, ivermectin
TRICHINOSIS (~1%) Trichinella spiralis	Albendazole	Thiabendazole (GI phase) Mebendazole (tissue larval phase)
GUINEA WORM Dracunculus medinensis	Metronidazole	Thiabendazole, Mebendazole

scratches, fingers become contaminated. Through contact, ova are spread among members of a family or play group. When contaminated objects or fingers are placed in the mouth, ova are swallowed. Ova hatch in the lumen of the GI tract where worms take up residence.

Trichinella life cycle (Pork roundworm). Hosts are infested by eating improperly cooked larvae-encysted pork.

DRUGS FOR INTESTINAL NEMATODES (TABLE 11.1 AND FIGURE 11.36)

1. *Mebendazole* (Vermox) is a broad-spectrum anthelmintic, useful against larval and adult stages of many kinds of nematodes, some trematodes and some cestodes. Ova of roundworms, hookworms and whipworms are killed.

    Pharmacokinetics (Figure 11.37). Mebendazole is poorly absorbed and rapidly inactivated by first pass metabolism, so its desired effect on worms within the intestine is accompanied by few adverse effects. The small amount absorbed is excreted via bile and urine.

Mechanism. Mebendazole binds to the β-subunit (dimer) of tubulin (in nematodes and mammals), inhibiting its assembly into tubulin **(Figure 11.42)**. Also, mebendazole inhibits glucose uptake by adult and larval stages, resulting in glycogen depletion, reduced ATP formation and ultimate immobility and death.

2. *Albendazole* (Zental) is a broad-spectrum anthelminthic that binds to tubulin **(Figure 11.42)**, like mebendazole, and is often used as an alternative to mebendazole.

   Pharmacokinetics (Figure 11.38). Albendazole is well-absorbed and has a long $t_{½}$. It is metabolized by monooxygenases in the liver and excreted in urine. However, it is quite safe, being associated with few side effects.

3. *Thiabendazole* (Mintezol) binds to tubulin and inhibits glucose uptake in nematodes, like mebendazole. Thiabendazole also inhibits fumarate reductase, thereby interfering with ATP formation in helminths **(Figure 11.42)**.

   Pharmacokinetics (Figure 11.39). Thiabendazole is rapidly absorbed and has high bioavailability. Consequently, it is associated with a greater incidence of side effects. Most is metabolized in liver to products that are excreted in urine.

4. *Piperazine* (Vermizine) is an alternative drug for Ascaris. Piperazine blocks nicotine receptors on nematode skeletal muscle, producing flaccid paralysis **(Figure 11.43)**. Live worms are expelled through normal peristaltic action.

   Pharmacokinetics (Figure 11.40). Piperazine has a high availability and is accordingly associated with a high incidence of side effects. Because of its neurotoxic potential, it should not be used in epileptics. However, it is not teratogenic and can be used in pregnancy.

5. *Pyrantel* (Antiminth) and *Oxantel* (Quantrel) are broad-spectrum anthelmintics **(Table 11.1)**. They produce spastic paralysis in nematodes via agonist activity at nicotine receptors

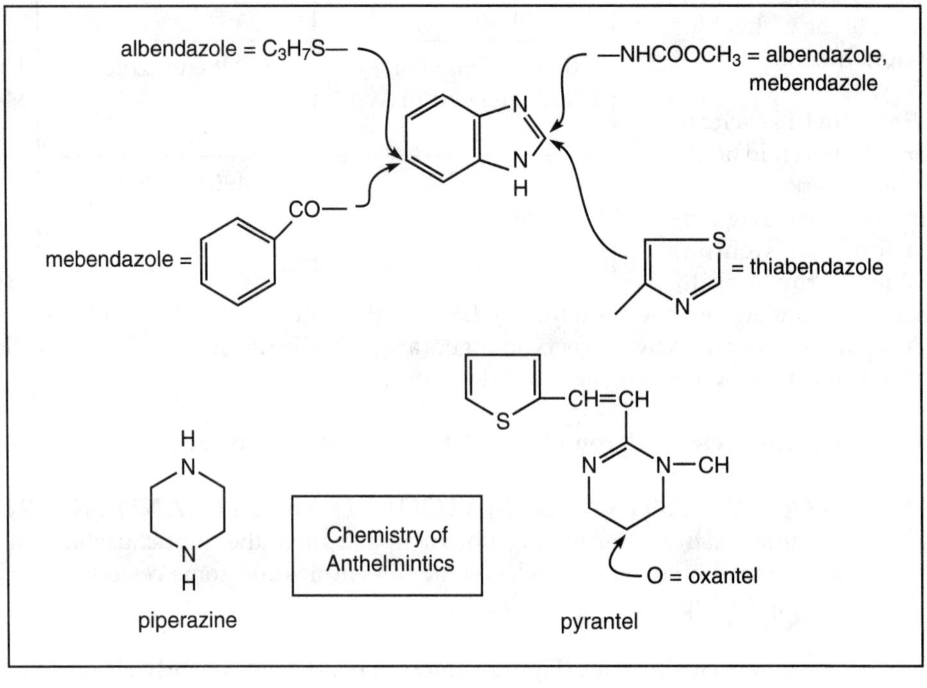

Figure 11.36

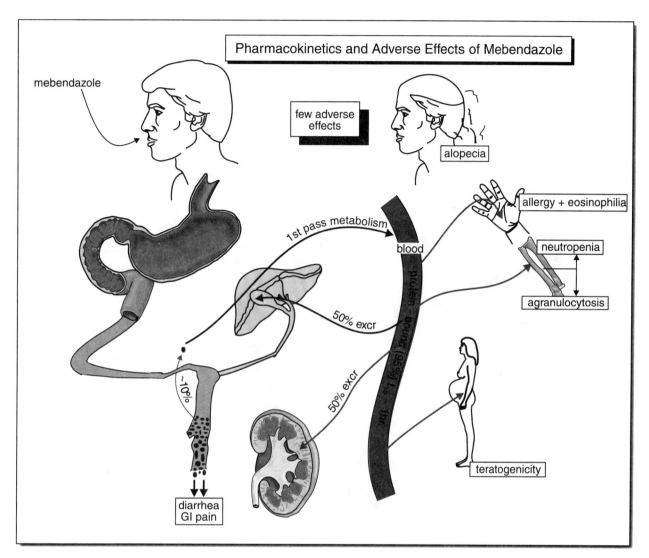

Figure 11.37

on skeletal muscle (i.e., succinylcholine-like) **(Figure 11.43)**. Cholinesterase is also inhibited. Live worms are expelled through normal peristaltic action. Use with piperazine is contraindicated, since piperazine blocks pyrantel/olantel action at nicotine receptors.

Pharmacokinetics (Figure 11.41). Only a small amount is absorbed, so there are few adverse effects. Parent drug and metabolites are excreted largely in urine.

B. ANTHELMINTICS FOR BLOOD AND TISSUE NEMATODES (FILARIASIS)

Filarial life cycle. Larvae enter the host via the vectors, mosquitoes (Bancrofti/Brugia) or black flies (Onchocerca). These develop and take up residence in lymphatic vessels and nodes, causing lymphangitis and lymphatic obstruction which leads to swelling of extremities and soft tissues like testes (Bancrofti causes elephantiasis) and eyes (Onchocerca causes River blindness).

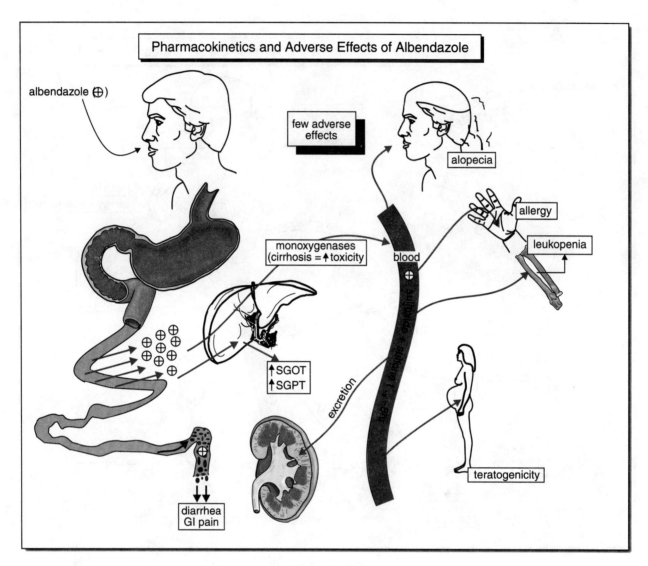

Figure 11.38

DRUGS FOR BLOOD AND TISSUE NEMATODES

1. *Diethylcarbamazine* (Hetrazan) produces spastic paralysis **(Figure 11.43)** and exposes surface protein of filariae, so filariae are made susceptible to host immune defenses **(Figure 11.42)**. Adult worms of B. malayi and Loa, loa are killed; W. bancrofti and Onchocerca are not.

   Pharmacokinetics. Diethylcarbamazine is rapidly absorbed and has a long $t_{1/2}$ (about 2 hours at urine pH is 5; about 12 hours at urine pH is 8). Damaged microfilariae produce Mazzotti reactions **(Figure 11.44)**, which can be attenuated by glucocorticoids.

2. *Ivermectin* (Mectizan) is a semisynthetic analog of a macrocyclic lactone that is naturally produced by *Strep. avermitilis*. It is broad-spectrum. Ivermectin kills microfilariae, even in the uterus of adult filariae **(Figure 11.45)**.

ANTIMICROBIAL CHEMOTHERAPY

Pharmacokinetics (Figure 11.45). Ivermectin is rapidly absorbed and largely excreted into bile. After being secreted with bile into the intestine, the drug can be absorbed again. This enterohepatic

Drugs Used to Eradicate Blood and Tissue Nematodes (Filariasis) (~15% globally) Table 11.2

Species	Drug of Choice	Alternate
Wuchereria bancrofti	Diethylcarbamazine	Ivermectin
Brugia malayi	Diethylcarbamazine	Ivermectin
Loa, Loa	Diethylcarbamazine	Ivermectin
Onchocerca volvulus	Ivermectin	Diethylcarbamazine + (later) Suramin

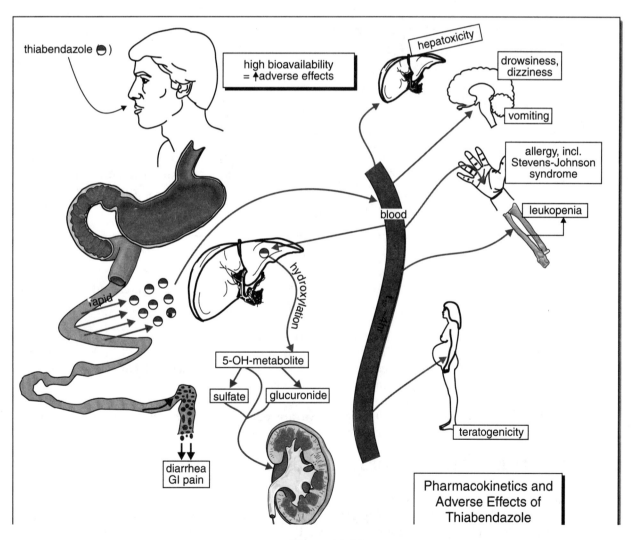

Figure 11.39

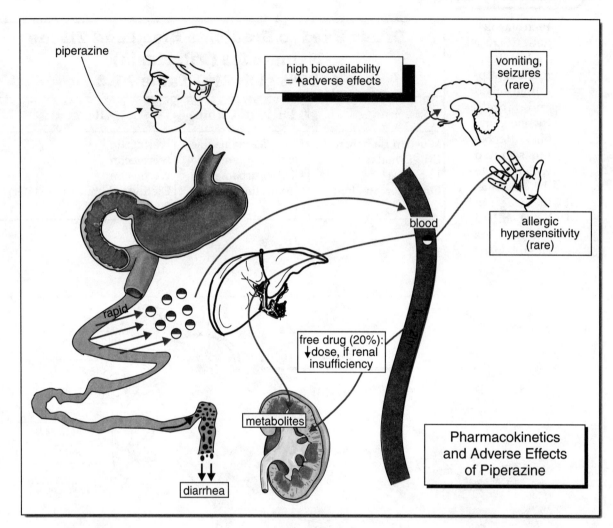

Figure 11.40

circulation extends the $t_{1/2}$ of ivermectin (about 12 hours). Most of the drug is excreted unmetabolized in feces.

Mechanism (Figure 11.43). Ivermectin is an agonist at GABA receptors and possibly enhances presynaptic GABA release. Consequent increased permeability to Cl⁻ results in hyperpolarization of filariae muscles and flaccid paralysis. Since ivermectin does not enter mammalian brain, mammalian GABA receptors are largely unaffected.

Therapeutic note. Since adult Onchocerca are not killed, ivermectin must be readministered annually to kill microfilariae that are produced by these worms.

C. ANTHELMINTICS FOR CESTODES (TAPEWORMS)

Life cycle of tapeworms. Tapeworm larvae develop in the above hosts and infect humans that eat improperly cooked larvae-infested food. Larvae develop into worms in human intestine and attach to the wall of the intestine, using contents of the GI tract for nourishment. Larvae may bore through the intestinal wall and lodge in internal muscles or organs (cysticerci) or brain (neurocysticerci), producing associated symptoms.

ANTIMICROBIAL CHEMOTHERAPY

Drugs for tapeworms

1. *Praziquantel* (Biltricide) **(Figure 11.46b)** kills adult worms and ova. Praziquantel increases the permeability to Ca^{++}, thereby producing muscle contractions and spastic paralysis of worms **(Figure 11.48)**. This causes them to be dislodged from their attachment sites. In high concentration, praziquantel causes blebbing of the tegmentum of worms, rendering them susceptible to host immune defenses **(Figure 11.48)**.

   Pharmacokinetics (Figure 11.47). Praziquantel is rapidly absorbed and undergoes first-pass metabolism. Non-absorbed drug acts on tapeworms within the intestine. Absorbed drug paralyzes flukes attached to the wall of mesenteric veins, causing their migration to the liver. Few adverse effects are normally produced. However, praziquantel should not be used to treat optic cysticercosis, since blindness could be produced.

2. *Niclosamide* (Niclocide) **(Figure 11.46b)** kills adult cestodes by preventing formation of ATP from ADP **(Figure 11.43)**. Ova are not killed, so there is a risk of producing cysticercosis with this drug when ova are released in the intestine from gravid proglotids. A cathartic should be given 3 to 4 hours after niclosamide, to flush ova from the intestine and thereby diminish the possibility of cysticercosis.

   Pharmacokinetics. Niclosamide is poorly absorbed, being largely retained in the intestine. Consequently, there are few adverse effects **(Figure 11.49)**.

D. ANTHELMINTICS FOR TREMATODES (FLATWORMS)

Life cycle of flukes. These enter the host by boring through the skin or via larvae in snails, fish or crayfish. Most larvae mature in

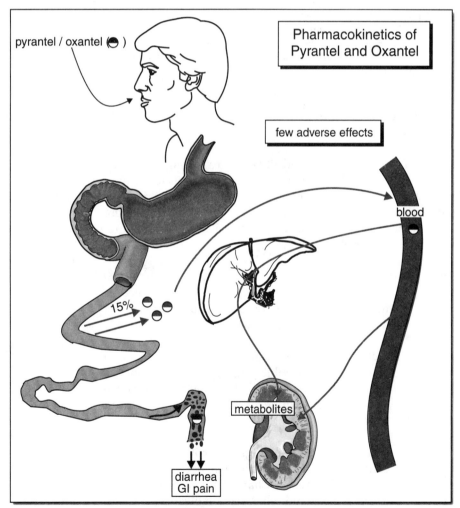

Figure 11.41

Drugs Used to Eradicate Cestodes (Tapeworms) (~3% global infection) Table 11.3

Worm Type	Drug of Choice	Alternate
BEEF TAPEWORM Taenia saginata	Praziquantel	Niclosamide, Dichlorophen, Mebendazole
PORK TAPEWORM T. solium Neurocysticercosis (larval stage)	Praziquantel Praziquantel	Niclosamide Albendazole
FISH TAPEWORM Diphyllobothrium latum	Praziquantel	Niclosamide, Albendazole, Dichlorophen
DWARF TAPEWORM Hymenolepsis nana	Praziquantel	Niclosamide
Cystic Hydatid Disease Echinococcus granulosus	Albendazole	Mebendazole
Alveolar Hydatid Disease E. multilocularis	Mebendazole	

the intestine (intestinal flukes). Some migrate and mature in the liver and bile ducts (liver flukes) or in the lungs (lung flukes).

Larvae of schistosomes (blood flukes) bore through the skin, then migrate through lymphatics and blood vessels to the liver where they mature. Adult male and female schistosomes migrate to the mesenteric vein and mate. Ova are able to bore through the intestine (S. mansoni, S. japonicum) or bladder (S. haematobium) for elimination.

Drugs for flukes

1. *Praziquantel* **(Figure 11.46b)** is the drug of choice for nearly all flukes. Actions of praziquantel were described in the section on tapeworms.

2. *Biothionol* (Bitin) **(Figure 11.46b)** kills flukes and some parasites by inhibiting respiration and energy production. Nearly half the patients experience adverse effects, often of a mild nature **(Figure 11.50)**.

3. *Oxamniquine* (Vansil) **(Figure 11.46b)** appears to be converted to a metabolite that decomposes to a carbonium ion **(see Chapter 12)** which alkylates parasite DNA and other macromolecules, thereby paralyzing the parasites. Flukes become dislodged from mesenteric veins and migrate to the liver.

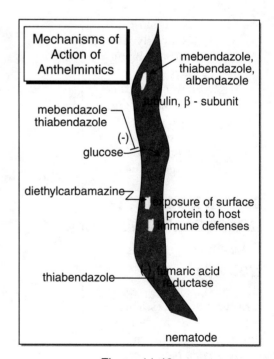

Figure 11.42

Drugs Used to Eradicate Trematodes (Flukes = nonsegmented flatworms) Table 11.4

	Species and % global infection	Drug of Choice	Alternate
BLOOD FLUKES (~5%) (SCHISTOSOMIASIS)	1. Schistosoma mansoni	Praziquantel	Oxamniquine
	2. S. hematobium	Praziquantel	Oxamniquine, Metrifonate
	3. S. japonicum	Praziquantel	Oxamniquine
	4. S. mekongi	Praziquantel	Oxamniquine
LIVER FLUKES (<1%)	1. Clonorchis sinensis	Praziquantel	Mebendazole, Albendazole
	2. Opisthorchis felineus, viverrini	Praziquantel	Mebendazole, Albendazole
	3. Fasciola hepatica (sheep)	Biothionol	Praziquantel, Dehydroemetine
LUNG FLUKES (<1%)	1. Paragonimus westermani	Praziquantel	Biothionol
	2. P. kellicotti	Praziquantel	Biothionol
INTESTINAL FLUKES	1. Heterophyes heterophyes	Praziquantel	Niclosamide
	2. Metagonimus yokogawai	Praziquantel	Niclosamide
	3. Fasciolopsis buskiPraziquantel Praziquantela	Praziquantel	Niclosamide

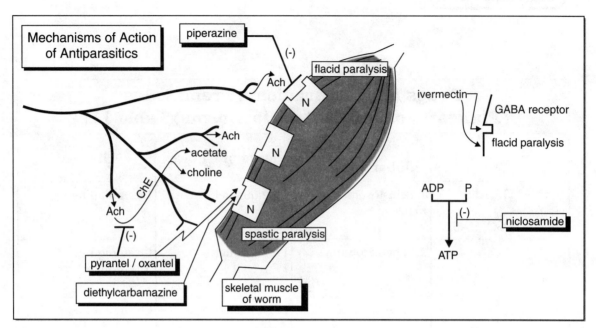

Figure 11.43

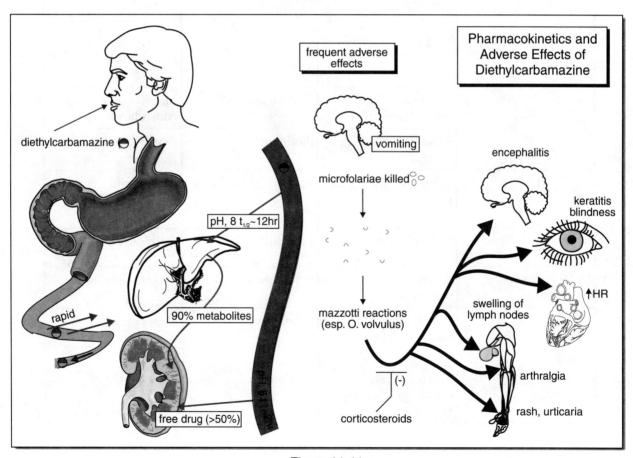

Figure 11.44

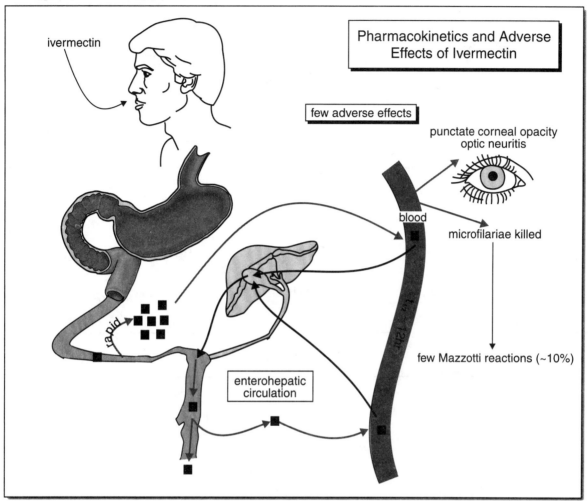

Figure 11.45

Male flukes accumulate more oxamniquine than females and most males are killed in the liver. Many females survive and reinfest the mesentery, but do not have mates and do not lay eggs.

Pharmacokinetics (Figure 11.51). Oxamniquine is rapidly absorbed and has high bioavailability. However, serious adverse effects are uncommon. Because this drug is metabolized in the liver and the gut to colored species that are excreted in urine, an orange-red urine is seen.

4. *Metrifonate* (Bilarcil) is an organophosphate **(Figure 11.46b)** that is metabolized to a product that inhibits cholinesterase (ChE) in the host (plasma and RBC ChE) and parasite (neuromuscular ChE). Paralysis is produced in parasites which detach from venous vessels in bladder. Flukes are transported to the lung where they become encased and die. Mature and immature stages, but not ova, are killed. Neuromuscular blockers should not be given within 48 hours of metrifonate.

Pharmacokinetics (Figure 11.52). Metrifonate produces few adverse effects. Symptoms of excess cholinergic activity may be seen, including vomiting, diarrhea and sweating.

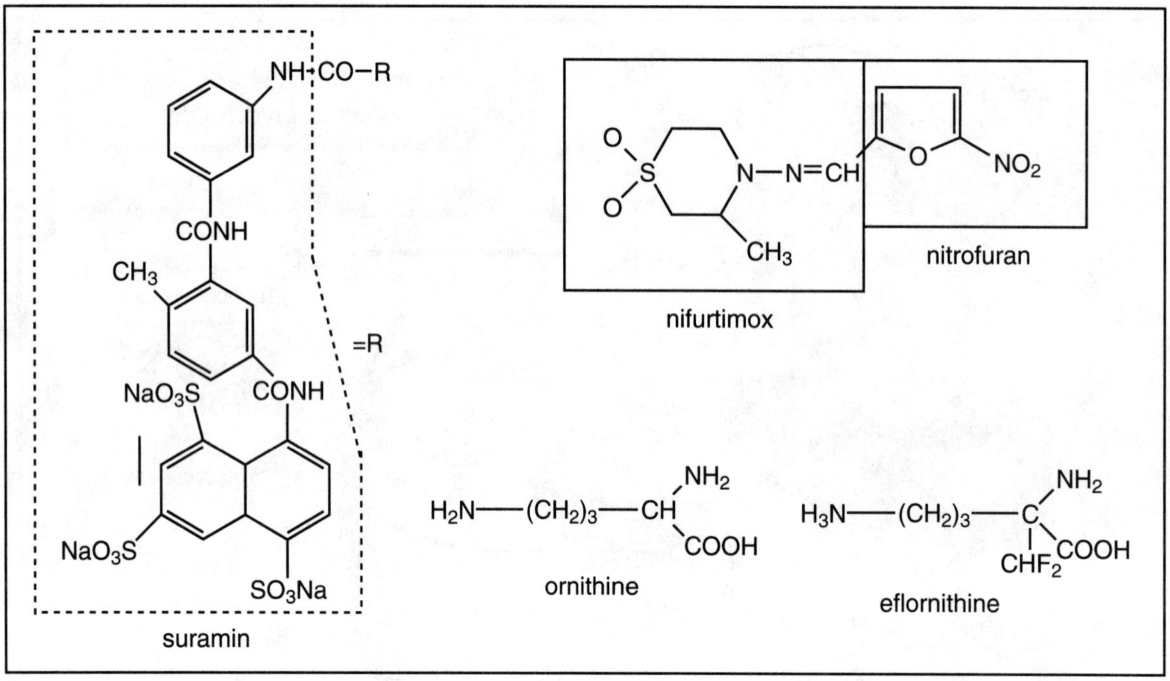

Figure 11.46a

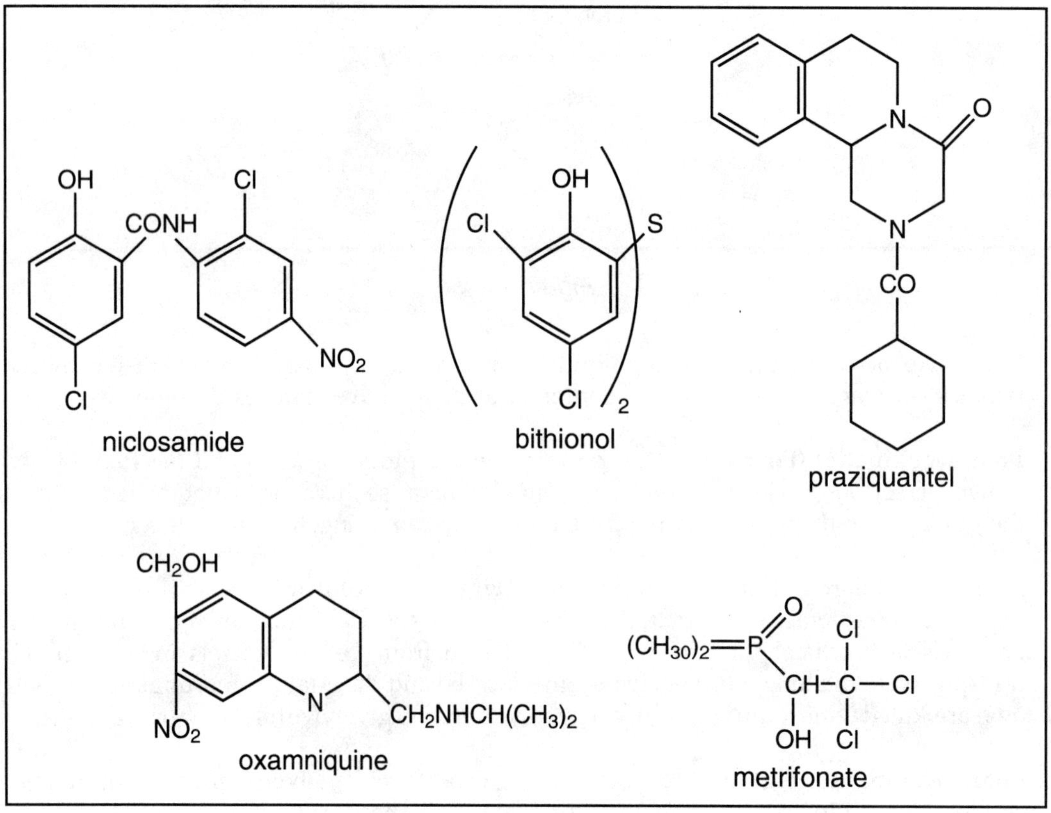

Figure 11.46b

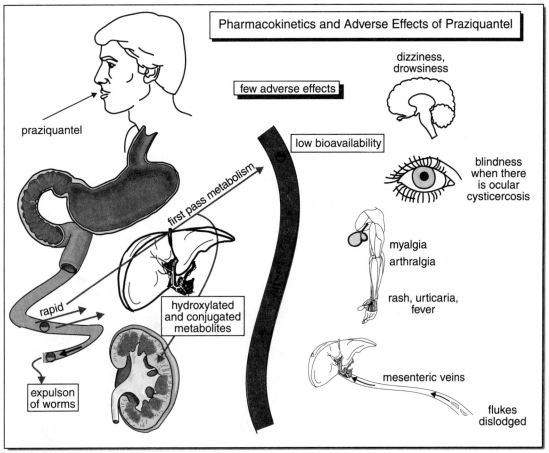

Figure 11.47

14. ANTIPROTOZOAL DRUGS

A. Treatment of Trichomonas Vaginalis

This occurs in women by the sexually-transmitted flagellated trophozoite of T. vaginalis. Male sex partners are asymptomatic carriers that must also be treated.

1. *Metronidazole* (Flagyl) is a lipophilic nitro compound (RNO_2) **(Figure 11.53)** that readily crosses lipid barriers. It is rapidly absorbed and readily crosses the blood-brain barrier to enter the CSF. Metronidazole enters all fluid compartments of humans **(Figure 11.54)**.

Mechanism. *Trichomonas* and other anaerobes readily reduce the metronidazole -NO_2 group to reactive species that interact with Trichomonas DNA and other macromolecules, particularly those with -SH and -S-S-

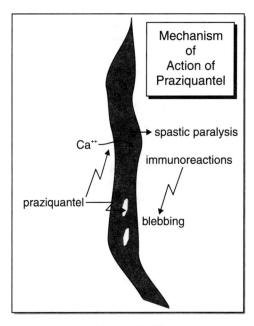

Figure 11.48

moieties **(Figure 11.53)**. Because of its inhibition of aldehyde dehydrogenase in humans, metronidazole has a disulfiram-like effect **(Figure 11.54)**.

B. Treatment of Giardiasis

Giardiasis represents the most common intestinal protozoan in the United States, usually caused when cyst-infested food or water is consumed.

1. *Metronidazole* is the drug of choice.
2. *Quinacrine* (Atabrine) is an alternative drug **(Figure 11.55)**. This is a yellow-colored lipophilic acridine analog that distributes throughout the body and is deposited in skin and under nails, causing yellow-colored skin (with urticaria causes exfoliative dermatitis) and blackened nails. Its protozoan toxicity is probably related to intercalation of DNA.

C. Treatment of Amebiasis

Amebiasis is usually caused by fecal-oral contact.

1. *Diloxide furoate* (Fumaride) is the drug of choice for asymptomatic and noninvasive intestinal forms. This drug is de-esterified in the intestine to furoic acid (causing diarrhea, flatulence) and diloxanol which is readily absorbed. The mechanism of its toxicity on trophozoites is not know.

2. *Paromomycin* (Humatin) is an aminoglycoside antibiotic that is an alternative to diloxanide. Paromomycin acts by killing intestinal bacteria on which intestinal trophozoites feed. Iodoquinol (Yodoxin) **(Figure 11.55)** is an alternative.

3. *Metronidazole* is the drug of choice for invasive amebic infections, particularly amebic dysentery and amebic abscesses.

4. *Chloroquine* **(Figure 11.59)** is an alternative drug for invasive amebic infection. Because chloroquine accumulates in the liver, systemic hepatic amebiasis is killed. Chloroquine is discussed in the section on malaria.

5. *Dehydroemetine* is an ipecac alkaloid that inhibits protein synthesis in amoeba, by preventing translocation of peptidyl-tRNA from acceptor to donor sites on amoeba ribosomes.

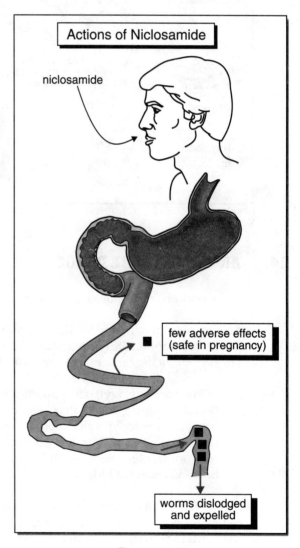

Figure 11.49

Drugs Used to Eradicate Protozoa (Table 11.5)		
Species	Drug of Choice	Alternate
TRICHOMONIASIS Trichomonas vaginalis	Metronidazole	
GIARDIASIS Giardia lamblia	Metronidazole	Quinacrine
AMEBIASIS Entamoeba histolytica a. asymptomatic b. noninvasive c. invasive(abscess, dysentery)	 Diloxanide Diloxanide Metronidazole	Quinacrine
LEISHMANIASIS 1. Leishmania tropica 2. L. mexicana, L. braziliensis 3. L. donovani	 Stibogluconate Stibogluconate Stibogluconate	Amphotericin B Pentamidine
TRYPANOSOMIASIS 1. Trypanosoma brucei rhodesiense 2. T. brucei gambiense 3. T. cruzi (Chagas' disease)	 Suramin Eflornithine Nifurtimox	Eflornithine, pentamidine Suramin, pentamidine, melarsoprol
PNEUMOCYSTOSIS Pneumocystis carinii	Bactrim	Pentamidine
MALARIA Plasmodium falciparum CHL-resistant P.f., prophylaxis CHL-resistant P.f., treatment P. malariae P. ovale P. vivax	Chloroquine (CHL) Pyrimethamine/ Chloroguanide Quinine Chloroquine Chloroquine Chloroquine	 +/-Doxycycline;Mefloquine +/-Doxycycline;Mefloquine +/-Doxycycline;Mefloquine Primaquine Primaquine

D. Treatment of Leishmaniasis.

There are three major forms of leishmaniasis transmitted by a sandfly vector:
  a. Cutaneous L. (Oriental sore) is a self-limiting infection, caused by L. tropica and L. mexicana.
  b. Mucocutaneous L. (espundia) is a severe form that causes erosion of the lips and nasal septum, leading to disfigurement. It is caused by L. braziliensis.
  c. Visceral L. (kala-azar) affects all internal organs and is caused by L. donovani.

1. *Stibogluconate* (Pentostam) **(Figure 11.55)** contains antimony ("stib") which interacts with protein sulfur groups in Leishmanias, thereby inactivating critical enzymes. Adverse effects resemble those after mercury poisoning, namely renal and hepatic failure.

2. *Amphotericin B* is an antifungal drug that is also useful for Leishmania.

3. *Pentamidine* (Pentam 300) accumulates in soft tissue (liver, spleen, kidneys) and is thought to exert its effect on protozoa by intercalating with DNA. Acute adverse effects are related to excess histamine release (decreased BP, increased HR, fainting, vomiting). Pancreatitis with hypo- or hyperglycemia is also produced.

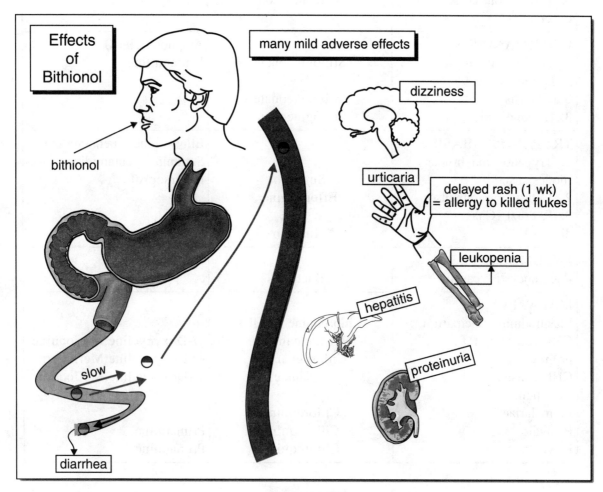

Figure 11.50

E. Treatment of Trypanosomiasis

African trypanosomiasis is caused by two species of Trypanosoma brucei:
1. T.b. rhodesiense produces a rapidly progressing disease, often fatal.
2. T.b. gambiense causes sleeping sickness.

The tsetse fly is the vector.

American trypanosomiasis is caused by amastigotes of T. cruzi (Chagas' disease). These cause destruction of myocardial cells and neurons in the myenteric plexus. There is no treatment. Nifurtimox suppresses the circulating form of the parasite. Blood-sucking reduviid bugs are the vector.

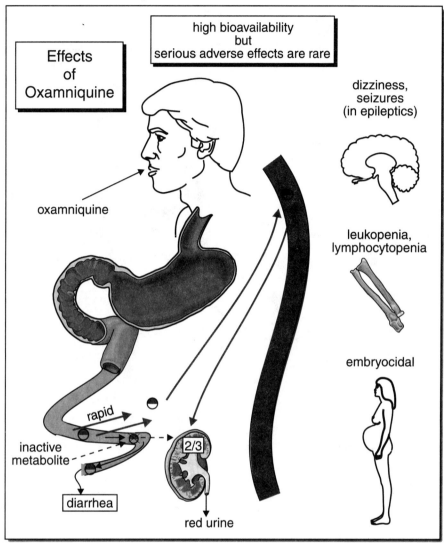

Figure 11.51

Drugs for Trypanosomiasis
1. *Suramin* (Germanin) **(Figure 11.46a)** is used for African trypanosomiasis. Its action is related to enzyme inhibition in trypanosomes, primarily glycerol phosphate oxidase and others involved in energy metabolism.

   Suramin is administered i.v. Therapeutic dosing extends effectiveness for months, making this drug useful for treatment and prophylaxis.

   Suramin produces many adverse effects **(Figure 11.56)**, with severity being related in part to initial health status. More sickly patients experience greater toxicity.

2. *Nifurtimox* (Lampit) **(Figures 11.46a, 11.55 and 11.57)** is used to treat Chagas' disease. This drug is converted to the nitro anion ($NO^\bullet$) which generates superoxide ($O_2^\bullet$) and peroxide (H_2O_2)-toxic oxidative species which damage the trypanosomes, since these are deficient in antioxidant enzymes.

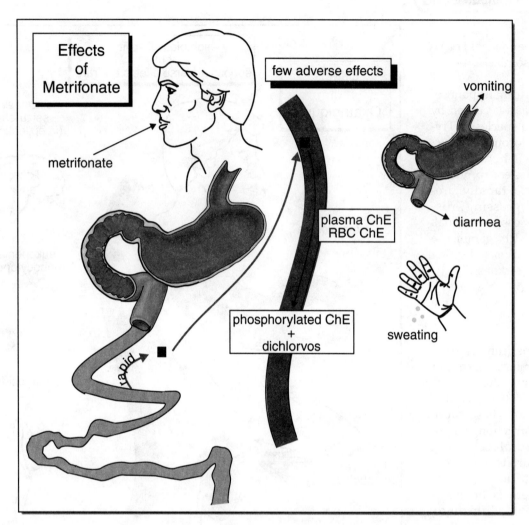

Figure 11.52

3. *Eflornithine* (Ornidyl) **(Figure 11.46a)** is administered i.v. Since it enters the brain, T.b. rhodesiense and gambiense are effectively treated. By irreversibly inhibiting ornithine decarboxylase (ODC) in trypanosomes, eflornithine prevents polyamine synthesis which is required for DNA, RNA and proteins. Selectivity is related to the fact that mammalian ODC has a $t_{1/2}$ less than 1 hour versus the trypanosome which has a $t_{1/2}$ of days, so that mammalian ODC is resynthesized rapidly. Since eflornithine sometimes rouses comatose sleeping sickness patients it has been called the "Resurrection Drug."

Toxic effects include those related to (a) CNS actions like seizures and vomiting, and those related to (b) inhibition of ODC in rapidly dividing cells like diarrhea, anemia, leukopenia, thrombocytopenia.

F. Treatment of Pneumocystosis

The antimicrobial Bactrim is the drug of choice; pentamidine is used mainly for prophylaxis of pneumocystosis.

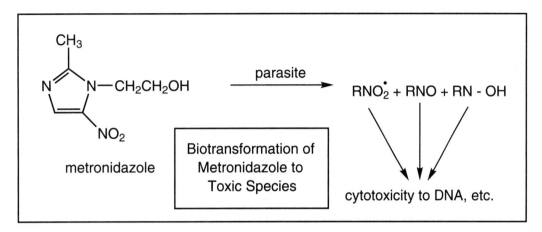

Figure 11.53

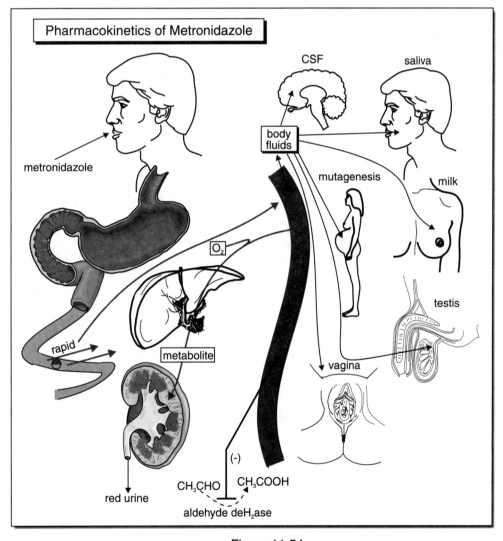

Figure 11.54

G. Antimalarials

Life Cycle of Malariae (Figure 11.58)

1. Infected Anopheles mosquitoes release sporozoites into blood vessels of humans. Sporozoites migrate to the liver where they develop during the next 6-12 days into merozoites. This is the primary exoerythrocyte stage.

2. *Merozoites* are released from the liver and enter erythrocytes (RBCs) where they asexually multiply and digest hemoglobin over the next 2 days (tertian: P.f., P.o., P.v.) or 3 days (quartan: P.m.). Merozoites, gametocytes and toxins are released from burst RBCs at the 2 day or 3 day interval, producing symptoms (fever, chills) and anemia. Merozoites may reinfest other RBCs. This is the erythrocyte stage.

3. When released from burst RBCs, merozoites may migrate to the liver and remain dormant for an indefinite time, even years. This is the secondary exoerythrocyte stage. Symptoms recur when merozoites exit the liver and reinfest RBCs, thereby restarting the erythrocyte stage.

4. Mosquitoes, sucking blood, become infected with gametocytes which reproduce sexually to sporozoites in the mosquito, thereby completing the cycle.

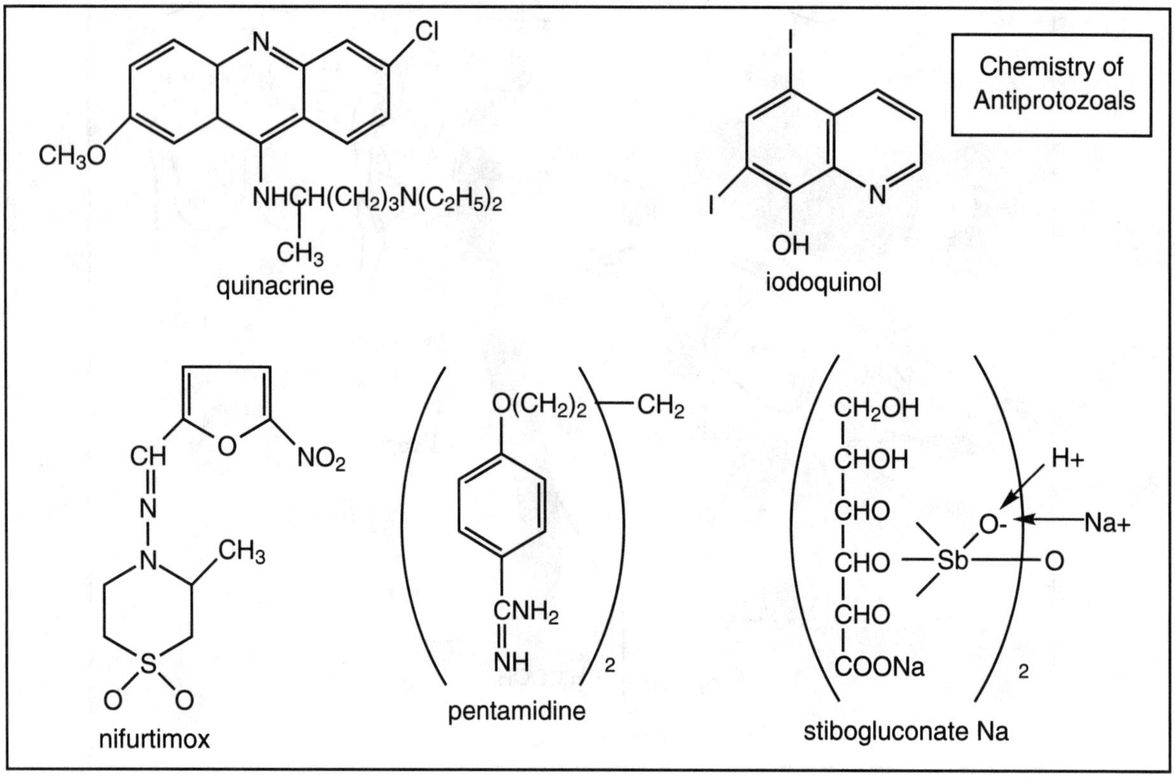

Figure 11.55

DRUGS USED TO TREAT MALARIA

Antimalarial act on one or more stages of the life cycle of Plasmodia, as shown in **Figure 11.58**.

A. Aminoquinolines **(Figure 11.59)**

1. Chloroquine (Aralen) **Pharmacokinetics (Figure 11.60)**. Chloroquine is well-absorbed and concentrates in soft tissues and melanin-containing sites like the eye. The N-desethyl ($-C_2H_5$) metabolite retains activity.

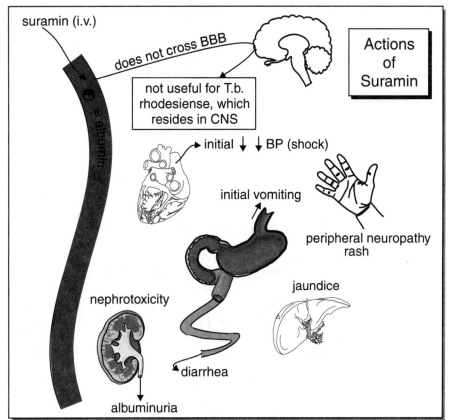

Figure 11.56

Mechanism. Chloroquine and other 4-aminoquinolines are actively accumulated by plasmodia. The alkalinity of 4-aminoquinolines is thought to alter the pH of plasmodia acidic vesicles, thereby interfering with lysosomal degradation of RBC hemoglobin.

Effect. Chloroquine and other 4-aminoquinolines only alter the erythrocytic stage of malariae. A radical cure (killing of all parasites) of P.f. and P.m. is produced, since these lack a secondary exoerythrocyte stage. Since chloroquine does not affect the exoerythrocyte stage, it only produces a clinical cure (suppression of erythrocyte stage and prevention of recurring symptoms) of P.v. and P.o.

Adverse effects. If administered at about 250 mg per day there are few adverse effects. Chloroquine is well-tolerated. When given in high dose for a long time, high amounts of chloroquine in the eye can cause retinopathy and blindness.

Acquired chloroquine-resistance is a problem with P.f. and is thought to be related to enhanced transport of chloroquine out of the parasite.

2. *Quinine* is recommended only for chloroquine-resistant P.f., although it has activity on all malarial schizonts.

Pharmacokinetics. Quinine is rapidly absorbed and largely bound to plasma proteins (about 75%). Most (about 75%) is metabolized in the liver ($t_{1/2}$ about 12 hours) and excreted in urine.

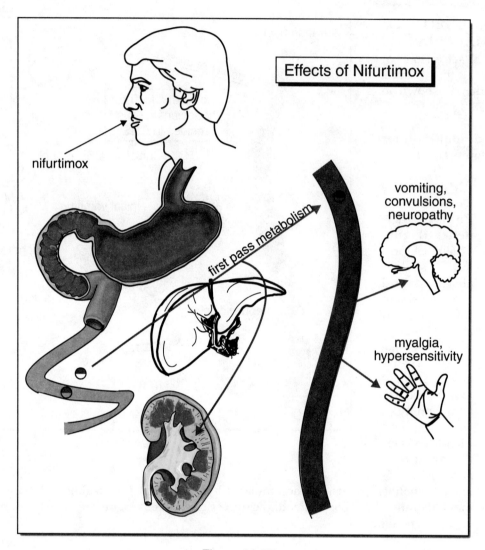

Figure 11.57

Mechanism. Quinine intercalates with Plasmodium DNA and interferes with transcription and protein synthesis.

Adverse effects. Quinine produces GI irritation with diarrhea and is associated with *cinchonism*: headache, dizziness, nausea, tinnitus and visual disturbance. Uncommon and even rare effects include hypoglycemia, marrow depression, hemolysis, optic atrophy with blindness and hypersensitivity.

3. *Mefloquine* (Lariam) has effects similar to quinine and should be used only for chloroquine-resistant malaria.

Pharmacokinetics. Mefloquine is handled by the body in a manner similar to chloroquine, but does not have an active metabolite and is excreted largely in feces.

Adverse effects. Mefloquine, like quinine, produces cinchonism and slows cardiac conduction and heart rate. Serious adverse effects can occasionally occur, such as leukocytosis, thrombocytopenia, psychosis and seizures (contraindicated in epileptics).

4. Primaquine is the only aminoquinoline that is a tissue schizonticide. It is used with an erythro-schizonticide like chloroquine to produce a radical cure of P.o and P.v. This is done only after a patient leaves the malarial geographic zone. Primaquine also is gametocidal and sporontocidal for all Plasmodia.

Pharmacokinetics and actions. Primaquine is well-absorbed and distributed, has a $t_{1/2}$ of about 6 hours and is metabolized to quinone intermediates that oxidize $Hb\text{-}Fe^{+2}$ causes $Hb\text{-}Fe^{+3}$ (methemoglobinemia) and hemolyze RBCs (especially in males with G-6-PD deficiency). Marrow suppression is a rare adverse effect.

B. Antimetabolites

1. Pyrimethamine (Daraprim)
   Pharmacokinetics. Pyrimethamine is slowly but completely absorbed. It accumulates in soft tissues (kidneys, liver, spleen, lungs) and has a $t_{1/2}$ more than 3 days.

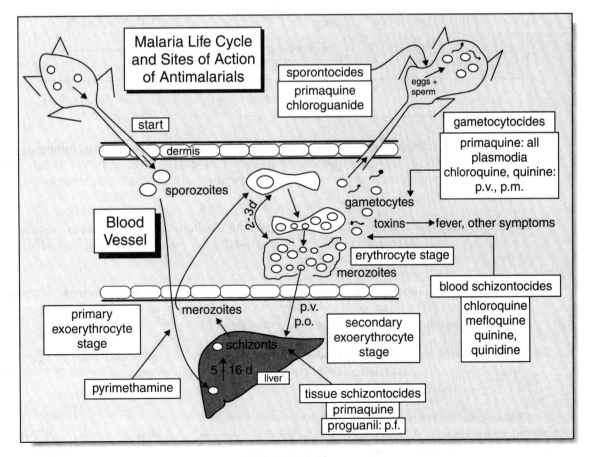

Figure 11.58

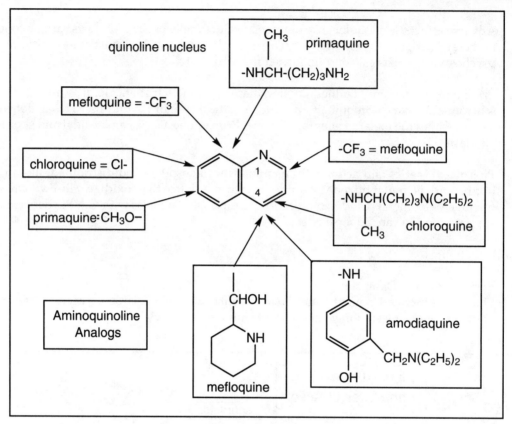

Figure 11.59

Mechanism. Pyrimethamine inhibits protozoal dihydrofolate reductase, thereby inhibiting de novo synthesis of folic acid. As a consequence, purine synthesis and ultimately DNA and RNA synthesis is inhibited. Mammals, but not protozoa, use dietary folic acid. This accounts for selectivity of this drug for Plasmodia.

Use. Pyrimethamine is almost always combined with a sulfonamide or sulfone in the treatment of acute attacks of chloroquine-resistant P.f. Because pyrimethamine and sulfonamides are slow acting, quinine is added for rapid action.

A common formulation is 25 mg pyrimethamine and 500 mg sulfadoxine to make Fansidar. The rationale is that these drugs each block separate steps in folic acid synthesis, so a super-additive effect is produced and the microorganism is much less likely to develop resistance.

Adverse effects. Rash and marrow depression may occur. The latter effect can be prevented by administering folinic acid, leucovorin. Hypersensitivity reactions are uncommon but can include Stevens-Johnson syndrome.

2. Chloroguanide (Paludrine)
   Pharmacokinetics. Chloroguanide, like pyrimethamine, is slowly but completely absorbed. Chloroguanide does not accumulate in tissue, so the $t_{1/2}$ of 18 hours is much less than that of pyrimethamine. About half the dose is excreted in urine as parent prodrug and an active metabolite.

Mechanism. Chloroguanide inhibits dihydrofolate reductase, like pyrimethamine.

Use. Chloroguanide is sometimes used in combination with chloroquine in the prophylactic treatment of P.f.

Adverse effects. High doses are associated with vomiting, diarrhea and possibly hematuria.

15. ANTISEPTICS AND DISINFECTANTS

An antiseptic ("anti" means against; "sepsis" means decay) is an agent that inhibits growth of microorganisms (bacteriostatic or bactericidal). An antiseptic is usually administered to biologic material (e.g., dermis).

A disinfectant kills infectious microorganisms. A disinfectant is usually used on non-biological material (e.g., prosthetic devices, surgical instruments).

These substances fall into the following categories:
1. Acidic
2. Lipophilic
3. Oxidative

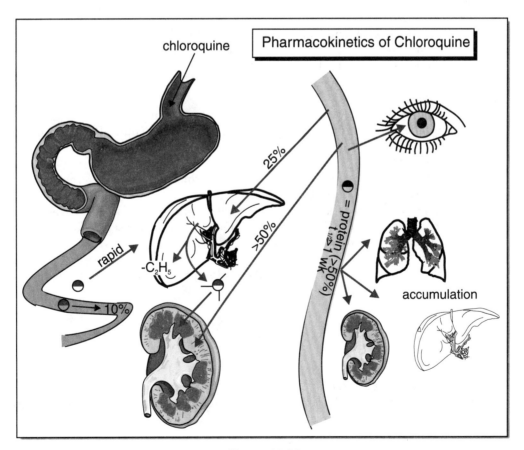

Figure 11.60

4. Detergent
5. Protein-denaturing

They all interact with microorganisms, often by damaging their cell wall, inhibiting growth or producing a cidal effect.

A. Acidic Substances

1. *Boric acid* (H_3BO_3)(5%) is used only as an ophthalmic ointment.
2. *Acetic acid* (CH_3COOH)(1%) is used to sterilize surgical dressings. Sometimes weak solutions are used as vaginal douches or for irrigating the ear or urinary tract.
3. *Salicylic acid*, undecylenic acid or other fatty acids are used topically as fungicides (e.g., athlete's foot).

B. Lipophilic Substances

1. *Ethanol* (CH_3CH_2OH)(70%) and *isopropanol* ($CH_3CHOHCH_3$)(90%) are used as antiseptics on skin (e.g., prior to needle puncture) and as disinfectants for medical instruments (e.g., thermometers). Alcohols precipitate protein and dissolve the lipid in microbial cell walls.

C. Oxidizing Substances

1. Halogens
   a. *Iodine* (I_2)(1:20,000) is used as an antiseptic to kill bacteria and spores.
   b. Iodine tincture (2% I_2 and NaI) is used to paint the skin prior to surgery.
   c. *Betadine* represents I_2 complexed with polyvinylpyrrolidone.
   d. *Bleach* or $HClO/Cl_2$ is used as a disinfectant.

2. Aldehydes
   a. *Formaldehyde* (HCHO)(over 1%) is used as a disinfectant.
   b. *Glutaraldehyde* (2% in 75% isopropanol) is used to sterilize optical and prosthetic instruments.

3. Oxygen generators
   a. *Peroxide* (H_2O_2)(3%) is an antiseptic, used on dermal abrasions, as a mouthwash and to clean contact lenses. Peroxide is also a disinfectant.
   b. *Benzoyl peroxide* is an antiseptic used as a keratolytic and antiseborrheic agent in acne.
   c. *Potassium permanganate* ($KMnO_4$)(1:10,000) is used as an antiseptic for dermal abrasions and topically in the treatment of eczema or ringworm.

4. *Nitrofurans* such as nitrofurazone (Furacin) are antiseptics for dermal abrasions and skin dressings.

D. Detergents

These substance would interact with the microbial cell wall.

1. Cationic agents increase the permeability of microbial cell walls and produce a cidal effect.
   a. *Benzalkonium chloride* (*Zephiran*) is used for dermal abrasions and to clean medical instruments.
   b. *Cetylpyridinium* chloride is infrequently used to clean medical instruments. It is not recommended for application to the dermis.

2. Anionic agents (soaps) should not be used with cationic agents, since each would inactivate the other.
   a. *Chlorhexidine* is used in soaps and mouthwashes and as a topical antiseptic to the skin.
   b. *Chloroxylenol* and trichlorocarbanilide are used in some soaps.

E. Protein-denaturing

1. *Phenols* (1-2%) include lysol, cresol, thymol, resorcinol and hexylresorcinol. These are used as antiseptics on the skin and to clean instruments.

2. *Hexachlorophene* is used in soaps for surgical scrubs. Until it was realized that substantial amounts can be absorbed through the dermis, hexachlorophene was widely used at one time for regular cleansing of infants. Hexachlorophene is a neurotoxin, producing spongy degeneration of white matter in the brain.

3. Heavy metals
   a. Silver salts precipitate microbial cell wall protein, producing a cidal effect. Silver compounds are frequently used for burn patients, since the silver residue is long-lasting. Also, the silver salt does not produce pain.
      i. Silver nitrate ($AgNO_3$)(1%) is an ophthalmic for newborns, exerting a cidal effect on gonococci. $AgNO_3$ (0.5%) is also used for burns.
      ii. Silver sulfadiazine is also used as an antiseptic for burns.
   b. Mercury salts are nearly obsolete as antiseptics and disinfectants.
      i. Mercuric chloride ($HgCl_2$) was an antiseptic for skin and instruments.
      ii. Mercurous ammonium chloride ($HgNH_2Cl$ ointment, 5%) was an antiseptic.
      iii. Thimerosal (Merthiolate) is used for dermal abrasions.
   c. Zinc salts are sometimes used as mild antiseptics for conjunctivitis, impetigo and ringworm. The main ingredient of calamine lotion is a zinc salt. Vaginal douches sometimes contain zinc salts.
   d. *Gentian violet* is a dye that precipitates protein. It is applied as an antiseptic to ulcerative lesions of the mouth (stomatitis).

CHAPTER 12
CANCER CHEMOTHERAPY

DEVELOPMENT OF CANCER
Cancer is an abnormal uncoordinated and uncontrolled growth of cells. The primary tumor invades and destroys the tissue of origin. Through metastasis, secondary tumors arise in many sites and destroy those tissue. These rapidly dividing cells compete with normal cells for the limited supply of body nutrients, thus compromising normal tissue growth and repair. In closed spaces, like the cranium, there are added problems. Excess tumor growth compresses the brain, increases intracranial pressure and compromises vascular flow. For these reasons an aggressive treatment of cancer is necessary for survival.

Substances in the environment promote cancer formation.
   Primary carcinogens directly cause cancer (e.g., epoxides, asbestos).
   Procarcinogens are metabolized to primary carcinogens (e.g., nitrosamines, aflatoxins, polynuclear aromatic hydrocarbons like benzanthrene).
   Cocarcinogens do not produce cancer but augment effects of primary carcinogens (e.g., tars in tobacco).

Specific viruses promote cancer. Some host-cell genes, oncogenes, are activated to promote uncontrolled cell proliferation.

There are a large number of drugs for treating cancer. Specific cancers are sensitive to some, not all of these drugs. This peculiarity is related to
   a. special properties of the drug: pharmacokinetics, mechanism of action and,
   b. special metabolism of different cancers.

Most of the drugs interfere with DNA replication. Some interact with DNA, producing breaks in DNA strands and thereby interfere with DNA replication, RNA synthesis and associated protein synthesis. Others inhibit production of nucleotides that are incorporated into newly synthesized DNA. As would be expected, drugs that inhibit DNA synthesis are especially cytotoxic to cells that are rapidly dividing, like cancer cells. The action of these drugs on normal cells, particularly in rapidly dividing tissue, produces predictable adverse effects as shown below:

Some antineoplastic drugs primarily affect protein synthesis. In some cases differences in cellular metabolism between normal cells and cancer cells can be used to an advantage, thus limiting the spectrum of side effects. Most anti-metabolites rely on such differences.

APPROACH TO TREATMENT OF CANCERS
Sometimes single drugs are used. Often, combinations of drugs are used, along with surgery or irradiation. Drug combinations are especially useful, since they can:

Antineoplastic Drug Effects	
Tissue	**Adverse Effect**
Bone marrow hematologic (myelogenous) system	Myelosuppression, agranulocytosis
Lymphatic system	Lymphosuppression, lymphocytopenia
GI epithelium	GI pain, diarrhea, vomiting
Reproductive organs (ovary, testes)	Sterility, mutagenicity
Any tissue	Carcinogenic effect

1. produce synergistic effects;
2. render a drug-resistant cancer susceptible to the drug combination;
3. impair development of drug-resistance;
4. reduce the overall incidence of adverse effects.

Drug dosage (mg) is frequently calculated on the basis of square meters of body surface per week.

One common drug combination used for Hodgkin's disease is MOPP therapy:
- M = Mechlorethamine (6 mg i.v., days 1 and 8)
- O = Oncovin (vincristine) (1.4 mg i.v., days 1 and 8)
- P = Procarbazine (100 mg oral, days 1 to 14)
- P = Prednisone (40 mg oral, days 1 to 14), for cycles 1 and 4

This regimen is given for 6 cycles, with 2 weeks drug-free between cycles.

The ABVD drug combination is used for advanced Hodgkin's disease:
- A = Adriamycin (Doxorubicin) (25 mg i.v.)
- B = Bleomycin (15 units i.v.)
- V = Vinblastine (4 mg i.v.)
- D = Dacarbazine (DTIC) (350 mg i.v.)

This regimen is given in 2 week cycles (with 2 weeks of rest) or alternated monthly with MOPP for 1 year.

VAMP is used for acute lymphocytic leukemia:
- V = Vincristine
- A = Amethopterin (Methotrexate)
- M = Mercaptopurine
- P = Prednisone

Anti-neoplastic drugs are ordinarily administered at the highest tolerable level, so that the cancer cell kill is as great as possible and the chance for development of drug resistance is reduced. The reductions in lymph cell count and leukocyte count are used as gauges of a tolerable dose. With this treatment approach the patient will feel sick, have many unpleasant side effects and will become susceptible to bacterial and viral infections. Patients must be monitored closely for symptoms. Laboratory testing is required, to assess liver function, marrow recovery, etc.

Several cancers can actually be sufficiently killed, so that the patient is cured. In many cases, the growth of the cancer is mostly suppressed by drugs, so that the patient can have a better quality of life for a longer time.

Cell-Cycle-Specific and Cycle-Nonspecific Drugs. *Cell-cycle-specific drugs* are those that act on specific portions of the cell cycle. These agents include:

Antimetabolites (MTX=S) (5-FU, cytarabine = S/G_1)

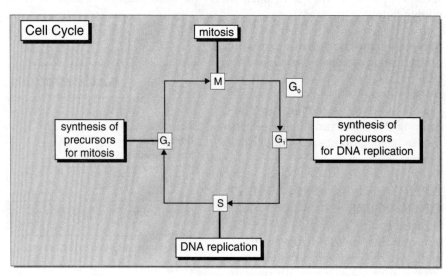

Figure 12.1

Bleomycin (G_2)
Podophyllin alkaloids (Etoposide, teniposide = S/G_2)
Vinca alkaloids (Vincristine, vinblastine = M)
Hydroxyurea (S)

Cycle-nonspecific drugs are those that act on all portions of the cell cycle. These agents include:

> **Alkylating agents** (Nitrogen mustards, busulfan, thiotepa, etc.)
> **Platinum agents** (Cisplatin)
> **Antibiotics** (Doxorubicin, dactinomycin, etc.)
> **Nitrosoureas** (BCNU, CCNU, Methyl-CCNU)

Alkylating Agents and/or their metabolites are strong electrophiles (love electrons) that generate highly reactive carbonium ion intermediates which form covalent alkyl bonds with nucleophiles (love protons).

Common nucleophiles include: -NHR, -SH, -S-S-, -OH, -COOH, -OPO_3.

The chemotherapeutic (i.e., cytotoxic) effect is most closely related to alkylation of DNA as illustrated in **Figure 12.3**. However, reactions with proteins and other small molecules would also interfere with normal cell function and would be detrimental.

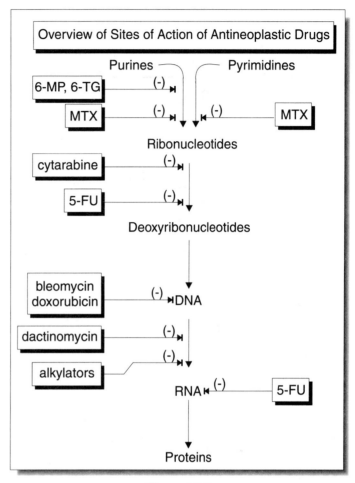

Figure 12.2

MOST COMMON DNA SITES ALKYLATED
Guanine N_7 much more than adenine N_1 and N_3, cytosine N_3, guanine O_6, -OPO_3 of DNA chains **(Figure 12.3)**.

Normally the purine guanine exists in the keto form and makes a hydrogen bonded base pairing with the pyrimidine cytosine (G···C). When carbonium ions interact, the enol form of guanine becomes preferred, so that (1) abnormal base pairings may form (e.g., G···T), leading to substitutions of A···T for G···C base pairs. The alkylated guanine (or adenine, etc.) may be (2) excised from the DNA chain or the purine/pyrimidine ring may be (3) cleaved. (4) Crosslinking of the alkylated purine with other DNA bases, proteins, etc. is a major reaction of bifunctional alkylating agents. Crosslinking within DNA may be intra- or inter-strand. All of these processes will lead to transcription errors that produce death or mutagenesis **(Figure 12.3)**. The chemotherapeutic effect is dependent on cytotoxic properties of the alkylating agents. Adverse effects are also related to this property.

BIFUNCTIONAL VERSUS MONOFUNCTIONAL ALKYLATING AGENTS
Monofunctional alkylating agents are able to bind with one purine/pyrimidine base of DNA. This interferes with transcription and can be lethal to the cell. However, the cell is able to use DNA repair enzymes to strip the alkylating agent from the DNA and thereby recover from such injury. Even when alkylated, the DNA may

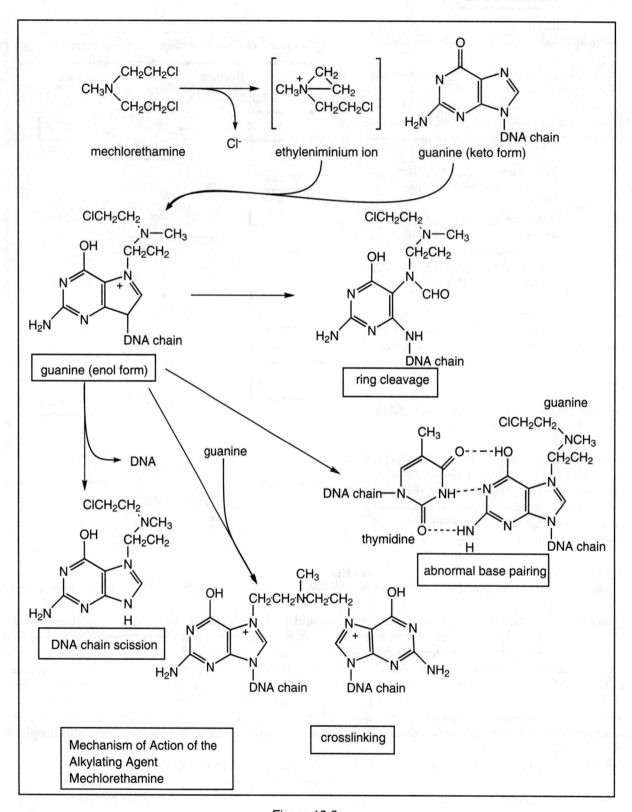

Figure 12.3

function in a reasonably normal manner except during mitosis when DNA replication errors would be introduced and progeny cells would be abnormal. For this reason, monofunctional agents tend to be mutagenic and carcinogenic.

Bifunctional alkylating agents (e.g., mechlorethamine) form crosslinks with DNA. These are much more difficult for DNA repair enzymes to strip away. Consequently, bifunctional agents are generally more *cytotoxic* than monofunctional agents.

DEVELOPMENT OF RESISTANCE TO ALKYLATING AGENTS
Acquired resistance to alkylating agents is fairly common, but slowly developing. There is some cross-resistance among alkylating agents. The following are possible mechanisms by which cancer cells become resistant to alkylating agents:

1. **Reduced permeation**.
    a. Mechlorethamine is transported inside cells via the choline transporter **(Figure 12.4)**. Melphalan, a phenylalanine analog, uses transporters for leucine and other neutral amino acids. Cells may synthesize a modified transporter protein that does not recognize these alkylating agents which then would be unable to enter the cell.

    b. In contrast to these agents, the nitrogen mustards are so lipid-soluble that they passively diffuse into cells. Resistance to nitrogen mustards cannot be based on changes in structure of a transporter protein.

2. **Increased production of nucleophiles**. Cells can make more molecules that have -SH, -NHR groups, etc. These would react with the alkylating agent, thereby reducing interactions with DNA.

3. **Increased production of DNA repair enzymes**. This may be especially important for monoalkylating agents, particularly in non-dividing cells.

4. **Increased rate of deactivation**. Cyclophosphamide is an agent that can be inactivated by liver enzymes. Induction of such enzymes in target tissue would represent resistance.

CELL CYCLE NONSPECIFICITY OF ALKYLATING AGENTS
Alkylating agents will react with DNA of any cell, regardless of the stage in the cell cycle. However, cells seem to be most sensitive during late G_1 or S phases, although toxicity is generally expressed when a cell enters the S phase. (During the S phase [DNA replication] portions of DNA become single-stranded. At such time DNA is more susceptible to alkylation**[Figure 12.5]**.) The cell cycle becomes blocked at the G_2 (premitotic) phase. These cells may continue to synthesize DNA and other cell constituents, leading to polyploid giant cells with an unbalanced cell growth.

Adverse effects of alkylating agents. As indicated above, these agents have their greatest effects on rapidly dividing cells. Accordingly, myelosuppression and immunosuppression are produced. These signs are often a gauge of dosing and effectiveness of these drugs. The monofunctional agents, in particular, tend to produce mutagenesis and sterility (effects on testes and ovaries); and carcinogenesis (e.g., leukemias) when administered for long periods.

Epithelial cells are also affected in the GI tract and this is manifested as diarrhea, severe nausea, and vomiting which can be controlled by anti-emetics. Effects on epidermal cells may be expressed as alopecia.

Uses of alkylating agents. Highly reactive compounds like the nitrogen mustard mechlorethamine must be administered i.v., to avoid severe GI effects. The prodrug cyclophosphamide is converted to an active slow-reacting species in the liver, so oral administration is practical.

Most alkylating agents are used for treating Hodgkin's disease, lymphocytomas and myelocytic leukemias. Some agents are useful for solid tumors.

SPECIFIC ALKYLATING AGENTS

A. Nitrogen Mustards

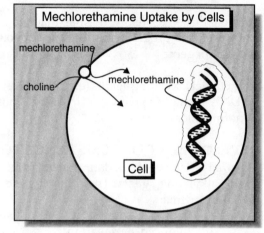

Figure 12.4

1. Mechlorethamine (Mustargen)
   Chemistry (Figure 12.9). This is a highly reactive, extremely unstable compound that must be freshly prepared immediately prior to i.v. injection. It reacts instantaneously with tissue. The irritant properties of this drug cause local tissue damage if extravasation occurs. Intra-arterial infusions are sometimes used in an effort to restrict actions of the drugs to the target tissue.

   Mechanism. Mechlorethamine is actively accumulated by cells via the choline uptake system **(Figure 12.4).** This is a bifunctional agent **(see Figure 12.1).**

   Use. Mechlorethamine is used primarily in MOPP therapy for Hodgkin's disease and for some solid tumors. Locally applied in dilute form to the dermis, this drug is effective for mycosis fungoides.

   Adverse effects. These include myelosuppression and immunosuppression. With this and other nitrogen mustards Herpes zoster and other latent viral infections become unmasked.

2. Cyclophosphamide (Cytoxan)
   Chemistry and activity. Cyclophosphamide is a *prodrug* that must be activated by hepatic cytochrome P_{450} enzymatic hydroxylation **(Figure 12.4).** Subsequently liberated phos-

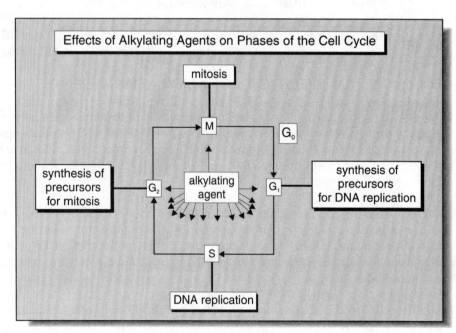

Figure 12.5

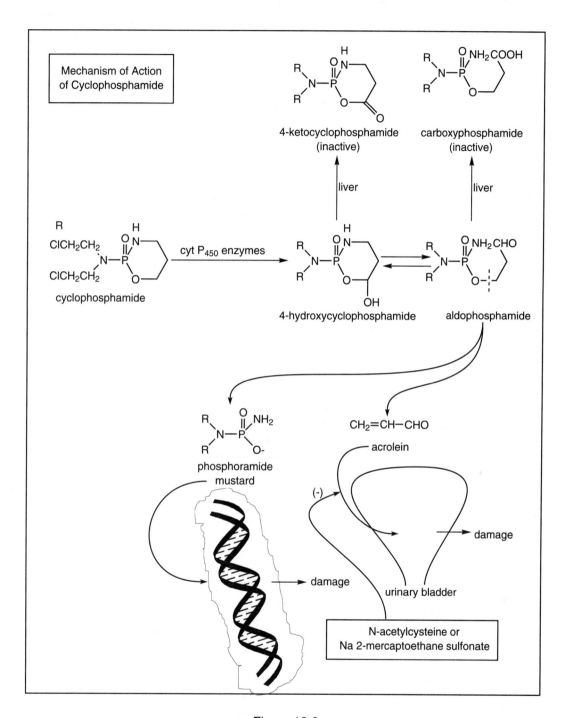

Figure 12.6

phoramide mustard interacts with DNA of tumor cells. The *undesired acrolein* is particularly damaging to urinary bladder, producing sterile hemorrhagic cystitis and *fibrosis of the bladder*. This effect can be prevented by hydration, diuresis (e.g., furosemide or mannitol) or by administration of the trapping agents N-acetylcysteine or Na 2-mercaptoethane sulfonate (MESNA).

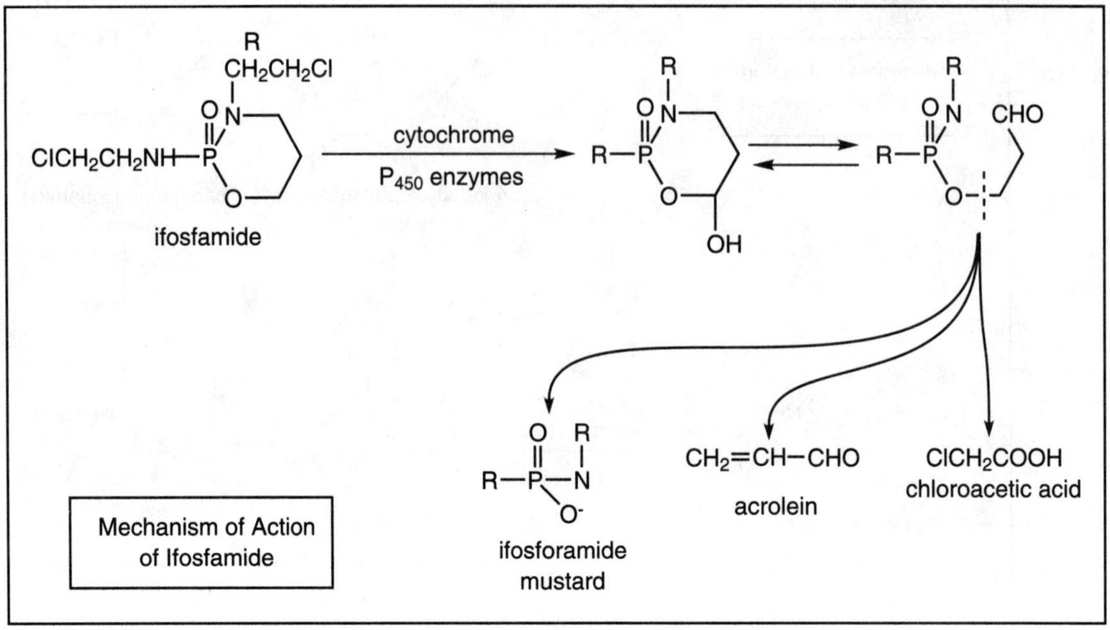

Figure 12.7

Pharmacokinetics. Cyclophosphamide is usually administered orally. Peak levels occur in about 1 hour; $t_{1/2}$ is about 8 hours. Liver cytochrome P_{450} enzymes hydroxylate the parent drug to a derivative that is transported to the target tissue where phosphoamide mustard is formed. Hepatocytes and other tissue cells also enzymatically inactivate the drug, thereby being partially protected from the adverse effects.

Uses. Cyclophosphamide is the most useful of the alkylating agents that are more toxic to granulocytes than lymphocytes. It is used alone or with other agents for a variety of neoplastic diseases that include myelomas, lymphomas (Burkitt's lymphoma, lymphoblastic leukemia) and some solid tumors (carcinomas of breast, ovary, cervix, neuroblastoma).

Adverse effects. These include the expected myelosuppression, ovarian and testicular damage. *Thrombocytes* and *megakaryocytes* are relatively resistant to damage by cyclophosphamide. Hair follicles are more sensitive to cyclophosphamide, so alopecia is common. In high dose, enhanced secretion of antidiuretic hormone (*ADH*) can be produced, leading to (a) water intoxication and (b) enhanced bladder toxicity of the acrolein metabolite.

3. Ifosfamide (Ifex)

Chemistry and activity. Ifosfamide is an analog of cyclophosphamide that is also activated by cytochrome P_{450} enzymes. The spectrum of actions and adverse effects are similar to those of cyclophosphamide, except that ifosfamide produces less myelosuppression but more adverse CNS effects including *seizures*. Hydration, diuretics or mesna are often routinely used to prevent bladder toxicity of the ifosfamide metabolites, acrolein and chloroacetic acid **(Figure 12.7)**.

Use. Ifosfamide is used for lymphomas and some solid tumors including ovarian and testicular cancer.

4. Melphalan (Alkeran)

Chemistry. Melphalan is a phenylalanine analog of mechlorethamine **(Figure 12.9)** that has properties of other nitrogen mustards. However, medphalan is not a vesicant.

Uses. Melphalan is used for myelomas, as well as solid tumors and melanoma.

Adverse effects. They are similar to that of other alkylating agents, except that alopecia, nausea and vomiting are minor.

5. Chlorambucil (Leukeran)

Chemistry. Because of the strong electron withdrawing capacity of the attached phenyl ring **(Figure 12.9)**, the rate of formation of carbonium ions is greatly reduced. This means that the drug can be

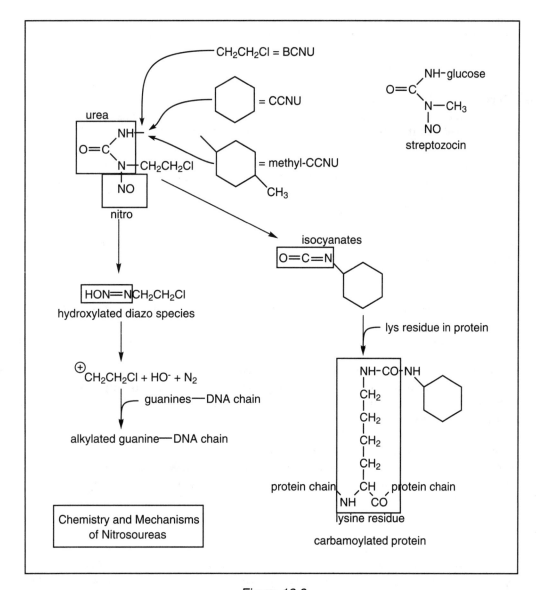

Figure 12.8

taken orally and that it can reach the desired target tissue before reacting with blood cells. Actions are similar to those of cyclophosphamide.

Uses. Chlorambucil is the drug of choice for chronic lymphocytic leukemia, but is also used for macroglobulinemia, Hodgkin's disease and other lymphomas or myelomas. Some solid tumors are treated with this drug. Chlorambucil has been used to stop the progression of rheumatoid arthritis.

Adverse effects. They are like those of other mustards, but *CNS effects* are more prominent with chlorambucil. Myelosuppression is less than with other mustards and recovery is rapid. Leukemias may be produced.

B. Nitrosoureas

    Carmustine   (BCNU = 1,3-bis-[2-chloroethyl]-1-nitrosourea)
    Lomustine   (CCNU = 1-[2-chloroethyl]-3-cyclohexyl-1-nitrosourea)
    Semustine   (Methyl-CCNU) and
    Streptozocin   (Streptozotocin)

Mechanism. These compounds spontaneously decompose **(Figure 12.8)** to
(a) *hydroxylated diazo* compounds that alkylate DNA and cause cross-links and
(b) *isocyanates* that carbamoylate lysine residues and alter proteins function.

Pharmacokinetics. Spontaneous non-enzymatic biotransformation to the active hydroxylated diazo species occurs in minutes. BCNU is administered i.v.; CCNU and methyl-CCNU are administered orally. Peak levels of metabolites of CCNU and methyl-CCNU occur in over 4 hours and have a $t_{1/2}$ less than 6 hours. Excretion is via the kidneys.

Uses. Being highly lipid-soluble, nitrosoureas readily cross the blood-brain barrier. Carmustine and lomustine are used primarily for solid tumors like brain tumors (e.g., gliomas, meningeal leukemias), lymphomas and myelomas.

Adverse effects. Delayed myelosuppression (usually at 1 month), nephrotoxicity (especially methyl-CCNU) and interstitial *pulmonary fibrosis* are seen with the nitrosoureas. Hepatotoxicity is more common with BCNU than CCNU. There is some cross-resistance with other alkylating agents.

Comparison of myelosuppression of nitrosoureas with nitrogen mustards. Mechlorethamine and cyclophosphamide produce maximal suppression on granulocytes in less than 2 weeks. With nitrosoureas this occurs in about 4 weeks. Recovery from cyclophosphamide takes about 3 weeks; with mechlorethamine and nitrosoureas, about 6 weeks.

Streptozocin (Zanosar) is a sugar-containing nitrosourea that destroys pancreatic β-cells and produces diabetes in animals, but not humans. Streptozocin is useful for insulin-secreting pancreatic carcinoma, carcinoid and sometimes in non-Hodgkin lymphomas. Bone marrow suppression is usually not produced, but hepato- and nephrotoxicity is common.

C. Ethylenimines

1. Triethylene thiophosphoramide (Thiotepa) and triethylenemelamine (TEM) **(Figure 12.10)** are used specifically for *bladder carcinoma*, often by direct instillation. They have also been used for lymphomas and solid tumors. Myelosuppression is one of the most serious adverse effects.

2. *Hexamethylmelamine* (Hexastat) is an ethylenimine that does not alkylate but is grouped here because of its chemical similarity to thiotepa. Hexamethylmelamine liberates formaldehyde (HCHO) and is often useful for tumors that are resistant to alkylating agents (e.g., in combination therapy of ovarian carcinoma, small cell lung carcinoma). Adverse effects include central and peripheral neuropathies.

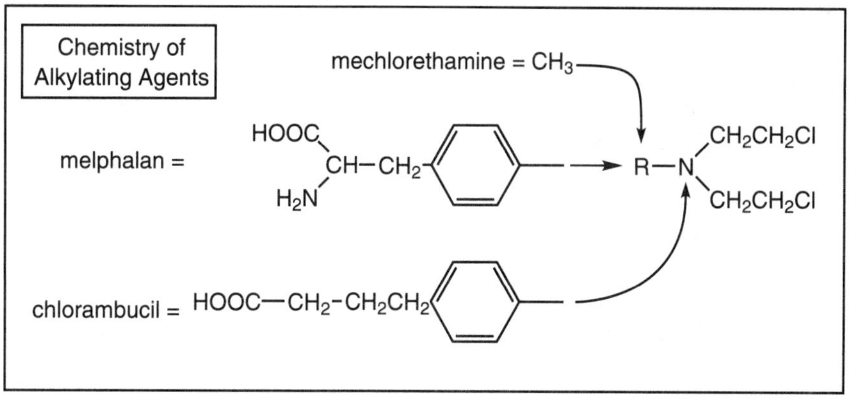

Figure 12.9

D. Busulfan (Myleran)

Chemistry and Pharmacokinetics. This sulfonic acid (**Figure 12.11**) is rapidly absorbed, has a $t_{1/2}$ of about 3 hours and is excreted in urine as methanesulfonic acid. This drug is bifunctional. It has little activity other than myelosuppression. Accordingly, adverse effects on the lymphatic system and GI tract do not occur.

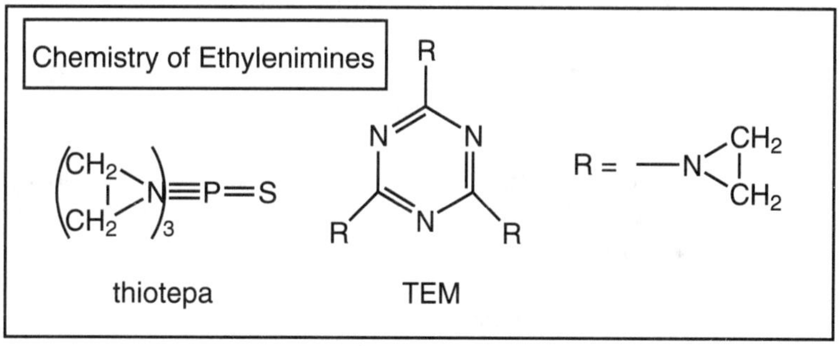

Figure 12.10

Suppressive activity on myelocytes is dose related as shown:

low dose → → → → → → → → → → high dose
granulocytopenia   thrombocytopenia   pancytopenia

Use. Busulfan is used for chronic granulocytic leukemia (about 90% remission) and other hematopoietic disorders such as polycythemia vera and myelofibrosis.

Adverse effects. Effects on reproductive organs, including teratogenesis, can occur; also, carcinogenesis. The release of large amounts of purines by destroyed blood cells results in *hyperuricemia* and gout. Allopurinol should be co-administered to inhibit uric acid formation **(see Chapter 8, Gout).**

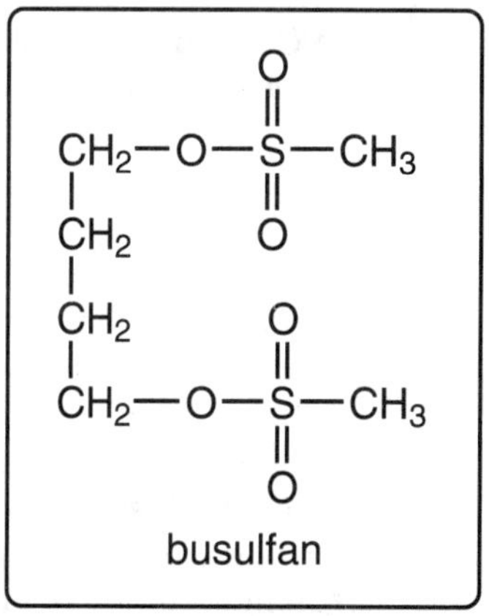

Figure 12.11

E. Triazenes

1. Dacarbazine (DTIC-Dome)

   Chemistry. Dacarbazine is a *hydrazine*, a chemical grouping that confers *MAO-inhibitory activity*. This drug is oxidatively demethylated by hepatic cytochrome P_{450} enzymes to a product that generates a reactive methylcarbonium ion that alkylates DNA and protein **(Figure 12.12)**.

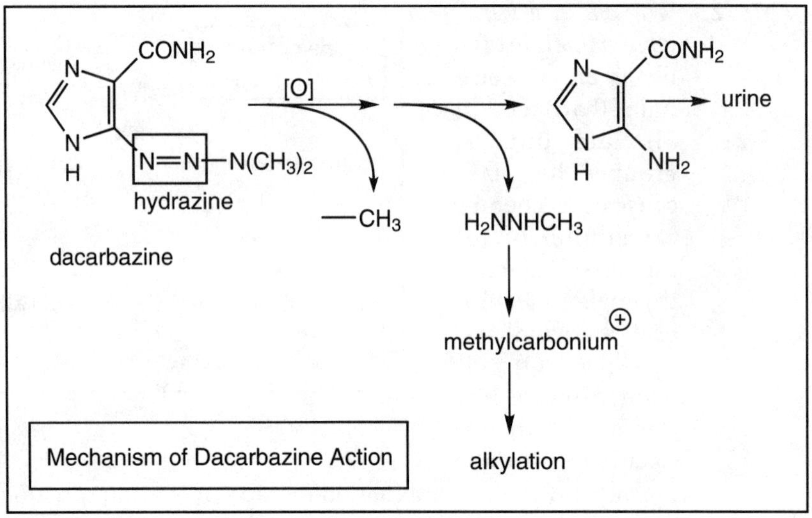

   Figure 12.12

   Pharmacokinetics. This drug is administered i.v., has a short $t_{1/2}$ (less than 6 hours) and is excreted partially unmetabolized (approximately 50%).

   Uses. Dacarbazine is often used in combination with other drugs for Hodgkin's disease, malignant melanoma and soft tissue sarcomas.

   Adverse effects. These include *hepatotoxicity*. Myelosuppression is not common. Since dacarbazine inhibits MAO, tyramine-containing foods and sympathomimetics should be avoided to prevent a hypertensive crisis.

2. Procarbazine (Matulane)

   Chemistry and Pharmacokinetics. Procarbazine, like dacarbazine, is a *hydrazide* with a short $t_{1/2}$, possessing MAO-inhibitory activity **(Figure 12.13)**. Procarbazine is highly lipid soluble, being orally absorbed and crossing the blood-brain barrier. Accordingly, this drug is useful for *brain tumors*. Metabolic activation produces reactive azoxy derivatives, hydrogen peroxide (H_2O_2) and free radicals which produce chromosome breaks.

   Uses. Procarbazine is used for Hodgkin's disease and is one of the principle drugs in MOPP therapy. Other uses include non-Hodgkin lymphomas, myelomas, melanoma, brain tumors and small cell carcinoma of the lungs. There is no cross-resistance between procarbazine and other antineoplastic agents.

   Adverse effects. Myelosuppressive, mutagenic and carcinogenic effects are risks. Procarbazine likewise inhibits aldehyde dehydrogenase (a disulfiram-like activity), so adverse effects are produced when alcohol is consumed **(see Chapter 9)**.

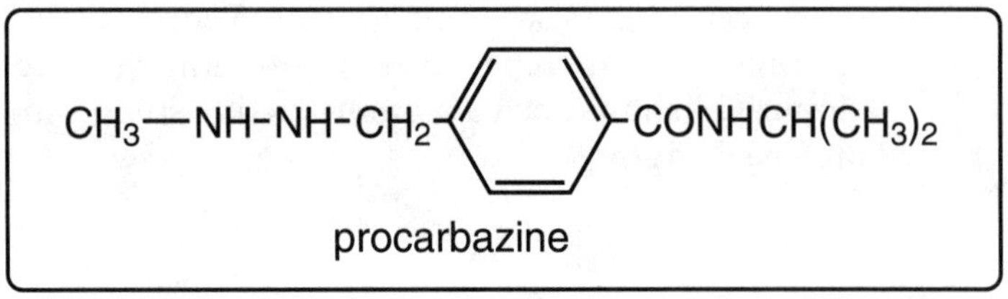

   Figure 12.13

F. Platinum analogs

Chemistry. *Cisplatin* (Platinol) and *carboplatin* (Paraplatin) are cationic compounds (+2 valence) that have 2 quaternary amino moieties ($-NH_4^+$) and other groupings complexed to platinum (Pt) in the cis-configuration **(Figure 12.14)**.

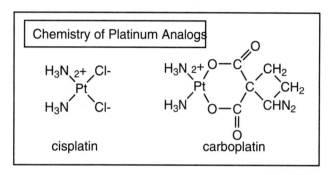

Figure 12.14

Pharmacokinetics. These are administered i.v. Cisplatin, but not carboplatin, becomes bound to plasma protein (greater than 90%). The $t_{1/2}$ of carboplatin is much less. Pt from these compounds concentrate in the liver, kidneys, GI tract and ovaries. This is related to the presence of high amounts of *metallothionein*, a protein which complexes heavy metals. These drugs do not cross the blood-brain barrier. Excretion is via the kidneys (which is the major site for toxicity).

Mechanism. These drugs enter cells by passive diffusion, then form tight bonds with the same nucleophilic groups to which alkylating agents bind. This binding is associated with removal of the Cl⁻ from cisplatin. Intra- and interstrand DNA crosslinks are produced.

Analogous to alkylating agents, all phases of the cell cycle are affected, but particularly G_1 and S phases. Drug resistance is related to increased metallothionein synthesis (in liver, especially), reduced uptake into cancerous cells and enhanced DNA repair.

Use. Cisplatin is used for bladder cancer (combined with vinblastine, bleomycin or etoposide), testicular (combined with bleomycin and vinblastine; 85% curative) and ovarian (combined with cyclophosphamide or doxorubicin) cancer, lung cancer, cancers of the head and neck, and squamous cell carcinoma (combined with 5-FU). *Carboplatin* is approved only for ovarian cancer.

Adverse effects
(a) *Myelosuppression* (carboplatin much greater than cisplatin), particularly thrombocytopenia.
(b) *Renal tubule damage* (cisplatin much greater than carboplatin), resulting from Pt^{+2} forming complexes with -SH and other groups in the distal tubule and collecting ducts. Consequent hypocalcemia and hypomagnesemia develop. Forced hydration and diuresis with mannitol or 3% saline diminishes renal toxicity.
(c) *Ototoxicity* (cisplatin much greater than carboplatin). This is related to accumulation of the Pt analogs in the middle ear. High frequency damage is most common, especially if aminoglycosides have been co-administered. Sodium thiosulfate prevents ototoxicity by virtue of its complexing Pt^{+2} and thereby diverting Pt^{+2} from -SH and other nucleophiles in the ear.
(d) *Peripheral neuropathies* (cisplatin much greater than carboplatin).

ANTIMETABOLITES

A. Folic Acid Analogs

1. Methotrexate (Mexate)
   Chemistry and Pharmacokinetics. Methotrexate (MTX) is a folic acid analog **(Figure 12.15)** that is orally absorbed. For high doses, however, i.v. administration is used. About 1/3 of MTX is bound

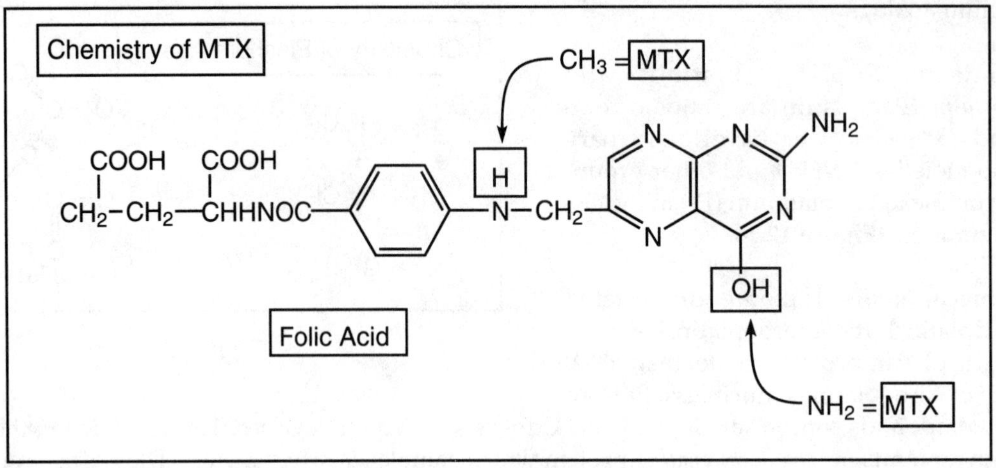

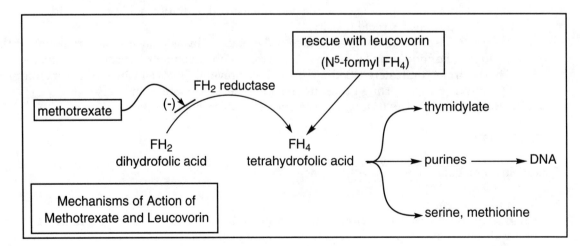

Figure 12.15

to plasma proteins, so interactions occur with salicylates, phenytoin, etc. Most of the drug is excreted in urine unchanged in less than 1 day. At least one metabolite is nephrotoxic.

Mechanism. MTX is taken into cells by active transport and *polyglutamated* to long-lived metabolites that are retained inside the cell. Parent and glutamated MTX competitively inhibit binding of folic acid to dihydrofolate reductase, thereby preventing formation of tetrahydrofolate and ultimate DNA synthesis **(Figure 12.16)**. As expected, this drug is S phase specific. Normal body cells can be rescued with leucovorin (folinic acid) which the malignant cells are unable to efficiently utilize. Resistance is related to reduced uptake of MTX, reduced affinity of MTX to FH_2 reductase and increased synthesis of FH_2 reductase.

Uses. MTX is used for lymphatic cancers (acute lymphoblastic leukemia, Burkitt's lymphoma); breast, ovarian and choriocarcinoma (with dactinomycin, 90% cure); lung carcinoma; and osteogenic sarcoma (high dose MTX with leucovorin rescue). It is also used for mycosis fungoides, psoriasis and rheumatoid arthritis.

Figure 12.16

Major adverse effects. These include myelosuppression, hepato- and nephrotoxicity. MTX is teratogenic.

PURINE ANALOGS

These substances produce metabolites that (a) compete with enzymes for native purines or (b) substitute for native purines in DNA or RNA strands. Enzymes become inhibited, DNA is unable to correctly transcribe, and RNA is unable to correctly translate the message for new protein synthesis.

1. *Mercaptopurine* (Purinethol; 6-MP) and *thioguanine* (TG; 6-TG), thio analogs of hypoxanthine, are activated by hypoxanthine-guanine phosphoribosyltransferase (HGPRT) to the products that inhibit purine interconversions **(Figure 12.17)**. Incorporation of purinethiols into DNA leads to transcription errors.

    The destructive effect on cancer cells results in degradation of DNA and release of large amounts of purine metabolites, leading to *hyperuricemia* **(see Chapter 8, Gout)**. *Allopurinol*, a xanthine oxidase (XO) inhibitor, is used to prevent conversion of purines to uric acid and thereby prevent drug-induced gout. Since 6-MP is inactivated by XO, the dose of 6-MP must be reduced by about 75% if allopurinol is used in combination. However, 6-TG is not inactivated by XO, so no dosage adjustment of 6-TG is needed for combined allopurinol therapy.

    6-MP and 6-TG (combined with cytarabine) are used primarily for acute myelogenous leukemia, chronic granulocytic leukemia and acute lymphocytic leukemia.

    Myelosuppression and hepatic necrosis are produced by higher doses of either agent. GI toxicity is much less with 6-TG.

2. *Deoxycoformycin* (Pentostatin) and *2-chlorodeoxyadenosine* interfere with DNA synthesis by inhibiting adenosine deaminase. These investigational drugs are effective in hairy cell leukemia.

PYRIMIDINE ANTAGONISTS

These substances produce metabolites that (a) compete with enzymes for native pyrimidines or (b) substitute for native pyrimidines in DNA or RNA strands. As with purine analogs, enzymes become inhibited, DNA is unable to correctly transcribe, and RNA is unable to correctly translate the message for new protein synthesis.

1. Fluorouracil (5-FU)
    Pharmacokinetics. 5-FU is erratically absorbed, so it is usually administered i.v. The $t_{\frac{1}{2}}$ is short, with metabolism occurring mainly in the liver.

    Actions. 5-FU is converted to 5-FdUMP, a metabolite that blocks DNA synthesis, by virtue of non-competitive inhibi-

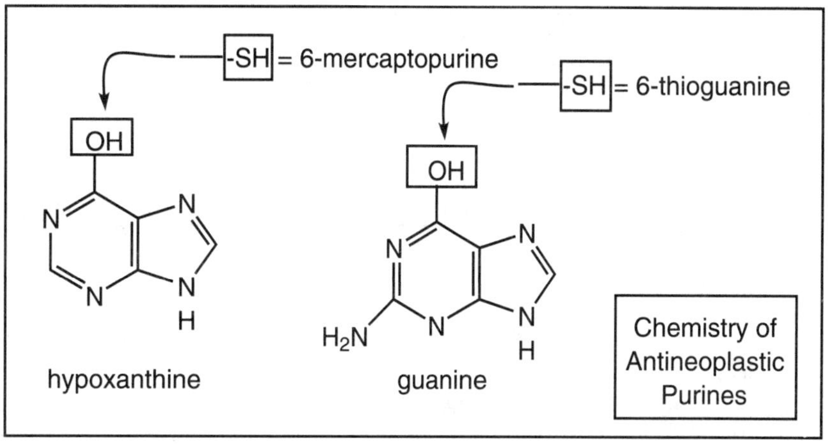

Figure 12.17

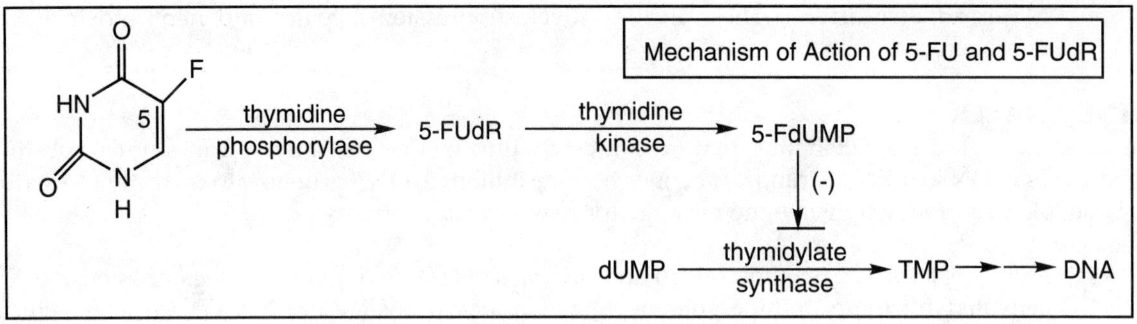

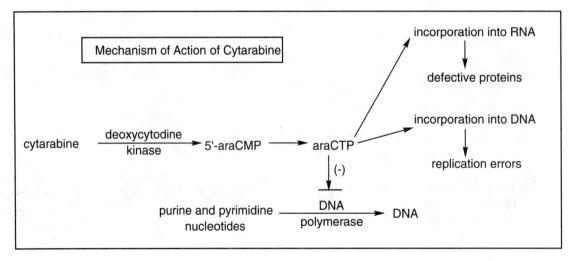

Figure 12.18

tion of thymidylate synthase **(Figure 12.18)**. As expected, this action is specific for the S phase in rapidly dividing cells. 5-FU is also incorporated into RNA, which would produce translation errors in protein synthesis. In non-proliferating cells this may be the major cause of death. When incorporated into DNA, 5-FU would produce added toxicity.

Uses
1. 5-FU is used for solid tumors such as colon, breast, head and neck cancers.
2. 5-FUdR (Floxuridine) is more rapidly metabolized to active compounds in the liver. This prep is administered directly into the hepatic artery to kill colon cancer cells that have metastasized to the liver.
3. A creme containing 5-FU (Fluoroplex) is used for superficial basal cell carcinoma.

Adverse effects. Acute myelosuppression is maximal at about 1 week, with gradual recovery over the next 2 weeks. *Cerebellar ataxia* occurs. GI toxicity is usually the dose-limiting effect.

2. Cytarabine (Cytosine arabinoside)
   Actions. Cytarabine is metabolized into araCTP which (a) inhibits DNA polymerase, thereby inhibiting DNA synthesis; (b) is incorporated into DNA, causing replication errors; and (c) is incorporated into RNA, causing defective proteins **(Figure 12.19)**. These actions result in cell death.

Figure 12.19

Predictably, this drug is S phase specific. Being schedule dependent, cytarabine must be administered continuously for several days or at selected intervals over several days.

Use. Cytarabine is used for acute myelogenous leukemia (67% remissions when combined with anthracycline). Myelosuppression is also the major adverse effect. Liver, GI and neurotoxicities can occur.

VINCA ALKALOIDS
These are naturally produced by the periwinkle plant.

Mechanism. These alkaloids bind selectively to tubulin in cells, damaging the microtubules and disrupting spindle formation during mitosis. Cell growth is inhibited in metaphase; these agents are M phase specific.

Pharmacokinetics. These are vesicants that must be administered i.v. The initial $t_{1/2}$ is about 5 minutes; the later $t_{1/2}$ is about 1-3 hours. Liver metabolism leads to fecal excretion of mostly metabolized drug. With liver impairment or with bilirubin levels greater than 3 mg%, dosage reductions should be made.

1. *Vincristine* (Oncovin) is a component of MOPP therapy and is used for Hodgkin's and non-Hodgkin's lymphoma. *Neurotoxicity* is the dose-limiting adverse effect. This is expressed as parasthesias, motor weakness and autonomic neuropathy. *ADH* secretion may be enhanced.

2. *Vinblastine* (Velban) is used for Hodgkin's disease and testicular cancer. Leukopenia (nadir at 1 week) is the dose-limiting adverse effect. Neurotoxicity is rare.

PODOPHYLLOTOXINS
These are naturally occurring alkaloids produced by the mayapple plant.

Mechanism. These agents induce scissions in DNA strands and block cells in the S to G_2 phases of the cell cycle.

1. *Etoposide* (Vepesid) is used for Hodgkin's disease, non-Hodgkin's lymphomas, testicular cancer and small cell cancer of the lung. Leukopenia is the major adverse effect.

2. *Teniposide* (Vumon) is an investigational drug used for neuroblastoma and retinoblastoma. Leukopenia is the major adverse effect.

ANTIBIOTICS
For cancer chemotherapy antibiotics are often obtained from the fungus, Streptomyces.

A. Anthracyclines

Mechanism. Anthracyclines are bulky molecules that (a) produce single and double strand scission of DNA and (b) intercalate in the DNA helix. These processes impair DNA replication and RNA synthesis and ultimate protein formation. These drugs (c) also interact with cell membranes, affecting membrane fluidity and ion transport. The reactive oxygens in anthracyclines (d) generate semiquinone radicals and oxygen radicals, which produce injury to virtually all elements of the cell, including energy production (respiratory transport chain) in mitochondria **(Figure 12.20)**. These combined actions produce a chemotherapeutic effect.

Pharmacokinetics. These drugs are administered i.v. and have a long duration of action. Liver metabolism produces both active and inactive metabolites, some of which undergo enterohepatic circulation and accordingly have a persistent action. Dosage reduction must be made if there is liver impairment. Most of the drug is eventually excreted in feces.

1. *Doxorubicin* (Adriamycin) is a major antineoplastic agent that is used for lymphatic leukemias (acute lymphocytic leukemia, Hodgkin's disease and non-Hodgkin's lymphomas), myelogenous leukemias, solid tumors (carcinomas of breast, ovary, endometrium, testes) and sarcomas (osteosarcoma, soft tissue sarcomas).

2. *Daunorubicin* (Cerubidine) and mitoxantrone are used mainly for acute myelogenous leukemia.

3. *Idarubicin* (Idamycin) is used for acute myeloid leukemia, often in combination with other agents.

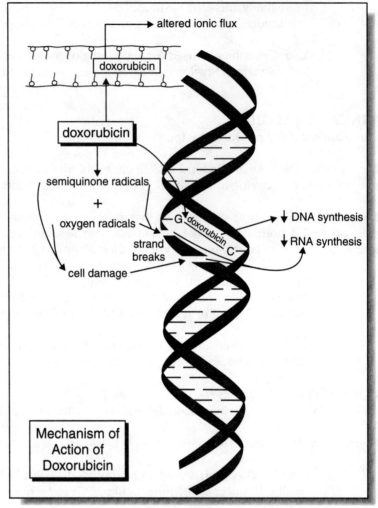

Figure 12.20

Adverse effects. These include transient myelosuppression and pronounced alopecia. These drugs also produce *cardiac toxicity* (conduction defects leading to arrhythmia) which is apparently related to generation of semiquinone and oxygen free radicals that injure cardiac cells.

B. Dactinomycin (Actinomycin D)

Mechanism. This drug *intercalates* in DNA between G-C and interferes with RNA synthesis and subsequent protein synthesis.

Pharmacokinetics. This drug is administered i.v. and has a long duration of action. It is excreted largely unchanged.

Uses. Dactinomycin is used for treating Wilm's tumor, choriocarcinoma and other neoplasms, often in combination with other agents.

Adverse effects. The dose-limiting adverse effects are leukopenia and thrombocytopenia, with the nadir about 2 weeks after treatment.

C. *Bleomycin* consists of several peptides isolated from Streptomyces.

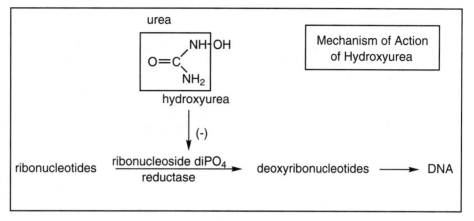

Figure 12.21

Mechanism. Bleomycin binds to DNA in a complex with Fe^{+2}, generating free radicals that produce single and double strand breaks in the DNA **(Figure 12.23)**. DNA replication becomes impaired. Cells are most labile in the G_2 phase of the cell cycle.

Pharmacokinetics. Bleomycin is administered by several parenteral routes, but not orally. It is short-lived, being excreted mainly by the kidneys. Bleomycin is schedule-dependent and should be administered continuously or at regular times over a fixed interval.

Uses. Bleomycin is used primarily for testicular cancer, often with vinblastine (90% cure); for squamous cell carcinoma; and several lymphomas.

Adverse effects. These include *hypersensitivity* reactions, like anaphylaxis—an expected effect, since bleomycin is a mixture of peptides. Dose-related fatal pulmonary fibrosis and hyperkeratosis/erythema are produced.

Advantages. Bleomycin has virtually no toxicity to bone marrow or the GI tract.

D. *Mitomycin* (Mitocin-C) is a mixture of quinones, carbamates and zairidine groups, isolated from Streptomyces.

Mechanism. Mitomycin is a substrate for cytochrome P_{450} enzymes, being converted by a reductase reaction to an agent that alkylates and crosslinks DNA.

Use. *Hypoxic tumor stem cells* are most efficient in bioactivating mitomycin, and accordingly are most susceptible to cytotoxic effects. Mitomycin is used in combination with other drugs for squamous cell carcinoma and adenocarcinomas.

Adverse effects. These include severe late-developing myelosuppression, probably related to effects on hematopoietic stem cells.

MISCELLANEOUS AGENTS

1. *Hydroxyurea* (Hydrea) is an inhibitor of ribonucleoside diphosphate reductase, the enzyme that converts ribonucleotides to deoxyribonucleotides in the sequence of DNA synthesis **(Figure 12.21)**. Hydroxyurea is S phase specific, so cell growth is arrested at the G_1-S interface. Since cells in the G_1 phase are particularly sensitive to irradiation, these two therapeutic approaches are synergistic.

Pharmacokinetics. Hydroxyurea is orally effective, has a $t_{½}$ of about 2 hours, crosses the blood-brain barrier and is excreted mostly in urine.

Uses. Hydroxyurea is used for granulocytic leukemia, polycythemia vera, thrombocytosis, melanoma and some solid tumors particularly of the genitourinary tract.

Adverse effects. They are mainly myelosuppression, GI toxicity and alopecia.

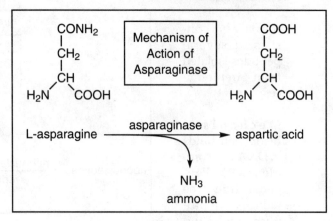

Figure 12.22

2. *Asparaginase* (Elspar) is an enzyme that converts L-asparagine to aspartic acid **(Figure 12.22)**. This enzyme is isolated commercially from bacteria and is used therapeutically to produced depletion of asparagine and glutamine—amino acids required for protein synthesis. Cancer cells are more susceptible than normal cells to depletion of these amino acids, because of their low levels of asparagine synthase, an enzyme that replenishes asparagine.

   Use. Asparaginase is used mainly for lymphatic cancers (e.g., acute lymphocytic leukemia).

   Adverse effects. It is not surprising that *hypersensitivity* (IgE-mediated) reactions, including anaphylaxis, occur subsequent to repeated treatments with the protein, asparaginase.

3. *Amsacrine* is an acridine analog that resembles doxorubicin in its actions. Amsacrine intercalates between DNA base pairs and produces crosslinks as well as single and double strand breaks in DNA. Cardiac toxicity and hepatic injury occur, and dosage adjustments must be made if there is liver impairment. Amsacrine is used for lymphomas, acute myelogenous leukemia and ovarian cancer.

   Amsacrine is useful when drug resistance develops to doxorubicin or cytarabine.

4. *Mitoxantrone* (Novantrone) is an anthracene that resembles anthracycline chemotherapeutic agents chemically and in its actions. Mitoxantrone can be used for anthracycline-resistant cancers. Like anthracyclines, mitoxantrone binds to DNA, producing strand breaks and thereby inhibiting DNA and RNA synthesis. Cardiac toxicity is one of the adverse effects. Uses are like those for anthracyclines.

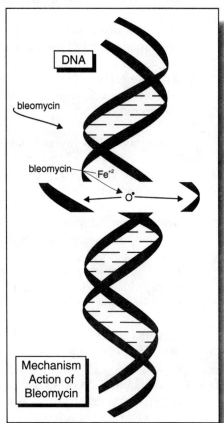

Figure 12.23

5. *Mitotane* (Lysodren) is an analog of DDT that inhibits adrenal steroid synthesis **(see Chapter 10, Adrenocortical Hormones)**. Accordingly, mitotane is used for producing tumor regression in *adrenal carcinoma*. Mental depression is one of the major adverse effects.

CHAPTER 13
TOXICOLOGY PRINCIPLES

Toxicology is the specialty involved in determining the toxic effects of chemicals and establishing safety measures to safeguard against these hazardous substances.

Some subspecialties of toxicology are listed:
1. **Occupational**. Defines safety standards in the workplace.
   TLV = Threshold Limit Value (ppm or mg/m^3) of the substance.
   TLV-TWA = Time-Weighted Average of the TLV for exposure by workers in a single workday or workweek
   TLV-STEL = the TLV Short-Term Exposure Limit. This represents the maximal exposure of workers to the hazardous material in a 15-min period.
   TLV-C = the TLV Ceiling of single exposure. This represents the maximal instantaneous exposure of workers.

2. **Environmental**. Establishes safe levels for plant, animal and human species in the environment.

3. **Ecotoxicology**. Determines the impact of a toxin on plant and animal species in defined environments. For example, if poisons kill the insects that lizards eat, then lizards will starve and die even if the poison does no direct harm to the lizard.

4. **Clinical toxicology**. Determines the chemical-induction of syndromes and disorders and establishes diagnostic criteria and treatment measures.

5. **Forensic**. Establishes whether drugs and chemicals are a cause of death.

6. **General**. Develops tests for establishing safety of new chemicals.

A toxin is a substance that, in small amounts, produces adverse effects. Substances are specifically classified on a scale from "practically non-toxic" to "supertoxic," according to the amount (mg/kg or g/kg) needed to cause injury.

In a mixture of two or more materials, toxicity may be additive (as with two similar toxins), superadditive (as with two toxins acting by different mechanisms in the same tissue) or subtractive (as with an antagonist or antidote).

Acute toxicity is an index of the immediate adverse effects following a single exposure or multiple exposures in a short time. Chronic toxicity is an index of eventual appearance of adverse effects following prolonged exposure (usually to subthreshold amounts). Delayed toxicity represents late-appearance of an adverse effect following acute exposure to a toxin. For example, liver necrosis after a high dose of acetaminophen may not appear until the second day.

A hazard represents the likelihood (risk) of a toxic substance causing injury in a specific setting. For example, the hazard to workers standing near a filling tank can be assessed by knowing the volatility of liquid, the distance of the workers from the tank, the effect of exhaust fans, etc.

Risk represents the calculated occurrence of adverse effects from exposure. In the previous example, the noted factors, as well as the classification of toxin (supertoxic or moderately toxic, etc.), would enter into the equation.

TOXIC AGENTS

A. Treatment of Heavy Metal Poisoning

Heavy metal poisoning is still prevalent in the United States. Many reports have surfaced in the past few years on the frequency of high lead levels in children in poor neighborhoods. In this chapter the toxicological properties of the most common heavy metals are outlined, as well as approaches toward treating acute and chronic heavy metal intoxication.

Chelating agents for eliminating heavy metals from the body. Chelating agents have two or more "ligands" (bi- or polydentate) that form a stable heterocyclic chelate ring with *heavy metals* (Me). This complex is then excreted, thereby eliminating the Me from the body. The objective in acute heavy Me poisoning is to quickly remove the Me from tissues before severe injury is produced. In chronic heavy Me poisoning the objective is to remove Me from its depot sites, thereby reducing the risk of injury to the depot tissue, as well as preventing mobilization of this Me to other tissues. According to the site in which Me is deposited, the chelating agent may be different. Useful chelating agents must have (a) higher affinity versus endogenous ligands for the Me and (b) relatively low affinity for Ca^{++}, so that hypocalcemia and tetany are not produced.

Cell injury is produced by Me when coordinate bonds (endogenous ligands contribute all electrons) are formed mainly with proteins that contain:

1. Oxygen   = -OH, -C=O, COOH, $-OPO_3H$
2. Nitrogen   = $-NH_x$
3. Sulfur   = -SH, -S-S-

This Me complex inactivates proteins thereby inhibiting enzymes, energy production (ATP) and other vital functions of cells. Injury and cell death is produced when cells are unable to maintain oxygen needs or ionic gradients.

Chelators
1. *Edetate Calcium Disodium* ($Ca•EDTA•Na_2$) (Trade name: Calcium disodium versenate) **(Figure 13.1)**. EDTA (without Na or Ca) is a good chelator, but it is poorly soluble and has high affinity for Ca^{++}. The solubility problem is solved when EDTA is complexed with 2Na. The potential for inducing hypocalcemia is averted by incorporating Ca^{++} into the chelator.

    The high solubility of $Ca•EDTA•Na_2$ limits penetrability through cell membranes, so this chelator is poorly absorbed by an oral route; similarly, extracellular (not intracellular) Me would be complexed by $Ca•EDTA•Na_2$. $Ca•EDTA•Na_2$ is mainly used for removing lead (Pb^{++}), as shown below. $Ca•EDTA•Na_2$ removes Pb^{++} mainly from bone. The resulting $Pb•EDTA•Na_2$ complex is filtered at the glomerulus and excreted in urine ($t_{½}$ of $Ca•EDTA•Na_2$ is one hour, so good renal function is essential). The major toxicity is at the renal tubules, possibly because these are rich in -SH groups which might combine with Pb in the $Pb•EDTA•Na_2$ complex during passage.

2. *Deferoxamine* (Desferal) **(Figure 13.2)** is a chelator with high affinity for iron ("fer" stands for Fe) and little affinity for Ca. However, deferoxamine does not remove Fe from hemoglobin or cytochromes—actions that would render the compound useless. Deferoxamine is given parenterally for acute

(usually children eating iron tablets) and chronic (hemochromatosis and hemosiderosis; also, thalassemia) Fe poisoning. Excretion of the dark red Fe-complex turns urine red. Adverse effects are mainly allergy related, including anaphylaxis. Renal, neuronal, visual and auditory effects also can occur.

Deferoxamine per se is prepared by removing Fe from the naturally occurring Fe-deferoxamine complex found in Streptomyces pilosus.

3. *D-Penicillamine* (Cuprimine) **(Figure 13.2)** is the drug of choice for acute and chronic (Wilson's disease: hepatolenticular degeneration) *copper (Cu) toxicity*. D-penicillamine is orally effective, metabolized by liver enzymes and excreted in urine and feces. Allergic reactions, from mild rash to agranulocytosis, have been associated with D-penicillamine, possibly from contamination with penicillin.

D-penicillamine is also used for Pb, mercury (Hg) and arsenic (As) toxicity. N-acetylpenicillamine has greater efficacy in Hg poisoning, possibly because of greater resistance to hepatic metabolism. D-Penicillamine has been used to treat arthritis **(see Chapter 8)**.

4. *Trientine* (Cuprid)**(Figure 13.3)** is used for Wilson's disease (chronic copper poisoning), in people intolerant to penicillamine. Since trientine complexes with Fe, Fe tablets must be taken as supplements (several hours after oral trientine).

5. *Dimercaprol* (BAL) **(Figure 13.3)** is administered parenterally, binds with Me^{++} in extracellular spaces and is excreted in urine in less than four hours. High amounts of dimercaprol are irritating. Pain is observed at non-injection sites (mouth,

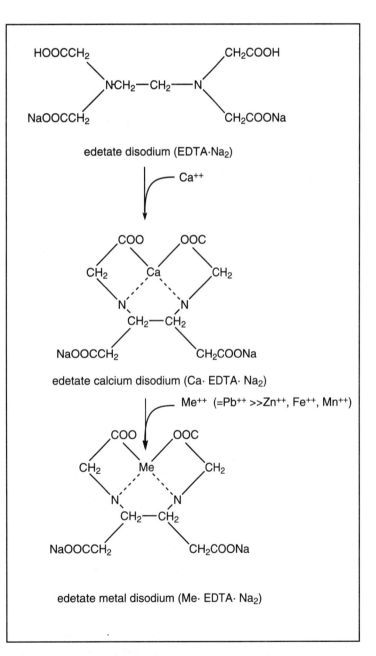

Figure 13.1

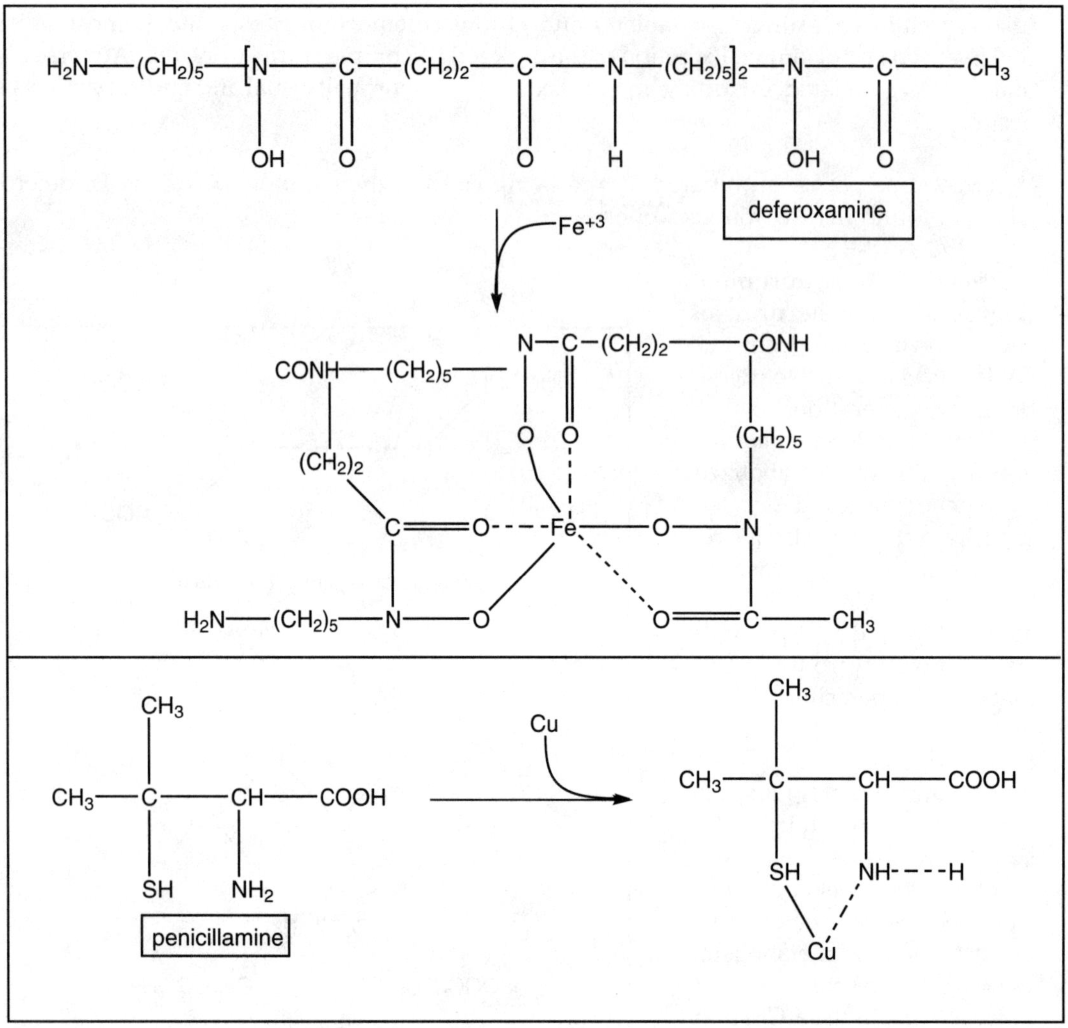

Figure 13.2

throat, penis). However, since dimercaprol-Me^{++} complexes can be cleaved in acid environs, the objective is to form the complex of dimercaprol-Me-dimercaprol and to alkalinize the urine, lest free Me^{++} be released in the nephron. Acute hypertension accompanied by tachycardia is observed shortly after dimercaprol administration.

Dimercaprol is the preferred agent for Hg and As intoxication. It is also used for Pb poisoning, in combination with CA•EDTA•Na$_2$.

6. *Succimer* (Chemet) **(Figure 13.3)** acts similarly to dimercaprol, but is orally effective and less toxic. Adverse effects include GI reactions (10%), (e.g., diarrhea) and allergic reactions (rash, stomatitis).

Succimer is recommended for childhood Pb intoxication and is useful for Hg and As poisoning.

Summary of treatments of heavy metal poisoning
  Pb: Oral D-penicillamine, dimercaprol (+Ca•EDTA•Na$_2$), oral succimer.
  Hg: Dimercaprol, oral succimer, penicillamine or N-acetylpenicillamine.

As: Dimercaprol, penicillamine, succimer.
Antimony (Sb): Effects similar to that of As and is treated similarly.
Cu: Penicillamine, Trientine.
Fe: Deferoxamine.
Cd: Ca•EDTA•Na$_2$.

Toxicological effects of heavy metals
1. *Inorganic lead* (Pb^{++}) is absorbed from the GI tract (10% or less in adults; 40% or less in children) and lungs (90% or less) **(Figure 13.4)**. Initially deposited in soft tissues (kidney, liver), Pb^{++} is eventually redistributed to bone. Although a substantial amount of Pb enters the brain of children, little enters the brain of adults. Adverse effects are shown in the table.

Pb: Acute Effects	Pb: Chronic Effects
GI pain and black stools (PbS)	Lead colic (pain)
Muscle weakness	Muscle weakness and lead palsy (wrist drop)
Hemolytic anemia	Hypochromic microcytic anemia (inh heme synthesis)
Renal damage	Interstitial nephropathy 1. hypertension 2. hyperuricemia and gout 3. proteinuria/hematuria
Brain edema, convulsions	(In children) Cerebral palsy, mental retardation, pallor and emaciation

Organic Pb^{++} poisoning, (e.g., tetraethyl lead, Pb(C$_2$H$_5$)$_4$) is accompanied by CNS effects in adults, as well as children. The effects include insomnia, delusions and mania.

2. *Inorganic mercury* (Hg$^+$, Hg^{++}) and organic Hg (RHg), but not elemental Hg (Hg$^\circ$), is well-absorbed from the GI tract. Hg$^+$ has little toxicity. However, Hg^{++} localizes in proximal tubules (which are rich in -SH groups) and damages the kidneys. Organic Hg (especially CH$_3$Hg) primarily affects the brain (erethism, tremor, mania: Remember the mad hatter in 'Alice in Wonderland!' Hg$^\circ$ was used for

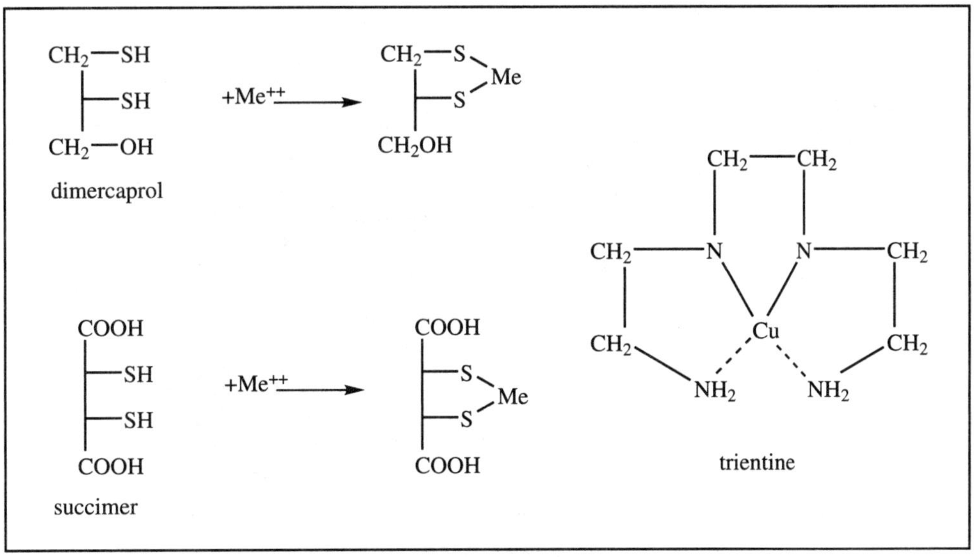

Figure 13.3

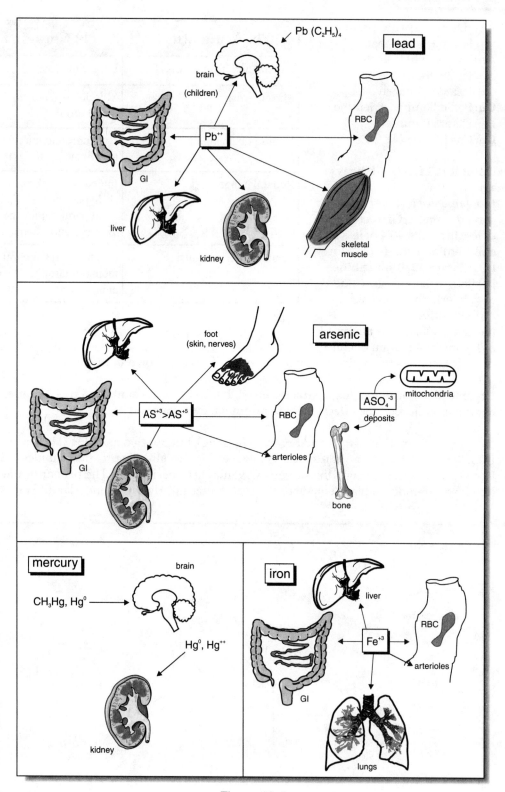

Figure 13.4

cleaning hats.) and does not injure the kidneys. Elemental Hg° is absorbed by the lungs and damages lung, kidney and brain (if chronic exposure). Gingivitis is often seen after any of the forms of Hg (Hg°, Hg$^+$, Hg^{++}).

3. *Arsenic* (As) produces the following effects **(Figure 13.4)**:

GI: As^{+3}>As^{+5}:	induction of GI cysts that rupture into the lumen: rice-water stools; later, bloody stools
Kidneys:	glomerular injury: proteinuria, hematuria; renal tubule damage
Blood:	anemia, leukopenia
CDV:	vasodilation: facial edema capillary permeability: gangrenous feet (blackfoot disease)
Liver:	fatty infiltration, centrolobular necrosis, cirrhosis, maybe cancer
Lungs:	cancer
Nerves:	neuropathy in extremities, encephalopathy (from mini-hemorrhages)
Skin:	vasodilation: milk (anemia) and roses (vasodilation) complexion hyperkeratosis of palms and soles
	deposition in hair and nails, since keratin is rich in -SH groups
	squamous and basal cell carcinoma

As^{+3} has high affinity for -SH groups, which accounts for its affinity for the kidney. As^{+5} forms AsO$_4^{-3}$, which appears to substitute for PO$_4^{-3}$ in the mitochondrial respiratory transport chain. The effect is an uncoupling of oxidative phosphorylation. AsO$_4^{-3}$ also deposits in teeth and bones, in place of PO$_4^{-3}$.

4. *Iron* (Fe) poisoning **(Figure 13.4)** usually occurs only in 1 to 2 year old children that eat iron tablets. Small amounts (1-10 g) are fatal. The following effects of acute iron toxicity are seen:

GI:	brown or bloody vomit, pain, bloody diarrhea
Respiration/acid-base balance:	acidosis and hyperventilation
Lungs:	pulmonary edema, cyanosis
CDV:	severe hypotension
Liver:	hepatotoxicity, failure

Deferoxamine is the treatment of choice when the plasma concentration of iron exceeds 3.5 mg/l.

5. *Cadmium* (Cd) is a contaminant of cigarette smoke and is present in low amounts in foods, particularly shellfish. Acute toxicity is primarily a reflection of the irritant property of cadmium. After eating Cd^{++} preps, GI pain, vomiting and diarrhea is seen. After breathing Cd^{++} fumes, pulmonary pain and edema are seen.

Chronic toxicity
Lung:	emphysema
Kidney:	glomerular damage, nephrotoxicity
Bone:	osteomalacia (found in "ouch-ouch" disease) or osteoporosis

Acute toxicity should be treated immediately with Ca•EDTA•Na$_2$, prior to Cd^{++} deposition at tissue sites that are inaccessible to. Increased renal toxicity has been reported after dimercaprol treatment.

B. Air Pollutants (Figure 13.5)

Air Pollutants are gaseous or airborne particulate contaminants that produce adverse effects in plants, animals and humans.

1. *Carbon monoxide* (CO) is formed by incomplete combustion of organic materials:

$$C \xrightarrow{O_2} CO \xrightarrow{O_2} CO_2$$

This colorless and odorless gas has about 200 times the affinity of O_2 for the O_2 binding sites of hemoglobin (Hb). Consequently, CO impairs the ability of Hb to transport O_2 in blood to tissues in the body. Cellular ischemia and anoxia (blue-tinged skin and lips) occurs if CO is present in high amount or if low amounts are breathed for prolonged periods. Adverse effects include confusion, tachycardia, shock, coma, respiratory failure and death.

$$Hb + CO \longrightarrow Hb \cdot CO = \text{carboxyhemoglobin}$$

O_2 should be administered to people acutely intoxicated with CO.

2. *Sulfur dioxide* (SO_2) is produced when sulfur (S)-containing materials are burned:

$$S \xrightarrow{O_2} SO_2 \xrightarrow{H_2O} H_2SO_3 \xrightarrow{O_2} H_2SO_4$$

When dissolved in water, as on moist surfaces in eyes, mouth, nose and throat, sulfurous (H_2SO_3) and sulfuric (H_2SO_4) acids are formed. Bronchiole and pulmonary irritation is associated with bronchoconstriction and actual pulmonary edema. Treatment is non-specific, directed towards reducing exposure and irritation.

3. *Nitrogen dioxide* (NO_2) is a brown gas produced by decomposing plants in silos, as an auto emission and when nitrogen (N)-containing materials are burned:

$$N \xrightarrow{O_2} NO_2 \xrightarrow{H_2O} HNO_3$$

When dissolved in water, as on moist surfaces in eyes, mouth, nose and throat, nitric acid (HNO_3) is formed. Bronchiole and pulmonary irritation is associated with bronchoconstriction and pulmonary edema. Alveolar type I cells are particularly sensitive to NO_2. "Silo-fillers disease" is the consequence of NO_2 effects. Treatment is non-specific as per SO_2 intoxication.

4. *Ozone* (O_3) is a bluish pungent gas that is produced naturally (electric storms) or by high-voltage equipment. It is also found in polluted air. This highly reactive oxygen species liberates free radicals

that react readily with any biological material. Airways and lungs become irritated; pulmonary edema can occur. Treatment is non-specific.

C. Solvents **(Figure 13.6)**

Aliphatic hydrocarbons (C_5-C_8), in small amounts, are mainly neurotoxic. This is due in part to toxic metabolites like 2,5-hexanediol. Larger chain aliphatics (larger than C_9) produce pneumonitis and a euphoric sensation if inhaled; GI pain, vomiting and diarrhea, if swallowed. Common household products in this category include turpentine, gasoline and kerosene.

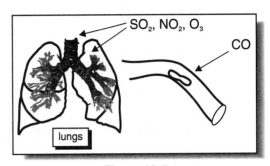

Figure 13.5

Treatment involves administration of O_2 under positive pressure, to displace residual hydrocarbons in the lung. If less than 1 mg/kg has been swallowed then emesis should not be attempted because of the inherent danger of aspiration into the lungs. If the hydrocarbon has other toxic products like insecticides or if less than 1 mg/kg has been swallowed, treatment consists of emesis or gastric lavage followed by a cathartic. If unconsciousness occurs, positive pressure O_2 must be continued.

Aromatic hydrocarbons. Benzene (C_6H_6) is the most widely used aromatic hydrocarbon and is accordingly the most common toxin in this class. CNS depression is the major acute effect. Aromatics also sensitize the heart to epinephrine and norepinephrine. Bone marrow depression and associated blood dyscrasias, including aplastic anemia, are the major chronic adverse effects. There is no antidote for benzene.

Toluene, methylbenzene (CH_3-C_6H_5), is found in detergents, perfumes, glues and many other household products. The acute toxicity of toluene is similar to that of benzene. However, chronic toluene exposure does not produce bone marrow depression.

Treatment involves gastric lavage or induced emesis.

Halogenated aliphatic hydrocarbons (R'CHX). Substances such as methylene chloride (CH_2Cl_2), chloroform ($CHCl_3$) and carbon tetrachloride (CCl_4) are capable of liberating free radicals ($Cl^\bullet$) *in vivo*. This highly reactive radical produces lipid peroxidation and

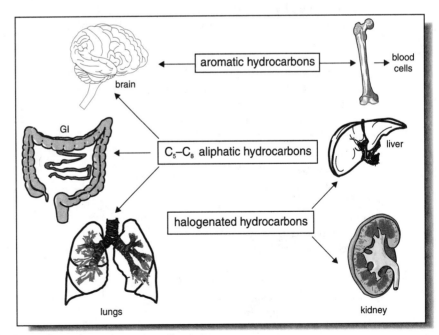

Figure 13.6

is particularly damaging to the liver and kidney. All of these products produce anesthesia, reflecting their CNS depressant action. In high concentrations or after long exposure, cardiotoxicity can be produced. There is no specific antidote or treatment.

D. Insecticides

1. *Cholinesterase-inhibitor* insecticides were described earlier **(Chapter 2)**.

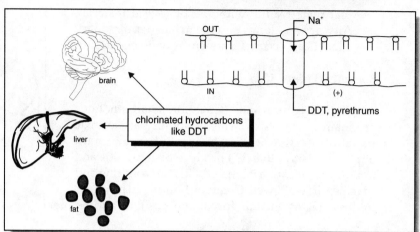

Figure 13.7

2. *Chlorinated hydrocarbon* insecticides include DDT (DichloroDiphenyl-Trichloroethane) and analogs **(Figures 13.7 and 13.8)**.

DDT prevents inactivation of the Na^+ channel of nerves, thereby producing repetitive nerve axon potentials (nerve firing), which is manifested as tremors and convulsions. Treatment consists of decontamination, maintenance of vital functions and administration of an anticonvulsant **(Figure 13.8)**.

Other chlorinated hydrocarbons (aldrin, dieldrin, lindane) enhance presynaptic neurotransmitter release, thereby producing CNS excitation which is expressed as convulsions. Liver toxicity is produced by some of these compounds (mirex, chlordecone). Poisoning is treated similarly to that described for DDT.

DDT and analogs are highly lipophilic and are absorbed through any biological membrane (skin, gut, lungs). These compounds are persistent in the environment; they likewise persist in the body in fat depots, a prominent storage site. DDT has limited use in the United States and Europe because of environmental consequences. In birds DDT causes thinning of the eggshell and reduces successful hatching.

3. Natural insecticides
   a. *Nicotine* is present in high amounts in tobacco and is readily absorbed

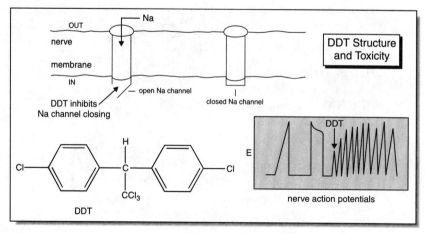

Figure 13.8

from mucous membranes and skin (base form of nicotine). Nicotine is an agonist at nicotinic receptors
   a. in sympathetic and parasympathetic ganglia: salivation, vomiting and diarrhea, increased blood pressure, etc.;
   b. at the neuromuscular junction (skeletal muscle): muscle fasciculations and
   c. in brain: tremors, convulsions and respiratory depression.

Actions of nicotine are described in **Chapters 2 and 3**. Treatment of nicotine toxicity includes gastric lavage plus activated charcoal and catharsis; administration of atropine (to block adverse muscarinic effects) and/or diazepam (for convulsions); maintenance of vital functions.

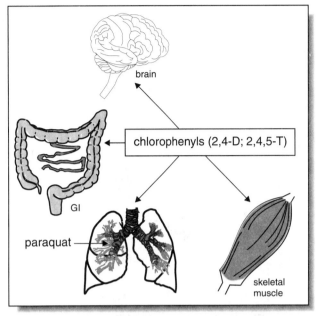

Figure 13.9

b. *Rotenone*, an inhibitor of cytochromes for oxidative phosphorylation, produces cytotoxicity by preventing adequate energy production (ATP) in cells.

c. *Pyrethrums* are the "instant kill" spray insecticides. They are associated initially with contact dermatitis and anaphylactic reactions; sometimes with convulsions. Pyrethrums, like DDT, prevent inactivation of Na channels of nerve membranes.

E. Herbicides **(Figure 13.9)**

1. *Clorophenyl analogs*, 2,4-D (2,4-dichlorophenoxyacetic acid) and 2,4,5-T (trichlorophenoxyacetic acid), act as growth hormones in plants. Human toxicity consists of GI pain, vomiting and diarrhea; bronchitis; muscle weakness (myotonia) accompanied by neuropathies (demyelination). The dioxin, TCDD (2,3,7,8-tetrachlorodibenzo-p-dioxin), has been a frequent contaminant in commercial preparations. TCDD produces chloracne and contact dermatitis.

2. *Paraquat* is a bipyridyl compound that is particularly toxic to humans, with the major danger being from delayed pulmonary damage. The mechanism of toxicity is illustrated in **Figure 13.10**.

1st day GI symptoms:	pain, nausea, vomiting, bloody diarrhea
2nd day:	renal impairment; eventual anuria
3rd day:	liver damage and oliguria
4th and subsequent days:	pulmonary edema and distress (hypoxia and cyanosis) accompanied by marked fibroblastic cell proliferation

Treatment should be vigorous and immediate, consisting of:
   a. Gastric lavage, then ingestion of Fuller's earth (kaolin), then a cathartic. The objective is to remove or inactivate as much paraquat as possible.

$$\text{Paraquat}^{+2} + 2\text{ Na-Clay} \longleftrightarrow \text{Paraquat-Clay} + 2\text{ Na}^+$$

b. Hemodialysis or hemoperfusion might be instituted, but this is not likely to be so effective since blood levels of paraquat are low.

c. Do not use hyperbaric O_2, since this potentiates paraquat toxicity in lungs. Instead, keep supplemental O_2 to a minimum, while ensuring that P_{O_2} levels are adequate.

3. *Diquat*, a less toxic analog of paraquat, produces GI, renal and hepatic effects analogous to that of paraquat. However, diquat does not have special affinity for the lung. Initially there may be CNS excitation and convulsions. Cataracts may occur after chronic exposure.

F. Fumigants

Hydrogen cyanide (HCN) is used as a fumigant in buildings, ships and at special food warehouses. Cyanide (CN^-) is an extremely toxic substance that produces death within minutes. Toxicity is due to the high affinity of CN^- for the Fe^{+3} in cytochromes. Since the complex is inactive, the mitochondrial respiratory transport chain is inhibited and cell respiration is compromised. Symptoms are a reflection of this effect: cyanosis, bright red venous blood (It contains O_2 that cells are unable to use.) and initial hyperventilation followed shortly by hypoxic convulsions.

Treatment of CN^- poisoning must be immediate, with diagnosis being the smell of bitter almonds. The treatment objective is to produce high amounts of non-cytochrome Fe^{+3} which will bind CN^- and thereby protect cytochrome Fe^{+3} from CN^- inactivation. This is usually accomplished with sodium nitrite ($NaNO_2$), as per **Figure 13.12**.

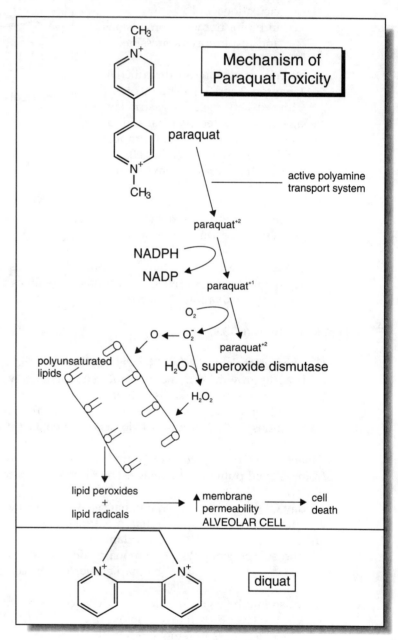

Figure 13.10

1. $NaNO_2$ oxidizes heme-Fe^{+2} to heme-Fe^{+3} (methemoglobin) which combines with CN^- to form cyanomethemoglobin **(Figure 13.12)**.

2. Sodium thiosulfate ($Na_2S_2O_3$) is next added to promote removal of CN^- from the complex, forming thiocyanate (SCN^-) which is readily excreted in urine.

3. Inadvertent excess methemoglobinemia that might result from $NaNO_2$ can be removed by giving i.v. methylene blue (which converts $HbFe^{+3}$ to $HbFe^{+2}$).

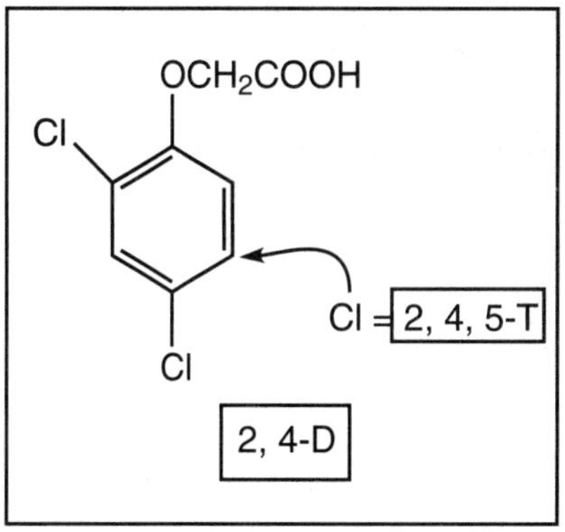

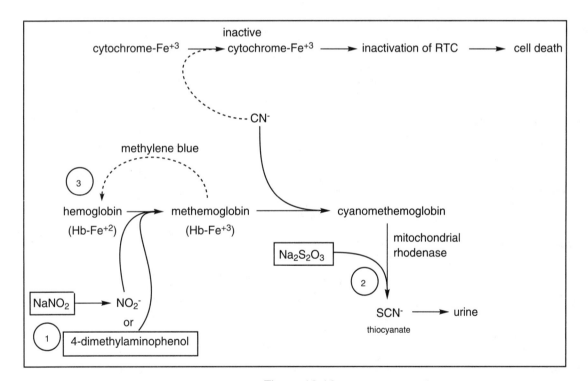

Figure 13.11

Figure 13.12

Index

Symbols

1,25(OH)$_2$D   271
1,25(OH)$_2$D$_3$ 268, 269, 270
15-methyl-PGF$_{2\alpha}$ 159
2-mercaptoethane sulfonate   357
2-PAM   54, 57
2,3,7,8-tetrachlorodibenzo-p-dioxin   383
2,4-D   383
2,4-dichlorophenoxyacetic acid   383
24,25(OH)$_2$D$_3$ 268, 269
2,4,5-T   383
25(OH)D$_3$ 268, 271
3-methylthiotetrazole   298, 299
4-dimethylaminophenol   385
5-aminosalicylate   291
5-FC   313, 314
5-FdUMP   365, 366
5-flucytosine   313
5-fluorouracil   285
5-FU   314, 352, 353, 363, 365, 366
5-FUdR   366
6-amidinopenicillanic acid   293
6-hydroxydopamine   59
6-mercaptopurine   171
6-MP   353, 365
6-TG   353, 365

A

α-adrenoceptors   40, 41, 43, 45, 97
α-antagonists   59–63, 65, 66
α-blockers   62–66, 97, 98, 101, 102 119
α-galactosidase   150
A-I   94–96
A-II   94–96
A-III   94–96
α-methyl-p-tyrosine   227
α-methyldopa   56, 58, 62
α_1-adrenoceptor agonists   58–64
α_1-adrenoceptor antagonists   63–71
α_1-adrenoceptors   43, 45,
α_2-adrenoceptor antagonist   67
α_2-adrenoceptors   43, 45
α_2-agonists   98, 99, 102
ABVD drug therapy   352
ACE inhibitors   95, 96, 101, 102, 119
Acebutolol   70, 97
Acetaminophen   160, 161, 177
Acetazolamide   85
Acetic acid   348
Acetohexamide   280
Acetylcholine   39, 41, 43, 44, 47, 48, 50
Acetylcholine, receptor   79–81
Acetylcholinesterase   42, 51–57
Acetylcysteine   160, 161, 357
Acetylsalicylic acid   161–163. See also Aspirin
Ach   48
Acne   304
Acrolein   358
Acromegaly   244

ACTH   181, 241, 242, 250–253
ACTH1-24   252
Actinomyces   303
Actinomycin   283–285
Actinomycin D   368
Actinomycosis   297
Acyclovir   318, 319, 321
Addiction   185, 236
Addison's disease   252, 253
Adenine arabinoside   321
Adenocarcinomas   369
Adenoviruses   320
ADH   83, 84, 95, 241, 242
ADH antagonists   84, 90, 91
ADHD   64, 207
Adrenal carcinoma   371
Adrenal medulla   43, 45
Adrenalin   60
Adrenaline. See Epinephrine
Adrenoceptors   58
Adrenocortical hormones   250–256
Adrenocorticotropic hormone   241, 242, 250-253
Adrenocorticotropin   241
Adrenolytics   68, 69, 71, 72, 99
Adriamycin   352, 368
Aerobic bacteria   301
AeroBid   138
Aflatoxins   351
Agonist, inverse   197
Agonists
    full   26
    inverse   28
    partial   26, 27
AIDS   319, 320, 322
Air pollutants   380
Albendazole   323, 324, 326, 330, 331
Albuterol   62, 64, 119, 137
Alcohol   237. See also Ethanol
Alcohol dehydrogenase   37
Aldactone   89, 94
Aldomet   62, 98
Aldosterone   83, 87, 90, 95, 96, 250–255
Aldosterone antagonist   84, 89, 90
Aldosteronism   90
Aldrin   382
Alfenta   193
Alfentanil   184, 193
Aliphatic hydrocarbons   381
Alkeran   359
Alkylating agents   353–363
Allergy   131–139, 158
    acute desensitization   134
    anaphylaxis   132, 133, 137
    asthma   131, 137–139
    conjunctivitis   135
    dermatitis   131
    food   131, 137
    hayfever   131, 135
    hives   131, 135
    prolonged desensitization   134
    respiratory   131, 137–139
    rhinitis   131, 135

Allopurinol   171, 365
Aloe   148
Alprazolam   193–197, 235
Alprostadil   159
Aluminum antacids   142, 145
Amantadine   219, 221, 317, 318, 320
Ambenomium   48
Ambien   197
Amdinocillin   293, 297
Amebiasis   336, 337
Amectine   79
Amethopterin   352
Amidate   192
Amikacin   300–302, 307, 312
Amikin   302
Amiloride   89, 90, 94
Aminocaproic acid   127, 129, 130
Aminocyclitols   300–302
Aminoglutethimide   252, 253
Aminoglycosides   79, 88, 283–285, 300–302
Aminophylline   139
Aminosalicylates   152
Aminosalicylic acid   307
Amitriptyline   177, 228–230
Amlodipine   106
Amobarbital   235
Amodiaquine   346
Amoxapine   231, 232
Amoxicillin   293, 296
Amphetamine   56, 58, 63, 64, 66, 207, 222, 227, 232, 233, 236, 237
Amphotericin B   283, 284, 312, 313, 317, 337, 338
Ampicillin   293–297
Amrinone   119
Amsacrine   370
Amyl nitrate   105
Anabolic steroids   257–258
Anaerobic infections   298, 299, 307
Anafranil   234
Analgesics   160–168
Anaphylaxis   132, 137
Anaquest   192
Ancobon   313
Androgens   250, 251, 257–258
Androstenedione   252
Anesthesia   183, 187–193
    balanced   187
    dissociative   192
    general   187–193
    Guedel's signs   187, 188
    stages   188
    surgical   187
Anesthetics   187–193. See Anesthesia
    general   187–193
    inhalant   189–192
    intravenous   192–193
Angel dust   239
Angina   103–107
    angiospastic   103
    classic   103

INDEX

unstable 103–104
variant 103–104
Angioneurotic edema 133, 258
Angiotensin 96
Angiotensin converting enzyme
  inhibitors 95, 96, 101, 102, 119
Angiotensin I 95
Angiotensin II 253
Ankylosing spondylitis 166
Antabuse 237
Antacids 141–147
  aluminum 142, 145
  calcium 142, 146
  magnesium 142, 144
  sodium bicarbonate 142, 143
Antagonists 27
  chemical 27
  competitive 29
  non-competitive 29
  physiological 27
  receptor 27
Anterior pituitary hormones 241–243
Anthelmintics 322–333
Anthracyclines 367–369
Anthrax 297
Anti-anginal drugs 103–107
Anti-diuretic hormone 241, 242
Anti-gout drugs 172–176
Anti-ulcer drugs 141–147
  antacids 141–147
  antibiotics 147
  cytoprotective agents 141–147
  H_2-blockers 141–147
  histamine H_2-blockers 141–147
  mucus stimulant 147
  PGE analogs 147
  proton pump inhibitor 141–147
Antiarrhythmic drugs 109–113
  calcium channel blockers 112
  class IA drugs 109–111
  class II drugs 111
  class III drugs 112
  class IV drugs 112
  potassium channel blockers 112
  sodium channel blockers 109–111
Antiasthmatics 137–139
Antibiotics 283. See also Antimicrobial chemotherapy
Anticholinergics. See Muscarinic blockers
Anticoagulants 125–130
Anticonvulsants 211–217
Antidepressants 228–235
Antidiarrheal drugs 150–152
Antidiuretic hormone 83, 84, 95, 241, 242
Antiemetics 135, 152–153
  antimuscarinics 152
  dopamine D_2-blockers 152
  histamine H_1-blockers 153
Antifolate drugs 287–291
Antifreeze 205
Antifungal drugs 312–317
Antihyperlipidemics 119–125
Antihypertensive drugs 93–102
  α-blockers 97
  $α_2$-agonists 98–99
  ACE-inhibitors 95–96
  adrenolytics 99
  angiotensin converting enzyme inhibitors 95–96

β-blockers 96, 97, 101, 102
calcium channel blockers 94–95
diuretics 93–94
ganglionic blockers 101
vasodilators 99–101
Antiinflammatory drugs. See NSAIDs
Antimalarials 342–347
Antimicrobial chemotherapy 283–349
  aminocyclitols 300–302
  aminoglycosides 300–302
  anthelmintics 322–333
  antifolate drugs 287–291
  antifungal drugs 312–317
  antimalarial drugs 342–347
  antiparasitic drugs 322–347
  antiprotozoal drugs 335–347
  antiseptics 347–349
  antiviral drugs 317–322
  basic principles 283–287
  beta-lactams 293–300
  broad spectrum antibiotics 283
  cephalosporins 297–299
  chloramphenicol 304–305
  disinfectants 347–349
  erythromycin 306
  fluoroquinolones 291–292
  macrolide antibiotics 306
  narrow spectrum antibiotics 283
  penicillins 293–297
  quinolones 291–292
  sulfonamides 287–292
  tetracyclines 302–304
  trimethoprim 289–292
  tuberculosis, treatment 307–312
  urinary tract antiseptics 292–293
Antimicrobials 283–349. See also Antimicrobial chemotherapy
Antiminth 324
Antimony 377
Antimony poisoning 377
Antimuscarinics 219, 221. See also Muscarinic blockers
Antineoplastics 351–371
  alkylating drugs 353–363
  cell cycle-specific drugs 352–355
  cycle-nonspecific drugs 352–363
Antiparasitic drugs 322–347
Antiparkinsonian drugs 218–221
Antiprotozoal drugs 335–347
Antipsychotics 221–225
Antipyretics 160–171
Antiseizure drugs 210–217
Antiseptics 347–349
Antithrombotics 127
Antiviral drugs 317–322
Anturane 174
Anxiety, treatment 193–198
Anxiolytics 193–198
Apomorphine 152
Apresoline 100
Ara-A 321
Aralen 169, 343
Aramine 64
Arfonad 73, 101
Aromatic hydrocarbons 381
Arrhythmia 107–113
  atrial fibrillation 118
  atrial flutter 118

  supraventricular 110–111, 118
  ventricular 110–111, 118
Arsenic 375, 378, 379
Arsenic poisoning 375–379
Artane 221
Arthritis 161–172
Asbestos 351
Ascaris 322, 323
Asendin 231, 232
Asparaginase 370
Aspergillosis 315
Aspirin 127, 128, 159, 161–166, 177
Astemizole 134, 136
Asthma 131, 137–139, 158
  extrinsic 137
  intrinsic 137
  treatment 137–139
Atabrine 336
Atarax 136, 200
Atenolol 70, 97
Ativan 193, 194, 197
Atovaquone 292
Atrial fibrillation 118
Atromid-S 124
Atropine 46, 47, 54–57, 119, 135, 150
Atrovent 55, 139
Attention deficit hyperactivity disorder 64, 207
Augmentin 296
Auranofin 169, 170
Aurothioglucose 169
Autacoid 155, 157
Autonomic nervous system 39–75
  metabolic effects 46, 47, 69–71
  neurotransmitters 41–45
  receptors 40, 41
  second messengers 45, 46, 70, 71
Aventyl 230
Axid 141
Azactam 300
Azaline 290
Azathioprine 170, 171, 175
Azithromycin 306
Azmacort 138
Azo gantanol 290
Azo gantrisin 290
AZT 320
Aztreonam 300

B

β-adrenoceptor agonists 138
β-adrenoceptors 40, 41, 43, 45
β-blockers 62, 65, 67–71, 96–97, 101, 102, 107
β-endorphin 181
β-lactams 283, 293–300
β-lipotropin 181
B. fragilis 298
$β_1$-adrenoceptor 43, 45
$β_1$-agonists 119
$β_2$-adrenoceptor 45
$β_2$-adrenoceptor agonists 138, 246
$β_2$-agonists 119, 137, 139
$β_3$-adrenoceptor 45

Bacitracin 283, 307
Baclofen 81
Bactericide 283
Bacteriostat 283
Bacteroides 297, 305
Bacteroides fragilis 307
Bactrim 290, 291, 337, 340
BAL 170, 375
Banthine 57
Barbiturates 192, 198–200, 235, 236, 238
Basal cell carcinoma 366
Basophilia 141
Basophilic leukemia 133
BCNU 353, 359, 360
Beclomethasone 138
Beclovent 138
Belladonna 57, 135
Benadryl 136, 153, 200, 221
Bendopa 218
Benemid 174
Benzalkonium chloride 348
Benzamides 222, 223
Benzanthrene 351
Benzathine penicillin G 296
Benzbromarone 174, 175
Benzedrine 63
Benzene 381
Benzisoxazoles 223
Benzocaine 178, 180
Benzodiazepines 193–198, 217, 236, 238
Benzonatate 185
Benzoyl peroxide 348
Benztropine 221
Beta-lactams 293–300
Betadine 348
Betamethasone 172, 252, 254
Betaxolol 65
Betazole 134
Bethanechol 47, 50
Biaxin 306
Biguanides 281
Bilarcil 333
Bilirubinemia 200
Biltricide 329
Bioavailability 16, 17
Bithionol 330, 331, 334, 338
Biotransformation 7, 33–37
   enzyme induction 34
   enzyme inhibition 34
   first pass metabolism 7, 8
   hydrolysis 33
   oxidation 33
   phase 1 reaction 33
   phase 2 reaction 33, 35
   reduction 33
Biphosphonates 273
Bisacodyl 148
Bismuth 141, 145, 146, 152
Bismuth subsalicylate 152
Bitin 330
Bitolterol 138
Black widow spider venom 46, 47

Bladder carcinoma 360, 363
Blastomycosis 315
Bleach 348
Bleomycin 352, 353, 363, 369, 370
Blood coagulation 125
Blood-brain barrier 16
Boric acid 348
Botulinum toxin 46, 47, 208, 209
Bradykinin 155, 157
Bran 147
Breast cancer 258, 263, 267, 358, 364, 366, 368
Brethaire 138
Brethine 62
Bretylium 112
Bretylol 112
Brevibloc 70
Brevital 192
Bricanyl 62, 246
Broad spectrum antibiotics 283
Bromfaromine 234
Bromocriptine 221, 238, 243, 244
Brucellosis 303, 305
Bumetanide 89, 94
Bumex 89, 94
Buprenex 185
Buprenorphine 185, 236
Bupropion 231, 232
Burkitt's lymphoma 358, 364
Burn sepsis 290, 291, 302
Buspar 197
Buspirone 197, 234
Busulfan 353, 361
Butazolidin 168
Butoconazole 317
Butorphanol 185
Butoxamine 70
Butyrophenones 218, 222, 223

C

C6 73
Ca-channel blockers 94, 95, 101, 102, 104–106, 112, 119
Cadmium 379
Cadmium poisoning 377–379
Ca•EDTA•Na$_2$ 374–377, 379
Cafergot 177
Caffeine 177, 207, 208, 236
Calcifediol 268
Calciferol 268, 269
Calcimar 270, 273
Calcitonin 270, 272, 273, 274
Calcitriol 268, 273, 275
Calcium antacids 142, 146
Calcium channel blockers 94, 95, 101, 102, 104–106, 112, 119
Calcium disodium versenate 374–377, 379
Calderol 268
Campylobacter 303
Cancer chemotherapy 351–371
Candida 299, 304
Candidiasis 313, 314–317

Capoten 95
Capreomycin 307, 312
Captopril 95, 119
Carafate 145
Carbachol 47, 50
Carbamates 51, 54
Carbamazepine 210–213, 235
Carbaryl 52
Carbenicillin 293, 296, 297
Carbenoxalone 147
Carbidopa 219, 220
Carbon monoxide 380–381
Carbon tetrachloride 381
Carbonic anhydrase inhibitors 83, 84, 85
Carboplatin 363
Carboprost 159
Carboxycellulose 147
Carboxypenicillins 297
Carcholine 50
Carcinogens 351
Carcinoid 133, 145, 244, 360
Carcinoma 358–371
Cardiovascular pharmacology 93–130
Cardizem 106
Carmustine 360
Cascara 148
Castor oil 148
Cat 236
Catapres 62, 98
Catechol-O-methyltransferase 42
Cathartics 147–149
CCNU 353, 359, 360
Cefaclor 297, 298, 299
Cefadroxil 298
Cefamandole 297, 298, 299
Cefazolin 297, 298, 299
Cefonicid 298
Cefoperazone 298, 299
Ceforanide 298
Cefotaxime 298, 299
Cefotetan 298
Cefoxitin 298, 299
Ceftazidime 298, 299
Ceftizoxime 298, 299
Ceftriaxone 298, 299, 303
Cefuroxime 298, 299
Cellulose 147
Celontin 215
Cephalexin 297, 298
Cephalosporins 89, 128, 283, 296–299
Cephalothin 298, 299
Cephamycins 298
Cephapirin 298
Cephradine 298
Cephulac 148
Cerebral edema 83, 91
Cerubidine 368
Cestodes 328–330
Cetylpyridinium 348
Chagas' disease 339
Chancroid 291
Chelating agents 374
Chemet 376

INDEX

Chemotherapy. *See* Antimicrobial chemotherapy; Cancer chemotherapy
Chlamydia 290, 303, 306
Chlamydia lymphogranuloma 303
Chlamydia psittacosis 303
Chlamydia trachoma 303
Chlamydia trachomatis 290
Chlamydia venereum 303
Chlor-trimeton 136
Chloral hydrate 201
Chlorambucil 171, 359, 360, 361
Chloramphenicol 283–285, 304–305
Chlordecone 382
Chlordiazepoxide 193–197
Chlorhexidine 349
Chlorinated hydrocarbon 382
Chlorodeoxyadenosine 365
Chloroform 188, 191, 381
Chloroguanide 345–347
Chloroquine 169, 336, 337, 343–347
Chlorothiazide 93
Chlorotrianisene 264
Chloroxylenol 349
Chlorpheniramine 136
Chlorpromazine 222, 223, 226, 227
Chlorpropamide 280, 281
Chlortetracyclin 303
Chlorthalidone 87, 94
Cholecalciferol 268
Cholera 303
Cholestipol 122, 123
Cholestyramine 122–124
Choline esters 47, 48
Cholinergic 39–42
Cholinesterase 42, 51–57
    carbamates 52, 54, 56
    organophosphates 53–55
    thiophosphates 56–57
Cholinesterase inhibitors 48, 49, 51–57, 382
Cholinomimetics 47, 48, 50, 51
Choriocarcinoma 364, 368
Chorionic gonadotropin 260
Chrysotherapy 169, 171
Cilastatin 300
Cimetidine 127, 141, 195
Cin-quin 110
Ciprofloxacin 285, 291, 292, 307, 312
Cirrhosis 85, 133, 141
Cisplatin 353, 363
Citalopram 234
Citanest 181
Clarithromycin 306
Claritin 136
Clavulanic acid 296
Clearance 11, 19, 21
Cleocin 307
Clindamycin 283, 284, 307
Clinoril 167
Clofazimine 307, 312
Clofibrate 124
Clomid 267
Clomiphene 267
Clomipramine 234

Clonazepam 197, 210, 211, 235
Clonazepine 210, 217
Clonidine 56, 59, 62, 98, 99, 177
Clorazepate 193, 194, 210, 211, 217
Clorophenyl 383
Clostridia 297, 304
Clostridium difficile 299
Clotrimazole 317
Cloxacillin 296
Cloxapen 296
Clozapine 223, 225, 227
Clozaril 223, 225
Cluster headache 178
CNS depressants 237, 238
CNS stimulants 206–208, 236, 237
CNS toxins 208
Cocaine 56, 58, 178, 179, 207, 235, 237
Cocarcinogens 351
Coccidioidomycosis 315
Codeine 150, 181–185, 235, 236
Cogentin 221
Coke 237
Colace 148
Colchicine 172–176. *See also* Glucocorticoids
Colestid 122
Colistin 307
Colloidal bismuth 146, 152
Colon cancer 366
Colyte 148, 150
Compliance 11
Congestive heart failure 83, 113–119
Conn's syndrome 90
Constipation 147–149
Contraception, oral 258, 263, 265–267
Contraceptives, oral 128, 265–267
Copper poisoning 375–379
Cor pulmonale 85
Corgard 69, 97
Corotrope 119
Corticosteroids 137, 138, 155
Corticotropin-releasing hormone 241, 242, 250, 251
Cortisol 137, 171, 250–255
Cortisone 171
Cosyntropin 252
Cough 185
Coumarin 126
Crack 237
Cresol 349
Cretinism 249
CRH 241, 242 250, 251
Cromolyn 137, 138, 139
Cryoprecipitate 129
Cryptococcal meningitis 313–315
Crystal 236
Crystodigin 118
Cuprid 375
Cuprimine 170, 375
Cushing's disease 250, 253, 267
Cutaneous fungal infections 317
Cyanide 384

Cyclizine 135, 136
Cyclo-prostin 159
Cyclophosphamide 171, 355–360, 363
Cyclopropane 188–191
Cycloserine 283, 284, 307, 311, 312
Cyproheptadine 136, 177
Cyproterone 257, 260
Cytadren 252, 253
Cytarabine 352, 353, 366, 370
Cytochrome P_{450} 16, 17, 33
Cytomegalovirus 320
Cytomel 249
Cytoprotective agents 145
Cytosine arabinoside 366
Cytotec 147, 159, 162
Cytovene 320
Cytoxan 171, 356

D

δ-9-THC 239
D-penicillamine 375
d-tubocurarine 46, 47, 77–79, 133, 187
Dacarbazine 352, 362
Dactinomycin 353, 364, 368
Dalmane 196
Danazol 257, 258, 260
Danocrine 257
Dantrium 81
Dantrolene 81
Daraprim 345
Darvon 183
Daunorubicin 368
DDT 371, 382
Decamethonium 79
Declamethasone 254
Declomycin 90, 91, 303
Deferoxamine 374–379
Dehydroemetine 331, 336
Delirium tremens 237
Demecarium 48, 54
Demeclocycline 91, 303, 304
Demerol 183
Deoxycoformycin 365
Depakene 215, 235
Depression 225–235
DES 264
Desferal 374
Desflurane 192
Desipramine 56, 58, 228–230
Desmethyldiazepam 193, 194
Desogestrel 265
Desyrel 232
Dexamethasone 172, 177, 252–254
Dextran 127
Dextran 40 128
Dextran 70 128
Dextran 75 128
Dextromethorphan 185
Dextropropoxyphene 236
DFP 46, 47, 54
DHE 45 177
Diabetes 276-281
    maturity-onset 276, 279-281

mellitus 276-279
  type I 276-279
  type II 276, 279-281
Diabetes mellitus 276
Diabinase 280
Diamox 85
Dianabol 257
Diarrhea 150-152, 159
  treatment 183, 184
Diazepam 56, 57, 81, 193-197, 211, 217
Diazoxide 100
Dibenzoxapines 223
Dibenzyline 67, 97
DichloroDiphenyl-Trichloroethane 382
Dichlorophen 330
Dicloxacillin 296
Didronel 270
Dieldrin 382
Diet pills 207
Diethylcarbamazine 326, 327, 330, 332
Diethylstilbestrol 264, 267
Digibind 119
Digitalis 110-119
Digitoxin 114, 115, 118
Digoxin 114, 115, 118, 142
Dihydroergotamine 177, 178
Dihydroindolones 223
Dihydropyridines 95, 104, 105
Dihydroxyphenylacetic acid 42, 219
Dihydroxyphenylalanine 219
Diisofluorophosphate 46, 47, 54
Dilantin 112
Diloxanide 337
Diloxide furoate 336
Diltiazem 94, 95, 106
Dimercaprol 170, 375-379
Dimethylphenylpiperazinium 46, 47
Dinoprost 159
Dinoprostone 159, 246
Diphenhydramine 136, 153, 200, 221
Diphenoxylate 56, 150, 183, 185
Diphenylhydantoin 211
Diphosphonate 270, 271
Diphtheria 306
Diprivan 193
Dipyridamole 127, 128
Diquat 384
Disinfectants 347-349
Disipal 221
Disopyramide 111
Disulfiram 204, 237, 238, 362
Diulo 87
Diuretics 83-93, 101, 102
Diuril 93
Diverticulitis 147
DMPP 46, 47
Dobutamine 61, 119
Dobutrex 61
Docusate 148, 150
Dolophine 183
Domperidone 219, 221
DOPAC 42, 219
Dopamine 42, 44, 60, 61, 119, 219-225, 243
Dopamine D_2 receptors 222-225
Dope 238
Dose-response 28-30
  graded 28, 30, 31
  quantal 28, 31
Doxepin 228, 230
Doxorubicin 352, 353, 363, 368, 370
Doxycycline 302-304, 337
Dronabinol 153, 239
Droperidol 193
Drug absorption 11-15
Drug abuse, withdrawal 185
Drug administration 7-10
Drug dependence 236
Drug deposition 16
Drug distribution 15-16
Drug effect 20-33
Drug excretion 17-19
Drug interactions 31
Drug metabolism 16-17, 31-37. See also Biotransformation
Drugs of abuse 235-239
DTIC 352, 362
DTIC-Dome 362
DTs 237
Ductus arteriosus 159
Dulcolax 148
Duodenal ulcers 141-147
Duracillin 296
Dwarfism 245
Dymelor 280
Dynorphin A 181
Dynorphin B 181
Dyrenium 90, 94
Dysmenorrhea 158, 159
  treatment 166-168

E

Echothiophate 48, 51, 52, 54
Econazole 317
Ecotoxicology 373
Ecstasy 236
ECT 234, 235
ED_{50} 29
Edecrin 89, 94
Edetate calcium disodium 374-377, 379
Edrophonium 48, 51, 52, 53
EETES 156, 158
Effexor 234
Efficacy 29
Eflornithine 334, 337, 340
Eicosanoids 155-160
Elavil 230
Eldepryl 221
Electroconvulsive therapy 234, 235
Elspar 370
Embolism 126
Emetics 152
Enalapril 95, 119
Endocarditis 296, 302, 306
Endocrine pharmacology 241-282
Endometrial carcinoma 263, 266
Endometriosis 260, 267
Enemas 147, 150
Enflurane 79, 188-192
Enkephalins 181
Enprostil 147
Enteritis 290
Enterobacteria 297, 298, 300
Enterobacteriaceae 310
Enterococci 299
Enterohepatic circulation 18, 20
Enthrane 191
Enuresis 55, 230
Ephedrine 63, 64, 66, 138
Epidermophyton infections 315
Epilepsy 208
  classification 209-211
  treatment 210-217
Epinephrine 40-47, 59, 60, 64, 137, 138, 179
Epinephrine-reversal 64-67
Epoxides 351
Equilin 262
Ergocalciferol 268
Ergomar 177
Ergonovine 177, 246
Ergot alkaloids 246, 238
Ergotamine 177, 178
Ergotrate 177, 246
Erythritol 105
Erythromycin 283-285, 306
Eserine 54
Esmolol 70
Esophageal candidiasis 315
Estrace 262
Estraderm 262
Estradiol 262
  benzoate 263
  cypionate 263
Estrogen 258-264, 273
Estrone 262
Estrovis 264
Ethacrynic acid 87, 89, 94, 301
Ethambutol 307, 308, 310
Ethanol 201-206, 237, 246, 348
Ether 188-192
Ethinyl estradiol 263, 265, 267
Ethionamide 307, 311, 312
Ethosuximide 210, 211, 215
Ethotoin 212
Ethyl alcohol. See Ethanol
Ethylene glycol 205
Ethylenimines 360, 361
Ethylestrenol 257
Ethynodiol 265
Etidronate 270, 273, 274
Etodolac 168
Etomidate 192
Etoposide 353, 363, 367
Eulexin 257
Euthroid 249
Ex-Lax 148

INDEX

F

Factrel 261
Famotidine 141
Fansidar 290, 291, 346
Felbamate 217
Feldene 167
Fentanyl 182–184, 193
Fetal alcohol syndrome 203
Fever 161–168
Fiber 147
Fibric acids 124, 126
Filariasis 325–328
Finasteride 257, 260
First pass metabolism 7, 8
Flagyl 335
Flatworms 329–333
Flaxedil 78
Floropryl 54
Flosequinan 119
Floxuridine 366
Fluconazole 315, 316
Flucytosine 283, 285, 313, 314
Fludrocortisone 254
Flukes 329–333
    blood 331
    intestinal 331
    liver 331
    lung 331
Flumadine 318
Flumazenil 193, 197, 198, 238
Fluocortisol 252
Fluoride 272, 275
Fluoroplex 366
Fluoroquinolones 283, 285, 291–292
Fluorouracil 365, 366
Fluothane 191
Fluoxetine 229, 233–235
Fluoxymesterone 257
Fluphenazine 223
Flurazepam 193, 194, 196
Flutamide 257, 260
Fluvoxamine 234
Folinic acid 364
Follicle stimulating hormone 241, 242, 255–257, 259–261
Forane 192
Formaldehyde 348
FSH 241, 242, 255, 258–260
Fumaride 336
Fumigants 384
Fungemia 314
Fungizone 312, 317
Funisolide 138
Furacin 348
Furadantin 292
Furosemide 87, 89, 94, 159, 270, 274, 301
Fusarium solani 317

G

γ-aminobutyric acid 81
γ-globulin 318
GABA 81

Gabapentin 217
Gallamine 78
Gallium nitrate 270, 274
Gamma globulin 317
Gancyclovir 319–321
Ganglionic blockers 72–74, 101, 102
Ganglionic stimulants 74, 75
Gantanol 290
Gantrisin 290
Garamycin 302
Gasoline 381
Gastric ulcers 141–147
Gastrointestinal drugs 141–153
Gemfibrozil 124
Gemonil 215
Genital herpes 319
Genital warts 322
Gentamicin 79, 300–302
Gentian violet 349
Gepirone 234
Germanin 339
GI ulcers 141–147, 159
Giardiasis 336
Glaucoma 83, 85, 91
    treatment 48, 50, 54, 59, 65, 68, 69
Glipizide 280, 281
Glucagon 276, 281, 282
Glucocorticoids 155–158, 171, 172, 176, 250–256, 272, 274
Glucotrol 280
Glucuronidation 35
Glutaraldehyde 348
Glyburide 280, 281
Glycerin 91
Glyceryl trinitrate 104
Glyrol 91
Glycopyrrolate 55
GnRH 241, 242, 255–257, 260, 261
Gold, therapy 169, 170
Gold sodium thiomalate 169
Golytely 150
Gonadorelin 259, 261
Gonadotropin-releasing hormone 241, 242, 255–257, 259–261
Gonorrhea 306
Gout 172–176
Gram-negative infections 298–304
Gram-positive 298, 302, 304, 309
Grass 239
Grave's disease 249
Grifulvin 315
Griseofulvin 315
Growth hormone 241–245
Growth hormone-releasing hormone 241, 242, 243
Guanethidine 56, 58, 67, 72, 99–102, 230
Guinea worm 323

H

Haemophilus influenzae 290, 291, 298, 302, 305, 310
H_1-blockers 131–136, 158, 221
    actions 132, 134–136
    toxicity 132, 135–136
H_2-blockers 141–147
Habitrol 236
Haemophilus vaginalis 290
Hairy-cell leukemia 322
Haldol 223
Half-life 19, 21
Hallucinogens 238
Halogenated hydrocarbons 381
Haloperidol 223, 227, 235
Halothane 79, 188–191
Hashimoto's thyroiditis 248
Hashish 239
Hayfever 131, 135
Hazard 373
Headache 177–178
Hemicholinium 46, 47
Hemochromatosis 375
Hemosiderosis 375
Hemostasis 125
Hemostics 127, 130
Heparin 125, 126, 127, 128
Hepatitis, viral 317, 321, 322
Hepatolenticular degeneration 375
Herbicides 383–384
Heroin 181, 182, 185, 235, 238
Herpes simplex 321
Herpes simplex 1 319
Herpes simplex 2 319
Herpes viruses 319–322
Herplex 321
HETES 156–158
Hetrazan 326
Hexachlorophene 349
Hexamethonium 73, 79
Hexamethylmelamine 361
Hexastat 361
Hexylresorcinol 349
Hismanal 136
Histalog 134
Histamine 79, 131–135, 158. See also H_1-blockers
    actions 131–134
    H_1-receptors 133–134
    H_2-receptors 133
    H_3-receptors 133
    storage 133
Histamine H_2-blockers 141–147
Histoplasmosis 315
HMG CoA reductase inhibitors 123–125
Hodgkin's disease 356, 360, 362
    lymphoma 352, 367, 368
Hog 239
Homovanillic acid 42, 219
Hookworm 323
HPETES 156–158
Human chorionic gonadotropin 260
Humatin 336
Humatrop 243
HVA 42, 219
Hydralazine 100, 119
Hydrea 369
Hydrochloroquine 169

Hydrochlorothiazide 87
Hydrocodone 181–183
Hydrocortisone 171
Hydrodiuril 87
Hydrogen cyanide 384
Hydrolysis reactions 34
Hydromorphone 181, 182, 185
Hydroxyamphetamine 64
Hydroxyprogesterone 264
Hydroxyurea 353, 369, 370
Hydroxyzine 136, 200
Hygroton 87, 94
Hypercalcemia 270–275
Hypercalcuria 270
Hypercholesterolemia 121, 122, 124
Hyperlipidemia 121
Hyperlipoproteinemia 124
Hyperparathyroidism 273, 274
Hypersensitivity 31
Hyperstat 100
Hypertension 83
   primary 159
   pulmonary 159
   treatment 93–102
Hypertensive crisis 63, 65, 72, 97, 100
Hyperthyroidism 71, 249–250, 275
Hypertriglyceridemia 121, 124
Hypervitaminosis A 271
Hypervitaminosis D 270, 273
Hypnotics. *See* Sedative-hypnotics
Hypoaldosteronism 255
Hypocalcemia 272, 275, 276
Hypogonadism 258
Hypomania 227, 235
Hypoparathyroidism 270, 275
Hypothalamic-pituitary axis 241–243
Hypothyroidism 248
Hypovitaminosis D 270, 275

I

Ibuprofen 166, 167, 177, 246
Ice 236
Idamycin 368
Idarubicin 368
Idiosyncratic reaction 30
Idoxuridine 319, 321
Ifex 358
IFNs 322
Ifosfamide 358
Imidazoles 314, 317
Imipenem 299
Imipramine 228–230
Imitrex 177
Imodium 150, 183, 185
Impotence 67
Imuran 170
Indapamide 87, 94
Inderal 68, 97
Indirect-acting sympathomimetic 64
Indocin 159, 167
Indomethacin 159, 167, 178
Infertility 261
Inflammation 155–176
   treatment 160–176
Influenza 298
Influenza A 318, 321
Influenza B 321
Inocor 119
Inorganic lead 377
Insecticides 51, 52, 54, 56, 57, 382–383
Insomnia, treatment 197–198
Insulin 276–279
   beef 277
   human 277
   Lente 279
   Lente, Ilentin I 279
   NPH Ilentin I 279
   NPH standard 279
   pork 277
   Ultralente Iletin I 279
   Ultralente standard 279
Insulin-like growth factor 243
Insulinomas 100
Intal 137, 138
Interferons 318, 322
Intermittent porphyria 200
Intrinsic activity 27
Intropin 60
Inverse agonist 197
Inversine 74
Iodides 247, 250
Iodine 348
Iodoquinol 336, 342
Iodotope-131 250
Ion trapping 19
Ipecac 152
Ipratropium 55, 139
Ipsapirone 234
Iron 374–379
Iron poisoning 374–379
Ismelin 99
Ismotic 91
Isocarboxazid 232
Isoflurane 79, 188–192
Isoflurophate 54
Isoniazid 283, 284, 307–309
Isopropanol 348
Isoproterenol 56, 58, 60, 137, 138
Isopto cetamide 291
Isosorbide 91, 103, 105
Isuprel 60
Itraconazole 315, 316
Ivermectin 323, 326–328, 332, 333

J

Juvenile arthritis 169

K

Kallikrein 155
Kanamycin 307, 312
Kaopectate 150
Kaposi's sarcoma 322
Keratolytics 166
Kernicterus 200
Kerosene 381
Ketalar 192
Ketamine 192, 239
Ketoconazole 314, 316
Klebsiella 297, 298

L

L-α-acetyl-methadol 238
L-DOPA 218–221
Labetalol 69, 97, 101
Labor
   induction 243, 244, 246
   inhibition 246
Lactulose 148
LAMM 238
Lamotrigine 217
Lampit 339
Lanoxin 118
Lariam 344
Larodopa 218
Lasix 89, 94
Laudanum 150
Laxatives 147–149
   contact agents 148
   fiber 147
   osmotic 147, 148
   saline 147, 148
   stool softeners 147, 148
LD_{50} 29
Lead 374–379
Lead poisoning 374–379
Legionnaires' diseases 306
Leishmaniasis 290, 337, 338
Lennox-Gastaut syndrome 217
Lesch-Nyhan disease 175
Leu-enkephalin 181
Leucovorin 364
Leukemia 356, 358, 360–370
Leukeran 171, 359
Leukotrienes 131, 134, 137, 138, 156–159
Leuprolide 259, 261
Leutinizing hormone 241, 242, 255–261
Levamisole 171, 323
Levodopa 218. *See also* L-DOPA
Levophed 59
Levopropoxyphene 185
Levothyronine 249
Levothyroxine 248
LH 241, 242, 255–261
Librium 194, 196, 197
Lidocaine 112, 119, 179–181
Lindane 382
Lioresal 81
Liotrix 249
Lipoxins 156
Lisinopril 95
Lithium 90, 91, 178, 233, 235
Loading dose 24, 25
Local anesthetics 155, 156, 178–181
Lodine 168
Lomotil 150, 183
Lomustine 360
Loniten 100
Loop diuretics 84, 88, 89, 94

INDEX

Loperamide 150, 183
Lopid 124
Lopressor 70, 97
Loratadine 136
Lorazepam 193–197, 211, 217
Lorelco 125
Losec 143
Lotrimin 317
Lovastatin 123, 124
Love pill 236
Loxapine 223, 232
Loxitane 223
Lozol 87, 94
LSD 235, 238
Ludiomil 232
Lugol's solution 250
Luminal 200
Lung cancer 363, 367
Lung carcinoma 361–364, 367
Lupron 261
Lupus erythematosus 170
Lyme disease 303
Lymphatic cacers 272, 273, 352, 356, 358, 360, 362, 364, 367–369
Lymphatic leukemias 352, 356, 358, 360, 362, 367–369
Lymphoblastic leukemia 358
Lymphocytic leukemia 352, 360, 370
Lymphocytomas 356, 358, 360, 362, 367, 369
Lymphogranuloma 291
Lymphomas 356, 358, 360, 362, 364, 367–369
Lysergic acid diethylamide 235, 238
Lysodren 253, 371
Lysol 349

M

M-blockers 220
MAC 189
Macroglobulinemia 360
Macrolide antibiotics 306
Mafenide acetate 290, 291
Magnesium 246
Magnesium antacids 142, 144
Malabsorption 270, 275
Malaria 290, 291, 337, 342–347
Malathion 51, 57
Malignant hyperthermia 81
Mandelamine 292
Mania 227, 235
Mannitol 91
Manoplax 119
MAO 42
MAO-inhibitors 59, 63, 220, 221, 227, 228, 230–234
Maprotiline 231, 232, 234
Marezine 135, 136
Marijuana 153, 235, 239
Marinol 153, 239
Marplan 232
Mastocytosis 133, 141
Matulane 362

Maxair 138
Maxibolin 257
Mazicon 197
MDA 235, 236
MDMA 235, 236
Measles 317
Mebaral 215
Mebendazole 323–325, 330, 331
Mecamylamine 74
Mechlorethamine 352–356, 359–361
Meclobemide 234
Meclofenamic acid 168
Meclomen 168
Mectizan 326
Medihaler 138
Mefloquine 344–346
Melanocyte stimulating hormone 251
Melanoma 359, 362, 370
Melarsoprol 337
Mellaril 223
Melphalan 355, 359, 361
Menaquinone 129
Menest 262
Meningitis 290, 296, 298, 299, 303, 305, 310
Menotropins 261
Meperidine 181–185, 232, 235, 238
Mephenytoin 212
Mephobarbital 214, 215
Meprednisone 171
Meprobamate 198, 201
Mepron 292
Mercaptopurine 175, 352, 365
Mercury 375, 377
Mercury poisoning 375–379
Mercury salts 349
Merthiolate 349
Mesantoin 212
Mescaline 235, 238
MESNA 357
Mesoridazine 226
Mestinon 54
Mestranol 263, 265
Met-enkephalin 181
Metals, heavy 374–379
Metapirone 252
Metaprel 138
Metaproterenol 138
Metaraminol 64, 66
Methacholine 47, 50
Methacycline 303
Methadone 181–185, 235, 238
Methamphetamine 64–66, 207, 235, 236
Methandrostenolone 257
Methanol 204–206
Methantheline 57
Metharbital 215
Methcathinone 236
Methemoglobinemia 385
Methenamine 283, 285, 290, 292
Methergine 246
Methicillin 293, 296
Methimazole 250

Methisazone 318, 322
Methohexital 192
Methotrexate 169, 170, 352, 353, 363–365
Methoxamine 61, 66
Methoxyflurane 188–192
Methsuximide 215
Methyl alcohol 204–205. See also Methanol
Methyl-CCNU 353, 359, 360
Methylcellulose 147, 152
Methyldopa 98, 99, 102
Methylene blue 385
Methylene chloride 381
Methylenedioxyamphetamine 235
Methylenedioxymethamphetamine 235
Methylergonovine 246
Methylphenidate 64, 207, 236
Methylprednisolone 171
Methylsalicylate 163, 165, 166
Methyltestosterone 257
Methylthiotetrazole 297
Methylxanthines 137, 139, 208
Methysergide 177, 178, 238
Metoclopramide 152
Metolazone 87, 94
Metoprolol 70, 97
Metrazol 209
Metrifonate 331, 333, 334, 340
Metrodin 261
Metronidazole 323, 335–337, 341
Metyrapone 252, 253
Mevacor 123
Mexate 363
Mexiletine 112
Mexitil 112
Mickey Finn 201
Micronase 280
Miconazole 315–317
Micronor 266
Microsporum 315
Microsulfon 290
Midamor 90, 94
Midazolam 193, 195, 196
Mifepristone 267
Migraine 177
Milontin 215
Milrinone 119
Miltown 201
Mineral oil 148, 150
Mini-pill 265
Minimum alveolar concentration 189
Minipress 67, 97
Minocycline 302–304
Minoxidil 100
Mintezol 324
Miochol 48
Mirex 382
Misoprostol 147, 159, 162
Mithramycin 270, 274
Mitocin-C 369
Mitomycin 369
Mitotane 253, 371

Mitoxantrone 370
Moban 223
Moclobemide 234
Molindone 223
Monistat 317
Monoamine oxidase 36, 42
Monoamine oxidase inhibitors 195, 221, 227, 228, 230–235
Mood disorder. *See* Depression; Mania
MOPP therapy 352, 356, 362, 367
Morning-after pill 267
Morphine 133, 181–185, 193, 235
Motion sickness 55, 135, 152
Motrin 166, 177, 246
Moxalactam 298, 299
MPP 218
MPTP 218, 238
MSH 181, 251
MTX 170, 352, 353, 363–365
Multiple myeloma 272
Murine typhus 303
Muscarine 50
Muscarinic blockers 54–58
Muscarinic receptor antagonists 139. *See also* Muscarinic blockers
   use in asthma 139
Muscarinic receptors 39–44
Muscle relaxants 77–81. *See* Skeletal muscle relaxants
Mustargen 356
Myalgia, treatment 161–172
Myambutol 310
Myasthenia gravis 52, 53
Mycifradin 302
Mycobacteria 307–312
Mycobacterium intracellulare 306
Mycoplasma pneumoniae 303, 306
Mycosis fungoides 356, 364
Mycostatin 317
Myelocytic leukemias 356
Myelofibrosis 361
Myelogenous leukemia 356, 358–360, 362, 365, 367, 370
Myelomas 358–362, 365, 367, 370
Myleran 361
Myocardial infarct 159, 164, 166
Myochrysine 169
Mysoline 215
Myxedema coma 249
Myxoviruses 320

N

N-acetylcysteine 357
N-acetylpenicillamine 375, 376
N. gonorrhoeae 298, 299
N. meningitidis 305
N_2O 191. *See also* Nitrous oxide
Nabumetone 167
Nadolol 69, 97
Nafarelin 259, 261
Nafcil 296
Nafcillin 296
Naftifine 317

Naftin 317
Nalbuphine 184
Nalidixic acid 291
Nalorphine 181, 182
Naloxone 181, 182, 185, 193, 238
Naltrexone 181, 182, 185, 237, 238
Naprosyn 166, 177
Naproxen 166, 177
Narcan 185, 238
Narcolepsy 207
Narcotic analgesics 181–185
Narcotic antagonists 181, 182, 185
Narcotics 235, 238
   addiction 185
   definition 181
Nardil 232
Narrow spectrum antibiotics 283
Natacyn 317
Natamycin 317
Navane 223
Nebcin 302
Nedocromil 137, 138
Nefazodone 234
Neisseria 309, 310
Neisseria gonorrhoea 303
Nematodes, intestinal 322–325
Neo-antergan 136
Neo-synephrine 61
Neomycin 302
Neostigmine 48, 52, 54
Nephron 83, 84
Nerve gas 51, 55
Nerve gas antidote 56
Netilmicin 300–302
Netromycin 302
Neuroblastoma 358, 367
Neurolepanalgesia 193
Neuroleptic malignant syndrome 224, 226
Neuroleptics 221–225
Neuromuscular blocking agents. *See* Skeletal muscle relaxants
Niacin 121, 122, 124
Nicardipine 105
Niclocide 329
Niclosamide 329–336
Nicoderm 236
Nicolar 121
Nicorette 236
Nicotine 47, 75–79, 236, 382, 383
Nicotine patches 236
Nicotine, receptor 79–81
Nicotinic acid 121, 122
Nicotinic receptors 40, 42, 44–48, 72–75
Nifedipine 105, 106
Nifurtimox 334, 337, 339, 342, 344
Nimodipine 105
Nipride 101
Nitrates 103, 119
Nitrites 103
Nitrofurans 348
Nitrofurantoin 283, 285, 292
Nitrofurazone 348
Nitrogen dioxide 380–381
Nitrogen mustards 353, 356–360

Nitroglycerin 104, 105
Nitroprusside 100, 101
Nitrosamines 351
Nitrosoureas 359, 360
Nitrous oxide 188–191
Nizatidine 141
Nizoral 314
N_M receptor 47, 49, 72–75
N_N receptor 46, 47, 49, 73
Nocardia 290, 291
Nolvadex 267
Nonsteroidal antiinflammatory drugs. *See* NSAIDs
Noradrenaline 39. *See* Norepinephrine
Noradrenergic 39–40
Norcuron 78
Nordazepam 193–194
Nordoxepin 228
Norepinephrine 39, 42–47, 58–62, 227–228
Norethindrone 265
Norethynodrel 264
Norfloxacin 285, 291, 292
Norfluoxetine 233
Norgestrel 264, 265
Norpace 111
Norplant 266
Nortriptyline 228–230
Norvasc 106
Nosocomial infections 299
Novantrone 370
Novocain 178
NSAIDs 155–168, 177
Nubain 184
Nydrazid 308
Nystatin 317

O

Obsessive compulsive disorder 234
OCD 234
Octreotide 243, 244
Ocular infections 290, 291
Ofloxacin 307, 312
Omeprazole 143–147
Oncovin 352, 367
Ondansetron 153
Opiates 235, 236, 238. *See also* Opioid
   for diarrhea 150, 151
Opioid 155, 156, 181–185
   receptors 183–185
Opioid antagonists 181–185
Opioid receptors
   delta 183–185
   kappa 183–185
   mu 155, 156, 183–185
   sigma 183–185
Opioids 193, 235, 236, 238
Opium tincture 150
Oral hypoglycemics 276, 279–281
Orap 223
Organophosphates 53–57
Orinase 280
Orlamm 238

INDEX

Ornidyl 340
Orphenadrine 221
Ortho-Novum 250, 251, 265
Ortho-Novum 7/7/7 265
Osmitrol 91
Osmotic diuretics 90, 91
Osmotic laxatives 147, 148
Osteoarthritis 166, 167
Osteomyelitis 310
Osteoporosis 258, 263, 270, 272
Otitis media 290, 291, 299
Ovarian cancer 358, 363, 364, 368, 370
Ovral 265
Ovrette 266
Oxacillin 296
Oxamniquine 330–334, 339
Oxantel 323, 324, 329, 332
Oxazepam 194
Oxidation reactions 33
Oxotremorine 50
Oxybarbiturates 199
Oxycodone 181–183
Oxymorphone 181, 182
Oxyphenbutazone 168
Oxypurinol 174–176
Oxytetracycline 303, 304
Oxytocin 241–246
Ozone 380–381

P

P. aeruginosa 300
Paget's disease 270
Pain 155–185
   description 155–158
   integumental 155
   treatment 160–185
   visceral 155
Pain and inflammation 155–185
Paludrine 346
Pancreatic carcinoma 360
Pancuronium 78
Panic attacks 235
Papaverine 177
Para-aminosalicylic acid 283, 285, 309, 311
Paracoccidioidomycosis 315
Paradione 217
Paramethadione 217
Paranoia 221
Paraplatin 363
Paraquat 383, 384
Paraquat+2 383
Parasympathetic nervous system 39–58
Parasympatholytics. See Muscarinic blockers
Parathion 51, 57
Parathyroid hormone 267–269
Paregoric 150
Parkinsonism 55, 136, 218
Parlodel 221, 243
Parnate 232
Paromomycin 336
Paroxetine 234

Partial seizures 217
Patent ductus arteriosus 167
Pavilon 78
Paxil 234
PCP 235, 239
Penicillamine 170, 376, 377
Penicillin G 293–296
Penicillin V 293, 296
Penicillins 283, 285, 293–297
Pentaerythritol 105
Pentagastrin 134
Pentam 300 338
Pentamidine 337, 338, 340, 342
Pentazocine 185
Pentobarbital 198, 235
Pentostam 338
Pentostatin 365
Pentothal 192, 200
Pentylenetetrazol 209
Pepanone 212
Pepcid 141
Pepto-Bismol 146, 152
Pergolide 219, 221
Pergonal 261
Periactin 136, 177
Permapen 296
Permax 221
Peroxide 348
Persantine 128
Pertrofrane 230
Pertussis 306
Peyote 235
Pharmacodynamics 25–26
Pharmacokinetics 19–25
Pharmacology 7
Phase 1 reactions 33
Phase 2 reactions 33
Phenazopyridine 290
Phencyclidine 235, 239
Phenelzine 177, 220, 232
Phenergan 136, 152, 225
Phenobarbital 127, 198–200, 210, 211, 214–217, 236, 238, 275
Phenolphthalein 148
Phenols 349
Phenothiazines 218, 222, 223
Phenoxybenzamine 56, 59, 67, 97
Phensuximide 215
Phentolamine 63, 67, 97
Phenylbutazone 168
Phenylephrine 56, 58, 61, 66
Phenyltrimethylammonium 46, 47
Phenytoin 112, 119, 210–212, 216, 217, 275
Pheochromocytoma 63–65, 97
Phethbarbital 200
Phospholine 54
Physical dependence 236
Physostigmine 48, 52, 54
Phytonadione 129
Picrotoxin 199, 208, 209
Pilocar 50
Pilocarpine 50

Pimozide 223
Pindolol 69, 97
Pinworm 323
Piperacillin 297
Piperazine 323, 324, 328, 332
Pirbuterol 138
Piroxicam 167, 168
Pitocin 244
Placental barrier 16
Plague 302, 303
Plaquenil 169
Platelet activating factor 156–158
Platelet aggregating factor 131
Platinol 363
Pleuropulmonary infections 299
Plicamycin 270, 274
Pneumocystis carinii 290, 291
Pneumocystosis 337, 340
Podophyllotoxins 367
Poliomyelitis 317
Polycystic ovarian disease 260
Polycythemia vera 361, 370
Polymixins 307
Polymyxin B 307
Polynuclear aromatic hydrocarbons 351
Pontocaine 178
Posterior pituitary hormones 241
Pot 239
Potassium channel blockers 112
Potassium permanganate 348
Potassium-sparing diuretics 89, 94
Potentiation 31
Pox-viruses 320, 321
Pralidoxime 53, 54, 56
Pravachol 123
Pravastatin 123
Praziquantel 329–331, 334, 335
Prazosin 67, 97, 119
Precocious puberty 257, 259, 260
Prednisolone 138, 171
Prednisone 138, 171, 252, 254, 352
Pregnyl 261
Premarin 262
Prilocaine 181
Primaquine 345, 346
Primaxin 300
Primidone 210, 211, 214, 215
Pro-banthine 57
Pro-opiomelanocortin 181, 251
Probenecid 172–175, 296
Probucol 125
Procainamide 110, 119
Procaine 178, 181
Procaine penicillin G 296
Procarbazine 352, 362
Procarcinogens 351
Procardia 106
Prochlorperazine 225
Prodynorphin 181
Proenkephalin A 181
Proenkephalin B 181
Progesterone 264, 265
Progestins 264–267

Proguanil 345
Progynon 262
Proklar 291
Prolactin 241, 243
Prolid 249
Promethazine 135, 136, 152, 225
Pronestyl 110
Propacil 250
Propantheline 57
Propofol 193
Propoxyphene 183
Propranolol 56, 59, 68, 97, 110, 177
Propylthiouracil 250
Prostacyclins 156–160
Prostaglandins 131, 134, 137, 138, 155–162
Prostate cancer 260, 263
Prostatitis 289–291
Prostep 236
Prostigmin 54
Prostin E_2 159, 246
Prostin $F_{2\alpha}$ 159
Prostin VR 159
Protamine sulfate 126
Proteus 297, 298
Proton pump inhibitor 142, 143
Protopam 54
Protriptyline 228, 229, 230
Protropin 243
Proventil 62, 138
Prozac 233
Pseud. aeruginosa 300
Pseudocholinestrase 42
Pseudomonas 297–299, 307
Psilocybin 238
Psoriatic arthritis 166
Psychedelics 238
Psychological dependence 236
Psychosis 221
Psyllium 147, 152
PTH 267–269
PTMA 46, 47
PTU 250
Pulmonary edema, treatment with morphine 184, 185
Purinethol 365
Pyrantel 323, 324, 329, 332
Pyrazinamide 307, 308, 310
Pyrethrums 383
Pyridostigmine 48, 54
Pyridoxine 219, 220
Pyrilamine 136
Pyrimethamine 289–291, 345, 346
Pyrophosphate 271

Q

Quantrel 324
Questran 122
Quinacrine 336, 337, 342
Quinidine 110, 111, 119, 231, 345
Quinine 337, 343–345
Quinistrol 264
Quinolones 291–292

R

Rabies 317
Ranitidine 141
Receptor desensitization 62
Receptors 26
Reduction reactions 34
Reflux esophagitis 141, 142
Regitine 67, 97
Reglan 152
Reitter's syndrome 166
Relafen 167
Renin 94, 95, 96, 97
Renoquid 291
Reserpine 56, 58, 67, 68, 71, 99, 101, 102, 218, 222, 227
Resorcinol 349
Restoril 196, 197
Resurrection drug 340
Retinoblastoma 367
Retrovir 320
Retroviruses 320
ReVia 237
Reye's Syndrome 161, 166
Rheumatoid arthritis 166
   treatment with aspirin 165–166
   treatment with glucocosticoids 171–172
   treatment with NSAIDs 166–168
   use of remittive drugs 168–171
Rheumatrex 170
Rhinitis 131, 135
Rhubarb 148
Ribavirin 318–321
Rickets 270
Rickettsia 303
Ridaura 169
Rifadin 309
Rifampin 283–285 307, 309, 322
Rimantadine 318, 320
Risk 374
Risperdal 223, 225
Risperidone 223, 225, 227
Ritalin 64
Ritodrine 62, 246
Robinul 55
Rocaltrol 268
Rocky Mountain spotted fever 303, 305
Rotenone 383
Roundworms 322–325
Routes of administration 7–10
   dermal 10
   enternal 7
   inhalation 9
   intestinal 7
   intramuscular 9
   intranasal 9
   intrathecal 10
   intravaginal 10
   intravenous 9
   oral 7, 8
   parenteral 9, 10
   rectal 8, 9
   subcuntaneous 9
   sublingual 7, 8
   topical 10
RU-486 253, 267
Rubella 318

S

Salicylate. *See* Aspirin
Salicylic acid 348
Salicylism 165
Saline laxatives 147, 148
Salmonella 290
Salsalate 162
Sandostatin 243
Sansert 177, 178
Sarcoidosis 271, 273
Sarcomas 362, 364, 368
Sarin 55
Schistosomes 330
Schizophrenia 221
Scopolamine 55, 57, 152, 187
Secobarbital 198, 200, 235
Seconal 200
Sectral 70, 97
Sedative-hypnotics 193–206, 236, 238
Seizures 209–211
   absence 210
   atonic 210
   clonic 210
   complex 209
   focal 209
   generalized 209
   Jacksonian 209
   myoclonic 210
   partial 209
   petit mal 210
   status epilepticus 197, 208, 210, 217
   tonic 210
   tonic-clonic 210
Seldane 136
Selegiline 219, 221
Semustine 360
Senna 148
Sepsis 299, 314
Sermorelin 242
Seromycin 312
Serotonin 227–228, 232
Serotonin reuptake inhibitors 228, 232–234
Serpasil 99
Sertraline 234
Serzone 234
Sevoflurane 189, 192
Shigella enteritis 290
Silo-fillers disease 380
Silvadene 290, 291
Silver nitrate 349
Silver salts 349
Silver sulfadiazine 290, 291, 349
Simvastatin 123
Sinemet 219
Sinequan 230
Skeletal muscle relaxants 77–81
   depolarizing 77–81
   leptocurares 77–81
   nondepolarizing 77–79
   pachycurares 77–79
Slow reacting substance of anaphylaxis (SRS-A) 158
Snow 237

Sodium bicarbonate 142, 143
Sodium channel blockers 109–111
Sodium nitrite 384
Sodium salicylate 162, 163
Solganal 169
Solid tumors 356, 358–371
Solvents 381
Soman 55
Somatomedin 241–244
Somatostatin 241–244
Somatrem 243
Spasmolytics 80
Spectazole 317
Spectinomycin 300
Speed 236
Spironolactone 87, 89, 90, 94, 260
Sporanox 315
Sporotrichosis 315
SRIH 241–244
SSRIs 228, 232–234
Stadol 185
Staph. aureus 294, 296, 302
Staphcillin 296
Staphylococcal infections 298, 304, 306, 310
Staphylococcal endocarditis 310
Staphylococcus aureus 294, 303, 309
Status epilepticus 197, 210, 211, 217
Stelazine 223
Stibogluconate 337, 338, 342
Stomach ulcers 141–147, 159
Stool softeners 147, 148
Strep pyogenes 296
Strep throat 290, 296
Strep. pneumoniae 290, 291, 296, 305
Strep. viridans 296
Streptococcal infections 291, 306
Streptokinase 127, 128, 130
Streptomycin 300–302, 307, 310
Streptozocin 359, 360
Streptozotocin 360
Strychnine 208
Sublimaze 183, 193
Succimer 376, 377
Succinylcholine 77–79, 187
Sucralfate 141, 142, 145, 146
Sufenta 193
Sufentanil 184, 193
Sulamyd 290
Sulbactam 296
Sulfacetamide 288–291
Sulfacytine 291
Sulfadiazine 288, 290, 291
Sulfadoxine 288, 291, 346
Sulfamerazine 288
Sulfamethazine 288
Sulfamethizole 288, 291, 292
Sulfamethoxazole 288–290
Sulfamylon 290, 291
Sulfanilamide 288
Sulfapyridine 291
Sulfasalazine 152, 290, 291
Sulfathalidine 290, 291

Sulfathiazole 292
Sulfinpyrazone 174, 175
Sulfisoxazole 288, 290, 291
Sulfonamides 283, 285, 287–292
Sulfonylureas 279–281
Sulfur dioxide 380–381
Sulindac 167
Sumatriptan 177, 178
Suramin 327, 334, 337, 339, 343
Surital 192
Symmetrel 221, 317
Sympathetic nervous system 39–47,
Sympathetics 207
Sympathomimetic amines 137
Sympathomimetics 58, 206, 236
  direct-acting 58–62
  indirect-acting 58, 62, 63
Synarel 261
Synthroid 248
Syntocinon 244
Syphilis 303, 306

T

Tabun 55
Tace 264
Tachyphylaxis 30, 64
Tagamet 141, 195
Talwin 185
Tamoxifen 267
Tapazole 250
Tapeworms 328–330
  beef 330
  dwarf 330
  fish 330
  pork 330
Tardive dyskinesia 223
TCADs 227–231
TCDD 383
Teebacin 311
Tegretol 212, 235
TEM 360
Temazepam 194, 196, 197
Teniposide 353, 367
Tenormin 70, 97
Tensilon 52, 53
TEPP 57
Terbutaline 58, 62, 66, 138
Terconazole 317
Terfenadine 134, 136
Terramycin 303
Testicular cancer 358, 363, 367–370
Testosterone 255–260
Tetanus 303, 306
Tetanus toxin 46, 47, 208, 209
Tetracaine 178, 181
Tetracyclines 142, 145, 152, 283–285, 302–304
Tetraethyl lead 377
Tetraethyl pyrophosphate 57
Tetrahydrocannabinoid 235
Tetrahydrocannabinol 239
Tetrahydrozoline 61
Tetraiodothyronine 247–250

TG 365
Thalassemia 375
THC 235
Theobromine 208
Theophylline 137, 139, 208
Theophylline ethylenediamine 139
Therapeutic index 29
Thiabendazole 323, 324, 327, 330
Thiamylal 192
Thiazide-like diuretics 87, 94
Thiazides 84–87, 93, 270, 273
Thimerosal 349
Thiobarbiturates 192, 198, 199
Thiocyanate 385
Thioguanine 365
Thionamides 250
Thiopental 192, 198–200
Thiophosphates 51, 52, 56
Thioridazine 223, 224, 226
Thiotepa 353, 360, 361
Thioxanthenes 223
Thorazine 223
Threadworm 323
Thrombolytics 127
Thrombosis 126
Thromboxanes 156, 158–160
Thymol 349
Thyroglobulin 249
Thyroid 249
Thyroid carcinoma 249
Thyroid hormone 246–250
Thyrotropin 241, 242, 246–249
Thyrotropin-releasing hormone 241, 242, 246–248
Thyrotropin-stimulating hormone 241, 242, 246–249
Ticarcillin 293, 296, 297
Tilade 137, 138
Timentin 296
Timolol 65, 68, 128
Timoptic 68
Tinactin 317
Tinea corporis 316
Tinea cruris 316
Tinea pedis 316
Tinea versicolor 316
Tioconazole 317
Tissue plasminogen activators 127, 128
Tobramycin 302
Tofranil 230
Tolazamide 280
Tolbutamide 280, 281
Tolectin 167
Tolinase 280
Tolerance 34, 236
Tolmetin 167
Tolnaftate 317
Toluene 381
Tomalate 138
Toxic adenoma 249
Toxicity 373
  acute 373
  chronic 373
  delayed 373

Toxicology 373–385
  clinical 373
  environmental 373
  forensic 373
  occupational 373
Toxin 373
Toxoplasma gondii 306
Toxoplasmosis 290, 291
tPA 127, 130
Trandate 69, 97
Tranexamic acid 127, 129, 130
Transcortin 253
Transient ischemic attacks 159, 166
Tranxene 193
Tranylcypromine 56, 59, 195, 232
Trazodone 231, 232–234
Trecator-SC 312
Trematodes 329–333
Treponema 303
Trexan 185, 238
TRH 241, 242, 246–249
Triamcinolone 138, 171
Triamterene 87, 89, 90, 94
Triazenes 362
Triazolam 193, 194, 195
Triazoles 314, 317
Trichinella 323
Trichinosis 323
Trichlorophenoxyacetic acid 383
Trichomonas vaginalis 335, 337
Trichomoniasis 335, 337
Trichophyton 315
Tricyclic antidepressants 224, 228–232
Tridione 217
Trientine 375, 377
Triethylene thiophosphoramide 360
Triethylenemelamine 360
Trifluoperazine 223
Trifluridine 319, 321, 322
Trihexyphenidyl 221
Triiodothyronine 247–250
Trimethadione 210, 217
Trimethaphan 46, 47, 73, 101
Trimethoprim 283, 285, 289–292
Trimipramine 228
Trimplex 290
Triphasil 265
Triple response of Lewis 133
Trisilicates 150
Trisulfapyrimidines 290
Trobicin 300
Trypanosomiasis 337, 339
TSH 247–250
Tubarine 78
Tuberculosis, treatment 307–312
Tularemia 302, 303
Tumors, solid 356–371
Turpentine 381
Tyicyclic antidepressants 58
Tylenol 160, 177
Typhoid fever 305
Tyramine 56, 58, 62, 231

U

Ulcerative colitis 147, 159, 290, 291
Ulcers 141–147, 159
Unasyn 296
Undecylenic acid 348
Urea 91
Ureaphil 91
Urecholine 50
Urex 292
Uricosurics 172–175
Urinary retention, treatment 50
Urinary tract antiseptics 292–293
Urinary tract infections 291, 292, 298, 299, 303, 310
Urinary urgency 55
Urofollitropin 261
Urogenital infections 306
Urokinase 127, 128, 130
Urticaria 133, 137
Uterine fibroids 260

V

Valium 81, 193–197
Valproate 210, 211, 216, 235
Valproic acid 215
VAMP therapy 352
Vancomycin 283, 297, 302
Vanillylmandelic acid 43
Vansil 330
Varicella-zoster 319, 322
Vasodilators 99–101, 119
Vasopressin 241, 242
Vasotec 95
Vasoxyl 61
Vecuronium 78
Velban 367
Venlafaxine 234
Vepesid 367
Verapamil 94, 95, 104, 106, 110, 112
Vermizine 324
Vermox 323
Versed 193, 196
Vertigo 152
Vibramycin 303
Vidarabine 319, 321
Vinblastine 352, 353, 363, 367, 369
Vinca alkaloids 353, 367
Vincristine 352, 353, 367
Virazole 320
Viroptic 322
Visine 61
Visken 69, 97
Vistaril 200
Vitamin B_6 219
Vitamin D 268–276
Vitamin D_2 268
Vitamin D_3 268
Vitamin K 129
Vitamin K antagonists 126
Vitamin K_1 129
Vitamin K_2 129
Vivactil 230

VMA 43
Volume of distribution 20
Vumon 367

W

Warfarin 126, 127, 129
Wellbutrin 232
Whipworm 323
Wilm's tumor 368
Wilson's disease 375
Withdrawal 236
  treatment 236–239

X

Xanax 196, 197
Xylocaine 112, 179

Y

Yodoxin 336
Yohimbine 67
Yohimex 67

Z

Zanosar 360
Zantac 141
Zarontin 215
Zaroxolyn 94
Zental 324
Zephiran 348
Zestril 95
Zidovudine 318, 320, 321
Zinc salts 349
Zithromax 306
Zn-insulin forms 278
  Humulin 278
  regular insulin 278
  Semilente insulin 279
  Velosulin 278
Zocor 123
Zofran 153
Zollinger-Ellison syndrome 134, 141, 145
Zoloft 234
Zolpidem 197, 198
Zovirax 319
Zyloprim 171, 174